Glycobiology

Nathan Sharon is one of the fathers of the science we now know as glycobiology. During his productive career, spent almost entirely at the Weizmann Institute in Rehovot, Israel, he has made ground-breaking discoveries in the area of protein–carbohydrate recognition. He is widely respected throughout the carbohydrate research community and as a mentor and friend to many younger glycoscientists. He is the author of over 450 peer-reviewed scientific publications and the acclaimed textbooks *Complex Carbohydrates* and *Lectins*.

We are delighted to be able to dedicate this book to Nathan, in celebration of his 80th birthday in 2005 and in recognition of his half-century of research at the carbohydrate–protein interface.

Clare Sansom and Ofer Markman, January 2007

Glycobiology

C. SANSOM & O. MARKMAN
Birkbeck College, UK   NutriCognia, Israel

© Scion Publishing Ltd, 2007

First published 2007

All rights reserved. No part of this book may be reproduced or transmitted, in any form or by any means, without permission.

A CIP catalogue record for this book is available from the British Library.

ISBN 978-1-904842-27-9

Scion Publishing Limited
Bloxham Mill, Barford Road, Bloxham, Oxfordshire OX15 4FF
www.scionpublishing.com

Important Note from the Publisher

The information contained within this book was obtained by Scion Publishing Limited from sources believed by us to be reliable. However, while every effort has been made to ensure its accuracy, no responsibility for loss or injury whatsoever occasioned to any person acting or refraining from action as a result of information contained herein can be accepted by the authors or publishers.

Typeset by Phoenix Photosetting, Chatham, Kent, UK
Printed by Replika Press Pvt. Ltd, India

Contents

Contributors	xiii
Abbreviations	xix

SECTION 1: INTRODUCTION — 1

Chapter 1
Half a century at the carbohydrate–protein interface — 2
Clare Sansom
    1. Protein–carbohydrate recognition: the first half century — 3
    2. A life with lectins — 3
    3. Lectins and beyond — 5
    4. Glycoscience comes of age — 6
    References — 6

Chapter 2
Glycobiology – discovery, diagnostics and drugs — 8
N. Zitzmann, T.M. Block, A.S. Mehta et al.
    References — 11

SECTION 2: SYNTHESIS OF CARBOHYDRATE AND GLYCOCONJUGATES — 13

Chapter 3
Sugar–lectin interactions: glycotargeting with synthetic glycoconjugates — 14
Michel Monsigny, Éric Duverger, Oruganti Srinivas et al.
    1. Introduction — 14
    2. Synthesis of neoglycoproteins and glycosylated polymers — 15
    3. Avidity effect — 16
    4. Endogenous lectins and drug targeting — 17
    5. Gene therapy — 18
    6. Conclusion — 21
    References — 22

Chapter 4
Recent developments in the synthesis and application of sialylmimetics — 24
Heike Busse and Hansjörg Streicher
1. Introduction — 24
2. Major applications of sialylmimetics — 25
3. Inhibition of viral sialidases — 27
4. Other sialylmimetics as inhibitors of viral and bacterial sialdases and parasitic trans-sialidases — 32
5. Sialyltransferases — 33
6. Outlook — 35
   References — 35

SECTION 3: ANALYSIS AND STRUCTURE — 37

Chapter 5
Structure and function of the chondroitin/dermatan sulfates — 38
Robert M. Lauder
1. Introduction — 38
2. Chondroitin/dermatan sulfate chain structure — 38
3. Chondroitin/dermatan sulfate chain function — 44
4. The future — 46
   References — 47

Chapter 6
Revealing the hidden dimensions of glycosaminoglycans in solution and protein complexes with saccharide libraries, FTIR, and SRCD — 49
E.A. Yates, T.R. Rudd, L. Duchesne et al.
1. Introduction — 49
2. Spectroscopic approaches to understanding higher order GAG structure — 52
3. Perspectives — 60
   References — 60

Chapter 7
Structural determination of N-linked glycans by matrix-assisted laser desorption/ionization and electrospray ionization mass spectrometry — 62
David J. Harvey
1. Introduction — 62
2. Mass spectrometry of intact glycoproteins and glycopeptides — 64
3. Release of N-glycans from glycoproteins — 66
4. Mass spectrometric profiling of N-glycans by mass spectrometry — 67
5. Structural information from released glycans — 70
6. Conclusions — 75
   References — 76

Chapter 8
Integrated approach to glycan structure–function relationships   78
Jonathan R. Behr and Ram Sasisekharan

  1. Introduction   78
  2. Analytical techniques for glycobiology   80
  3. Bioinformatics platforms for glycomics   86
  4. Applications   88
  5. Conclusions   90
  References   91

SECTION 4: CELL BIOLOGY OF CARBOHYDRATES AND GLYCOPROTEINS   93

Chapter 9
Some observations on the biology of cell surface heparan sulfate proteoglycans   94
John T. Gallagher

  1. Introduction   94
  2. The molecular structure of heparan sulfate   95
  3. Saccharide conformation and orientation of sulfate groups   98
  4. The biosynthesis of heparan sulfate   98
  5. "Editing" the structure of heparan sulfate   99
  6. Cell surface heparan sulfate proteoglycans   100
  7. Summary   104
  References   104

Chapter 10
Galectins: effective modulators of cytoskeletal organization and cellular growth – focus on galectin-8 as a model system   107
Yehiel Zick

  1. Introduction   107
  2. Characteristic features of the galectin family   107
  3. Galectin-8 – a tandem repeat galectin   108
  4. Galectin-8 as a mediator of cell adhesion   109
  5. Galectin-8 as a mediator of cell growth   115
  6. Future perspectives   119
  References   120

Chapter 11
Intracellular lectin involvement in glycoprotein maturation and quality control in the secretory pathway   123
Gerardo Z. Lederkremer

  1. Introduction   123
  2. Glycoprotein biosynthesis and maturation   123
  3. ER quality control of glycoproteins   126
  4. ER-associated degradation of glycoproteins   127
  5. Lectins in glycoprotein maturation   128

6.	Lectins in ER quality control and ERAD	130
7.	Concluding remarks	132
	References	132

SECTION 5: CARBOHYDRATE-BINDING PROTEINS   135

Chapter 12
The diversity of sialic acids and their interplay with lectins   136
Roland Schauer

1.	Introduction	136
2.	The distribution of sialic acids in nature and their chemical diversity	137
3.	Biosynthesis of sialic acids	139
4.	Degradation of sialic acids	141
5.	Sialic acids regulate molecular and cellular interactions	142
6.	Conclusion	148
	References	148

Chapter 13
Structure and carbohydrate specificity of ß-prism I fold lectins   150
K. Sekar, A. Surolia and M. Vijayan

1.	Introduction	150
2.	Examples	150
3.	Concluding remarks	157
	References	158

Chapter 14
Protein–glycan interactions in immunoregulation: galectins as tuners of the inflammatory response   160
Gabriel A. Rabinovich and Marta A. Toscano

1.	Introduction	160
2.	Protein–glycan interactions in health and disease	160
3.	Galectins as novel immunomodulatory agents: biochemistry and cell biology	162
4.	Galectins as tuners of the adaptive immune response	163
5.	Galectins as modulators of innate immune responses	168
6.	Galectins in immunopathology	169
7.	Conclusions and future directions	174
	References	175

Chapter 15
Recognition of bacterial glycoepitopes by pulmonary C-type lectins as an arm of lung innate immunity   178
Itzhak Ofek, Hany Sahly, Yona Keisari and Erika Crouch

1.	Introduction	178
2.	Glycoepitopes of *Klebsiella* and their interrelationship with the innate immune response	180

  3. General overview and concluding remarks 185
  References 187

Chapter 16
Selectin and integrin recognition of ligands under shear flow: affinity and beyond 190
Ronen Alon, Revital Shamri and Sara Feigelson
  1. Introduction 190
  2. Selectin recognition of ligands under shear stress: affinity, avidity and beyond 191
  3. Cytoskeletal associations of L-selectin regulate mechanical stabilization of adhesive bonds under shear flow 193
  4. VLA-4 and LFA-1 integrins also utilize cytoskeletal associations to stabilize their adhesive bonds under shear flow 194
  5. Why do selectin-mediated adhesions fail to arrest rolling leukocytes? 196
  6. Conclusions 197
  References 197

SECTION 6: CARBOHYDRATE-MODIFYING ENZYMES 201

Chapter 17
Complex carbohydrate-modifying enzymes 202
Karen J. Loft and Spencer J. Williams
  1. Introduction 202
  2. Carbohydrate sulfation 203
  3. Carbohydrate phosphorylation 209
  4. Carbohydrate carboxylic acid esters 212
  5. Conclusion 214
  References 215

Chapter 18
Glycosyltransferases specific for the synthesis of mucin-type O-glycans 217
Inka Brockhausen
  1. Introduction 217
  2. Glycosyltransferase structures 222
  3. Polypeptide GalNAc-transferases 224
  4. Core 1 β3-Gal-transferase 226
  5. Core 3 β3GlcNAc-transferase 227
  6. Synthesis of cores 2 and 4 228
  7. Extension and termination of *O*-glycans 230
  8. Conclusions 232
  References 232

Chapter 19
Glycosylation in development — 235
Shaolin Shi and Pamela Stanley
1. Introduction — 235
2. Glycosylation and oogenesis — 235
3. Glycosylation and spermatogenesis — 236
4. Glycosylation and fertilization — 237
5. Glycosylation and preimplantation development — 237
6. Glycosylation and implantation — 238
7. Glycosylation and postimplantation development — 238
8. Molecular bases of roles for glycosylation in development — 243
References — 244

SECTION 7: GLYCOBIOLOGY AND MEDICINE — 247

Chapter 20
Glycobiology of mucins in the human gastrointestinal tract — 248
Anthony P. Corfield
1. Mucins and glycobiology — 248
2. Mucins in the gastrointestinal tract — 252
3. Mucins and mucosal protection — 255
4. The mucous barrier and the intestinal microflora — 256
5. Future perspectives of intestinal mucin glycobiology — 258
References — 259

Chapter 21
Mucosal glycoconjugates in inflammatory bowel disease and colon cancer — 261
Barry J. Campbell, Lu-Gang Yu and Jonathan M. Rhodes
1. Mucosal glycosylation changes in inflammatory bowel disease and colon cancer — 261
2. Mechanisms underlying altered glycosylation — 262
3. Interaction of mucosal glycoconjugates with dietary lectins — 264
4. Interactions between intestinal epithelia and microbial lectins — 266
5. Implications for inflammatory bowel disease and colon cancer — 269
6. Interaction of mucosal glycoconjugates with endogenous galectins — 270
References — 271

Chapter 22
Selectin-mediated metastasis of tumor cells: Alteration of carbohydrate-mediated cell–cell interactions in cancers induced by epigenetic silencing of glycogenes — 274
Reiji Kannagi, Keiko Miyazaki, Naoko Kimura and Jun Yin
1. Introduction — 274
2. Induction of sialyl Lewis$^{a/x}$ expression in early cancers through the incomplete synthesis mechanism — 275

3.	Tumor hypoxia and enhancement of sialyl Lewis[a/x] in locally advanced cancers	280
4.	Sialyl Lewis[a/x] in hematogenous metastasis of cancers	283
	References	286

Chapter 23
Glycosphingolipids in health and disease — 288
Swetlana A. Boldin-Adamsky and Anthony H. Futerman

1.	Introduction	288
2.	Biological functions of glycosphingolipids	288
3.	The pathophysiology of glycosphingolipid storage disorders	289
4.	Treatment of glycosphingolipid storage disorders	293
5.	Perspectives	295
	References	295

Chapter 24
Pathogen–host interactions in *Entamoeba histolytica* — 297
David Mirelman and William A. Petri Jr.

1.	Introduction	297
2.	Virulence factors	298
	References	305

SECTION 8: INDUSTRIAL GLYCOBIOLOGY — 307

Chapter 25
The development of iminosugars as drugs — 308
Bryan Winchester

1.	Introduction	308
2.	Classes of iminosugars	309
3.	Potential applications	313
4.	DNA and purine and pyrimidine metabolism	320
5.	Conclusion	322
	References	322

Chapter 26
Carbohydrate biosensors — 325
Raz Jelinek and Sofiya Kolusheva

1.	Introduction	325
2.	Deciphering carbohydrate structures	325
3.	Lectin-based biosensors	327
4.	Glycoprotein and glycosylation biosensors	328
5.	Pathogen detection assays	328
6.	Cancer diagnostics	334
7.	Carbohydrate nano-biosensors	335
8.	Biosensors utilizing protein–carbohydrate interactions	336
9.	Concluding remarks	337
	References	337

Chapter 27
Glycoanalysis on a lectin array: applications to the development of
    biopharmaceuticals and life science research   340
Ruth Maya, Yehudit Amor, Revital Rosenberg et al.
    1.   Introduction   340
    2.   The technology   341
    3.   The fingerprints   341
    4.   Fingerprint deconvolution   344
    5.   Glycoanalysis of human milk lactoferrin glycovariants   345
    6.   Glycoanalysis of complex protein mixtures   346
    7.   Discussion   350
    References   351

Chapter 28
Lectin chips and CarboDeep™: a rapid carbohydrate fingerprinting
    technology and its application for the food scientists   352
Gabriel Faiman and Ofer Markman
    1.   Introduction   352
    2.   Technological applications   356
    3.   Specific markets   358
    4.   Conclusions   360
    References   360

Afterword   363
Nathan Sharon
Index   369

Contributors

Alon, Ronen Weizmann Institute of Science, Rehovot 76100, Israel. E-mail: ronalon@weizmann.ac.il

Amor, Yehudit Procognia Ltd, Habosem St. 3, Ashdod 77610, Israel. E-mail: yehudit.amor@procognia.com

Bangio, Haim Procognia Ltd, Habosem St. 3, Ashdod 77610, Israel. E-mail: haim.bangio@procognia.com

Behr, Jonathan 16-720, Massachusetts Institute of Technology, 77 Massachusetts Avenue, Cambridge, MA 02139, USA. E-mail: jbehr@mit.edu

Block, Timothy Pennsylvania Biotechnology Center, Drexel Institute for Biotechnology and Virology Research, 3805 Old Easton Rd, Doylestown, PA 18901, USA. E-mail: tim.block@drexel.edu

Boldin-Adamsky, Swetlana Department of Biological Chemistry, Weizmann Institute of Science, Rehovot 76100, Israel. E-mail: sweta.adamsky@gmail.com

Brockhausen, Inka Queen's University, Department of Medicine, Etherington Hall, Kingston, Ontario K7L 3N6, Canada. E-mail: brockhau@post.queensu.ca

Burton, Dennis Department of Immunology, The Scripps Research Institute, 10550 North Torrey Pines Rd, La Jolla, CA 92037, USA. E-mail: burton@scripps.edu

Busse, Heike Fachbereich Chemie, Universität Konstanz, D-78457 Konstanz, Germany. E-mail: heike.busse@gmail.com

Butters, Terry Glycobiology Institute, Department of Biochemistry, University of Oxford, South Parks Rd, Oxford, OX1 3QU, UK. E-mail: terry.butters@bioch.ox.ac.uk

Byk-Tennenbaum, Tamara Procognia Ltd, Habosem St. 3, Ashdod 77610, Israel. Current address: Evogene Ltd, PO Box 2100 Rehovot 76121, Israel. E-mail: tamara.byk@evogene.com

Campbell, Barry Liverpool University, School of Clinical Sciences, Division of Gastroenterology, Liverpool L69 3GA, UK. E-mail: b.j.campbell@liverpool.ac.uk

Clarke, DT CCLRC, Daresbury Laboratory, WA4 4AD, UK. E-mail: d.t.clarke@dl.ac.uk

Corfield, Anthony Clinical Science at South Bristol, Bristol Royal Infirmary, Level 7, Marlborough Street, Bristol BS2 8HW, UK. E-mail: tony.corfield@bristol.ac.uk

Crouch, Erika Department of Pathology and Immunology, Washington University School of Medicine, St Louis, USA. E-mail: crouch@path.wustl.edu

Duchesne, L Centre for Nanoscale Science, School of Biological Sciences, University of Liverpool, Liverpool, L69 7ZB, UK. E-mail: lduchesn@liv.ac.uk

Duverger, Eric Glycobiologie CNRS-CBM, 1, rue Charles-Sadron, 45071 Orléans, cedex 2, France. E-mail: duverger@cnrs-orleans.fr

Dwek, Raymond Glycobiology Institute, Department of Biochemistry, University of Oxford, South Parks Road, Oxford OX1 3QU, UK. E-mail: raymond.dwek@exeter.ox.ac.uk

Faiman, Gabriel The Institute for Standardization and Control of Pharmaceuticals, Pharmaceutical Administration, Ministry of Health, 9 Eliav St, Jerusalem 91342, Israel. E-mail: gabriel.faiman@eliav.health.gov.il

Fajac, Isabelle Lab. Physiol. Respir., Faculté Cochin, 24, rue du Faubourg St-Jacques, 75679 Paris, cedex 14, France.
E-mail: isabelle.fajac@cochin.univ-paris5.fr

Feigelson, Sara Weizmann Institute of Science, Rehovot 76100, Israel. E-mail: sara.feigelson@weizmann.ac.il

Fernig, David Centre for Nanoscale Science, School of Biological Sciences, University of Liverpool, Liverpool L69 7ZB, UK. E-mail: dgfernig@liv.ac.uk

Futerman, Anthony Department of Biological Chemistry, Weizmann Institute of Science, Rehovot 76100, Israel. E-mail: tony.futerman@weizmann.ac.il

Gallagher, John Cancer Research UK, Department of Medical Oncology, University of Manchester, Christie Hospital NHS Trust, Wilmslow Road, Manchester M20 4BX, UK. E-mail: jgallagher@picr.man.ac.uk

Harvey, David Glycobiology Institute, Department of Biochemistry, University of Oxford, South Parks Road, Oxford OX1 3QU, UK. E-mail: david.harvey@bioch.ox.ac.uk

Jelinek, Raz Department of Chemistry, Ben Gurion University, Beersheva 84105, Israel. E-mail: razj@bgu.ac.il

Kannagi, Reiji Department of Molecular Pathology, Aichi Cancer Center, 1-1 Kanokoden, Nagoya 464-8681, Japan. E-mail: rkannagi@aichi-cc.jp

Keisari, Yona Department of Human Microbiology, Sackler Faculty of Medicine, Tel Aviv University, Ramat Aviv, Israel. E-mail: ykeisari@post.tau.ac.il

Kimura, Naoko Department of Molecular Pathology, Aichi Cancer Center, 1-1 Kanokoden, Nagoya 464-8681, Japan. E-mail: nkimura@aichi-cc.jp

Kolusheva, Sofiya Ilse Katz Centre for Nanotechnology, Ben Gurion University, Beersheva 84105, Israel. E-mail: kolushev@bgu.ac.il

Lauder, Robert Department of Biological Sciences, I.E.N.S., Lancaster University, Lancaster LA1 4YQ, UK. E-mail: r.lauder@lancaster.ac.uk

Lederkremer, Gerardo Department of Cell Research and Immunology, The George S. Wise Faculty of Life Sciences, Tel Aviv University, Tel Aviv 69978, Israel. E-mail: gerardol@tauex.tau.ac.il

Loft, Karen Bio21 Molecular Science and Biotechnology Institute and School of Chemistry, University of Melbourne, Parkville, VIC 3010, Australia. E-mail: k.loft@pgrad.unimelb.edu.au

Markman, Ofer NutriCognia Ltd, PO Box 2230, Hahadarim St. 2, Heavy Ind. Estate, Ashdod 77613, Israel. E-mail: ofer.markman@nutricognia.com

Maya, Ruth Ben-Yakar Procognia Ltd, Habosem St. 3, Ashdod 77610, Israel. E-mail: ruth.maya@procognia.com

Mehta, Anand Pennsylvania Biotechnology Center, Drexel Institute for Biotechnology and Virology Research, 3805 Old Easton Rd, Doylestown, PA 18901, USA. E-mail: anand.mehta@mail.tju.edu

Mirelman, David Department of Biological Chemistry, Ullman Building, Weizmann Institute of Science, 76100 Rehovot, Israel. E-mail: david.mirelman@weizmann.ac.il

Miyazaki, Keiko Department of Molecular Pathology, Aichi Cancer Center, 1-1 Kanokoden, Nagoya 464-8681, Japan. E-mail: kmiyazaki@aichi-cc.jp

Monsigny, Michel Glycobiologie, CNRS-CBM, 1, Rue Charles-Sadron, 45071 Orléans, cedex 2, France. E-mail: monsigny@aol.com

Nichols, R.J. Centre for Nanoscale Science, School of Biological Sciences, University of Liverpool, Liverpool L69 7ZB, UK. E-mail: nichols@liv.ac.uk

Ofek, Itzhak Department of Human Microbiology, Sackler Faculty of Medicine, Tel Aviv University, Ramat Aviv 69978, Israel. E-mail: aofek@post.tau.ac.il

Olender, Roberto Procognia Ltd, Habosem St. 3, Ashdod 77610, Israel. E-mail: roberto.olender@procognia.com

Oruganti, Srinivas Glycobiologie, CNRS-CBM, 1, Rue Charles-Sadron, 45071 Orléans, cedex 2, France. E-mail: srinivas_o@hotmail.com

Petrescu, Stefana Institute of Biochemistry, Splaiul Independentei 296, 77700 Bucharest, Romania. E-mail: stefana.petrescu@biochim.ro

Petri, William Jr. Division of Infectious Diseases, MR4 Building, Room 2115, Lane Rd, P.O. Box 801340, University of Virginia Health System, Charlottesville, VA 22908-1340, USA. E-mail: wap3g@virginia.edu

Platt, Frances Department of Pharmacology, University of Oxford, Mansfield Rd, Oxford, OX1 3QT, UK. E-mail: frances.platt@pharm.ox.ac.uk

Rabinovich, Gabriel Instituto de Biología y Medicina Experimental, Consejo Nacional de Investigaciones Científicas y Técnicas de Argentina, Vuelta de Obligado 2490, Buenos Aires, C1428ADN, Argentina. E-mail: gabyrabi@ciudad.com.ar

Rhodes, Jonathan Liverpool University, School of Clinical Sciences, Division of Gastroenterology, Liverpool L69 3GA, UK. E-mail: j.m.rhodes@liverpool.ac.uk

Roche, Annie-Claude Glycobiologie, CNRS-CBM, 1, Rue Charles-Sadron, 45071 Orléans, cedex 2, France. E-mail: roche@cnrs-orleans.fr

Rosenberg, Revital Procognia Ltd, Habosem St. 3, Ashdod 77610, Israel. E-mail: revital.rosenberg@procognia.com

Rosenfeld, Rakefet Procognia Limited, Habosem Street 3, Ashdod 77610, Israel. E-mail: rakefet.rosenfeld@procognia.com

Rudd, Pauline NIBRT, Conway Institute, University College Dublin, Belfield, Dublin 4, Ireland. E-mail: pauline.rudd@nibrt.ie

Rudd, T.R. Centre for Nanoscale Science, School of Biological Sciences, University of Liverpool, Liverpool L69 7ZB, UK. E-mail: trudd@liv.ac.uk

Sahly, Hany Institute for Infection Medicine, Faculty of Medicine, University of Kiel, Kiel, Germany. E-mail: sahly@infmed.uni-kiel.de

Samokovlisky, Albena Procognia Ltd, Habosem St. 3, Ashdod 77610, Israel. E-mail: albena.samokovlisky@procognia.com

Sansom, Clare Department of Crystallography, Birkbeck College, Malet Street, London WC1E 7HX, UK. E-mail: c.sansom@bbk.ac.uk

Sasisekharan, Ram Biological Engineering Division, Center for Biomedical Engineering, Massachusetts Institute of Technology, 77 Massachusetts Ave., 16-561, Cambridge, MA 02139, USA. E-mail: ram@mit.edu

Schauer, Roland Christian-Albrechts-Universität zu Kiel, Biochemisches Institut in der Medizinischen Fakultät, Eduard-Buchner-Haus, Otto-Hahn-Platz 9, 24118 Kiel, Germany. E-mail: schauer@biochem.uni-kiel.de

Sekar, K. Bioinformatics Centre, Supercomputer Education and Research Centre, Indian Institute of Science, Bangalore 560 012, India. E-mail: sekar@physics.iisc.ernet.in

Shamri, Revital Weizmann Institute of Science, Rehovot 76100, Israel. Current address: Division of Allergy and Infectious Diseases, Dana 613, Beth Israel Deaconess Medical Center, 330 Brookline Avenue, Boston, MA 02215, USA. E-mail: rshamri@bidmc.harvard.edu

Sharon, Nathan Weizmann Institute of Science, Rehovot 76100, Israel. E-mail: nathan.sharon@weizmann.ac.il

Shi, Shaolin Division of Nephrology, Department of Medicine, Mount Sinai School of Medicine, 1 Gustave L. Levy Place, Box 1243, New York, NY 10029, USA. E-mail: shaolin.shi@mssm.edu

Skidmore, M.A. Centre for Nanoscale Science, School of Biological Sciences, University of Liverpool, Liverpool L69 7ZB, UK. E-mail: m.a.skidmore@liv.ac.uk

Stanley, Pamela Department of Cell Biology, Albert Einstein College of Medicine, 1300 Morris Park Avenue, New York, NY 10461, USA. E-mail: stanley@aecom.yu.edu

Streicher, Hansjörg Department of Chemistry, University of Sussex, Falmer, Brighton BN1 9QJ, UK. E-mail: h.streicher@sussex.ac.uk

Surolia, Avadhesha Molecular Biophysics Unit, Indian Institute of Science, Bangalore 560 012, India. E-mail: surolia@mbu.iisc.ernet.in

Toscano, Marta Instituto de Biología y Medicina Experimental, Consejo Nacional de Investigaciones Científicas y Técnicas de Argentina, Vuelta de Obligado 2490, Buenos Aires, C1428ADN, Argentina. E-mail: martalitos@yahoo.com.ar

Vijayan, M Molecular Biophysics Unit, Indian Institute of Science, Bangalore 560 012, India. E-mail: mv@mbu.iisc.ernet.in

Williams, Spencer Bio21 Molecular Science and Biotechnology Institute and School of Chemistry, University of Melbourne, Parkville, VIC 3010, Australia. E-mail: sjwill@unimelb.edu.au

Wilson, Ian Department of Molecular Biology & Skaggs Institute for Chemical Biology, BCC206, The Scripps Research Institute, 10550 North Torrey Pines Rd, La Jolla, CA 92037, USA. E-mail: wilson@scripps.edu

Winchester, Bryan Biochemistry, Endocrinology and Metabolism Unit, Institute of Child Health at Great Ormond Street Hospital, University College London, 30 Guilford Street, London WC1N 1EH, UK. E-mail: b.winchester@ich.ucl.ac.uk

Wormald, Mark Glycobiology Institute, Department of Biochemistry, University of Oxford, South Parks Rd, Oxford, OX1 3QU, UK. E-mail: mark@glycob.ox.ac.uk

Yates, E.A. Centre for Nanoscale Science, School of Biological Sciences, University of Liverpool, Liverpool L69 7ZB, UK. E-mail: e.a.yates@liv.ac.uk

Yin, Jun Department of Molecular Pathology, Aichi Cancer Center, 1-1 Kanokoden, Nagoya 464-8681, Japan. E-mail: yinjun@aichi-cc.jp

Yu, Lu-Gang Liverpool University, School of Clinical Sciences, Division of Gastroenterology, Liverpool L69 3GA, UK. E-mail: lgyu@liverpool.ac.uk

Zick, Yehiel Department of Molecular Cell Biology, Weizmann Institute of Science, Rehovot 76100, Israel. E-mail: yehiel.zick@weizmann.ac.il

Zitzmann, Nicole Glycobiology Institute, Department of Biochemistry, University of Oxford, South Parks Rd, Oxford, OX1 3QU, UK. E-mail: nicole.zitzmann@bioch.ox.ac.uk

Abbreviations

ß-GalT	ß-D-galactosyltransferase
ΔUA	4,5-unsaturated hexuronic acid (4-deoxy-α-L-*threo*hex-4-enepyranosyluronic acid)
2-AA	2-aminobenzoic acid
2-AB	2-aminobenzamide
2-AP	2-aminopyridine
6S/4S/3S/2S	*O*-ester sulfate group on C-6/C-4/C-3/C-2
AFM	atomic force microscopy
APP	amyloid precursor protein
Ara-A	arabinosyl-adenine
Ara-AMP	9-ß-D-arabinofuranosyl-adenine 5′ monophosphate
ASK1	apoptosis signal regulating kinase 1
A-SMase	acid sphingomyelinase
AT	antithrombin
BCA	bicinchoninic acid
BCR	B-cell receptor
bFGF or FGF2	basic fibroblast growth factor
BMP	bone morphogenetic protein
BSA	bovine serum albumin
b-SiA	bound sialic acid
CBE	conduritol-B-epoxide
CCK	cholecystokinin
CCSD	Complex Carbohydrate Structures Database
CD	circular dichroism
CD22	siglec-2
CDK	cyclin-dependent kinase
CDKI	cyclin-dependent kinase inhibitors
CE	capillary electrophoresis
CEACAM	carcinoembryonic antigen cell adhesion molecule
CF	cystic fibrosis
CFG	Consortium for Functional Glycomics
CFTR	cystic fibrosis transmembrane conductance regulator
CHO	Chinese hamster ovary
CIA	collagen-induced arthritis

CID	collision-induced decomposition
CMT	chaperone-mediated therapy
CMV	cytomegalovirus
CNS	central nervous system
con A	concanavalin A
CPS	capsular polysaccharide
CP	cysteine proteinase
CRD	carbohydrate recognition domain
CS	chondroitin sulfate
CSR	composite sulfated region
CT	cholera toxin
CTL	cytotoxic T-lymphocyte
DAB1	1,4-dideoxy-1,4-imino-D-arabinitol
DAEC	diffusely adherent *E. coli*
DAF	decay accelerating factor
DANA	2-deoxy-2,3-didehydro-*N*-acetylneuraminic acid
DC	dendritic cell
DC-SIGN	dendritic cell-specific ICAM grabbing non-integrin
DHB	dihydroxybenzoic acid
DHPE	1,2-dihexadecyl-*sn*-glycero-3-phosphoethanolamine
DMDP	2,5-dideoxy-2,5-imno-D-mannitol
DRB	1,4-dideoxy-1,4-imino-D-ribitol
DS	dermatan sulfate
EAE	experimental autoimmune encephalomyelitis
EAMG	experimental autoimmune myasthenia gravis
EAU	experimental autoimmune uveitis
ECM	extracellular matrix
ELISA	enzyme-linked immunosorbent assay
Endo H	endoglycosidase H
Eph	ephrin
ER	endoplasmic reticulum
ERAD	ER-associated degradation
ERGIC	ER-to-Golgi intermediate compartment
ERQC	ER-derived quality control compartment
ERT	enzyme replacement therapy
ESI	electrospray ionization
ESI-MS	electrospray ionization–mass spectrometry
FAK	focal adhesion kinase
FGF	fibroblast growth factor
FGly	formylglycine, 2-amino-3-oxopropanoic acid
FISH	fluorescence *in situ* hybridization
FPA	fluorescence polarization assay
FRET	fluorescence resonance energy transfer
FTICR-MS	Fourier transform ion cyclotron resonance mass spectrometry
FTIR	Fourier transform infrared

Fuc	fucose
GAG	glycosaminoglycan
Gal	galactose
GalNAc	2-deoxy-2-*N*-acetylamino-D-galactose
GBP	glycan-binding protein
gCOSY	gradient-selected correlation spectroscopy
GDNF	glial cell-derived neurotrophic factor
Glc	glucose
GFP	green fluorescent protein
Glc	glucose
GlcA	D-glucuronic acid
GlcCer	glucosylceramide
GlcNAc	*N*-acetylglucosamine
GlcNS	*N*-sulfated glucosamine
GMP	glycomacropeptide
GPC3	glypican 3
GPCRs	G protein-coupled receptors
GPI	glycosylphosphatidylinositol
GSDs	glycosphingolipid storage disorders
GSK3β	glycogen synthase kinase 3 beta
GSL	glycosphingolipid
GT	glycosyltransferase
GVHD	graft versus host disease
HA	hyaluronan
HBP	heparan binding protein
HBV	hepatitis B virus
HCV	hepatitis C virus
H-D antigen	Hanganutziu–Deicher antigen
Hex	hexose
HexNAc	*N*-acetylaminohexose
HGF/SF	hepatocyte growth factor/scatter factor
HIF	hypoxia inducible factor
HIQ	1-amino-*iso*-quinoline
HIV-1	human immunodeficiency virus type 1
HMBC	heteronuclear multiple bond correlation
HME	hereditary multiple exostosis
hmLF	human milk lactoferrin
HMQC	heteronuclear multiple quantum correlation
HPLC	high-performance liquid chromatography
HPRG	histidine proline-rich glycoprotein
HS	heparan sulfate
HSPGs	heparan sulfate proteoglycans
HSV	herpes simplex virus
ICAM	intercellular adhesion molecule
IdoA	α-L-iduronic acid
IFNδ	interferon gamma

IgA1	immunoglobulin A1
IgG	immunoglobulin G
IL-8	interleukin 8
IP$_3$R	inositol 1,4,5-triphosphate receptor
ISFET	ion selective field effect transistor
ITIM	immunoreceptor tyrosine-based inhibition motif
JNK	C-jun N-terminal kinase
Kdn	2-keto-3-deoxy-nononic acid
KS	keratan sulfate
LAT	latex agglutination test
LC/MS	liquid chromatography/mass spectrometry
Lewis Le^a	Galß-3[Fucα-4]GlcNAc,
Lewis Le^x	Galß-4[Fucα-3]GlcNAc
LMWHs	low molecular weight heparins
LPG	lipophosphoglycan
LPPGs	lipophosphopeptidoglycans
LPS	lipopolysaccharides
LSDs	lysosomal storage disorders
M6P	mannose-6-phosphate
mAb	monoclonal antibody
MAG	siglec-4, myelin-associated glycoprotein
mAGP complex	mycolylarabinogalactan-peptidoglycan complex
MALDI	matrix-assisted laser desorption/ionization
Man	mannose
M6P-R	mannose-6-phosphate receptor
MAPK	mitogen-activated protein kinase
MBP	myelin-basic protein
MDP	muramyldipeptide
MHC	major histocompatibility complex
MIP-1α	macrophage inflammatory protein 1 alpha
MM	molecular modeling
MOG	myelin oligodendrocyte glycoprotein
MPS	mucopolysaccharidoses
MPT	α-D-mannosylphosphate transferase
MR	mannose receptor
MS	mass spectrometry
MSD	multiple sulfatase deficiency
MX	α-mannosidase-IIx
NCAM	neural cell adhesion molecule
NDST	N-deacetylase/N-sulfotransferase
Neu	neuraminic acid
Neu2en5Ac	2-deoxy-2,3-didehydro-N-acetylneuraminic acid
Neu5,9Ac$_2$	N-acetyl-9-O-acetylneuraminic acid
Neu4,5Ac$_2$	N-acetyl-4-O-acetylneuraminic acid
Neu5Ac	N-acetylneuraminic acid
Neu5Gc	N-glycolylneuraminic acid

NIGMS	National Institute for General Medical Sciences
NLS	nuclear localization signal
NMR	nuclear magnetic resonance
NO	nitrous oxide
NPD-A	Niemann–Pick A
NP-HPLC	normal-phase HPLC
oN	oligonucleotide
OST	oligosaccharyltransferase
PAGE	polyacrylamide gel electrophoresis
PAK	p21-activated protein kinase
PAPS	3'-phosphoadenosine-5'-phosphosulfate
PCNA	proliferating cell nuclear antigen
PCTA-1	prostate carcinoma tumor antigen 1
PDA	polydiacetylene
PDB	Protein Data Bank
PEI	polyethylene imine
PET	photoinduced energy transfer
PF4	platelet factor 4
PG	proteoglycan
PHA	phytohemagglutinin
PI3K	phosphatidylinositol 3-kinase
PKB	protein kinase B
PLP	myelin proteolipid protein
PM	plasma membrane
PNA	peanut agglutinin
PNGase	peptide:N-glycanase or protein N-glycosidase
PNP	purine nucleoside phosphorylase
PPD	polarized photometric detection
PSA	polysialic acid
PSD	post-source decay
PSGL-1	P-selectin glycoprotein ligand-1
PtdIns(4,5)P$_2$	phosphatidylinositol 4,5-bisphosphate
Q	quadrupole
QCM	quartz crystal microbalance
QFIR	quantitative fingerprint interpretation relation
QIT	quadrupole ion trap
Q-TOF	quadrupole-TOF
RT-PCR	reverse transcriptase polymerase chain reaction
SBA	soya bean agglutinin
SDS	sodium dodecyl sulfate
SEA	sea-urchin sperm protein, enterokinase, and agrin
SEC	size exclusion chromatography
SERCA	sarco(endo)plasmic reticulum Ca^{2+}-ATPase
Sia	sialic acid
Siglec	sialic acid binding immunoglobulin (Ig)-like receptor
siRNA	small interfering RNA

SLeA	sialylated Lewis antigen
SM	sphingomyelin
SPR	surface plasmon resonance
SRCD	synchrotron radiation circular dichroism
SRP	signal recognition particle
SRT	substrate reduction therapy
SSEA-1	stage-specific embryonic antigen 1
TCR	T-cell receptor
TDM	trehalose dimycolate
TF	Thomsen Friedenreich
TFA	trifluoroacetic acid
TGFα	transforming growth factor alpha
THAP	2,4,6-trihydroxyacetophenone
TLC	thin-layer chromatography
TM	transmembrane domain
TNBS	2,4,6-trinitrobenzene sulfonic acid
TNFα	tumor necrosis factor alpha
TOCSY	total correlated spectroscopy
TOF	time-of-flight
UCE	uncovering enzyme
UGGT	UDP-glucose glycoprotein glucosyltransferase
u-PA	urokinase-type plasminogen activator
UPR	unfolded protein response
VCD	vibrational circular dichroism
VEGF	vascular endothelial growth factor
Vg1	vitellogenin 1
VNTR	variable number of tandem repeat
Xyl	xylose

SECTION 1
Introduction

CHAPTER 1
Half a century at the carbohydrate–protein interface

Clare Sansom

This book has its origin in a coincidence. Back in the summer of 2004 I was first asked by my collaborator (and now co-editor), Ofer Markman, if I would be able to help him plan the celebrations for the 80th birthday of his mentor, Nathan Sharon. That was soon after I was first approached by Jonathan Ray about editing a book for Scion Publishing. Putting the two together seemed an obvious move, and the list of invited speakers at the anniversary symposium held in Rehovot in November 2005 proved an excellent starting point for soliciting contributions. The fact that we have been able to assemble such a distinguished list of contributors to this book is, above all, a tribute to the exceptional standing of Professor Sharon in the glycoscience community. The final list of contributors includes many of his current and former colleagues and collaborators, ranging from early colleagues including David Mirelman, a student in the group in the 1960s, and Itzhak Ofek, to his last post-doc, Hansjörg Streicher.

Sharon's long and exceptionally productive scientific career has been spent almost completely at the Weizmann Institute in Rehovot, Israel. He started work as a research assistant at the Dairy Research Laboratory of the Agricultural Research Station in Rehovot in the year that the Weizmann was founded, 1949, and moved to that world-famous institute 5 years later. He is the author of over 450 scientific publications and has received numerous honors and awards, among them FEBS' Datta Medal, an honorary doctorate from the University of Paris, and the Israel Prize in Biochemistry and Medicine. He is a member of the Israel Academy of Sciences and Humanities and of the European Molecular Biology Organization (EMBO) and an honorary member of the American Society of Biochemistry and Molecular Biology and the American Society of Microbiology. Even now, as he enters his ninth decade, he continues publishing [1–3]. He is best known for his groundbreaking work on proteins that bind to and recognize carbohydrates – lectins – although he has research interests in many other aspects of pure and applied carbohydrate chemistry. It is not surprising that the area of protein–carbohydrate recognition has been covered particularly comprehensively in this book.

Glycobiology (C. Sansom and O. Markman, eds.)
© Scion Publishing Limited, 2007

1. PROTEIN–CARBOHYDRATE RECOGNITION: THE FIRST HALF CENTURY

The first of the proteins that are now known, collectively, as lectins were discovered many decades before anything about their molecular function was understood [4]. By the end of the nineteenth century it was generally known that some plant proteins were able to agglutinate erythrocytes. These proteins became known as hemagglutinins. Paul Ehrlich, one of the fathers of immunology, used them to establish some of the basic principles of that science. The first hemagglutinin to be isolated came from the seeds of the castor tree and was named ricin after the Latin name of that tree, *Ricinus communis*. In contrast to most other hemagglutinins, ricin is extremely toxic; it holds a notorious place in twentieth-century political history as the poison used in the so-called "umbrella murder" of the Bulgarian dissident, Georgi Markov, in 1978. More recently, the general public will have become aware of hemagglutinin, probably without realizing it, as the "H" in the names of influenza subtypes including the feared H5N1. An inhibitor of influenza virus hemagglutinin, named Fludase®, is under development by San Diego-based biotechnology company NexBio [5]; it is scheduled for clinical trials in late 2006.

The exquisite molecular specificity of the hemagglutinins was first hinted at in the 1940s, when two researchers, William Boyd and Karl Renoken, discovered simultaneously that different hemagglutinins agglutinate erythrocytes from different blood groups. This discovery was one of the keys to the understanding, developed soon afterwards, that it was sugars bound to the cell surfaces that gave the erythrocytes their specificity (and that group A specificity was conferred by alpha-*N*-acetyl-D-galactosamine and group H(O) specificity by alpha-L-fucose) [6]. This was the first realization of the important role that carbohydrates and glycoconjugates on cell surfaces play in molecular and cellular recognition. The name "lectins" – from the Latin *legere*, meaning to pick out or choose – was first applied to plant proteins with this ability to distinguish between different blood groups in 1954 [7]. That same year the young Sharon, newly awarded his PhD from the Hebrew University, Jerusalem, moved across Rehovot from the Dairy Institute to the Department of Biophysics at the Weizmann Institute: ironically, intending to work on proteins. His interest in carbohydrates had, however, already been aroused by his PhD research into the aldose–amino acid reaction between the milk sugar lactose and proteins found in milk. This topic, chosen prosaically enough because of its importance to the dairy industry, gave Sharon his first publication on carbohydrates [8].

2. A LIFE WITH LECTINS

In his first years at the Weizmann Institute, Sharon worked primarily on the synthesis and characterization of peptides and proteins. His interest in carbohydrates was rekindled when, by chance, during studies of bacterial protein synthesis, he found a novel polysaccharide in *Bacillus subtilis* [9]. Subsequently, he isolated an unusual diamino sugar from this polysaccharide, which he characterized and named bacillosamine [10] and which is now attracting considerable attention. In 1962, he began work on his first agglutinating protein, soybean agglutinin (SBA). This protein is a hemagglutinin, although it is not itself blood group specific. Sharon and his long-time colleague, Halina Lis, later proposed that the term "lectin" be extended to encompass all non-immune proteins that were sugar specific and could agglutinate cells [4,11]. Thus began a long association with lectins that Sharon has described as "an intellectual adventure, exciting and rewarding" [4].

The biological importance of these proteins, all of which show specific binding to, and recognition of, carbohydrates, first began to be understood in the 1960s. Peter Nowell of the University of Pennsylvania was the first to recognize that lectins could stimulate mitosis. Lectins thus became useful tools for the investigation of the events that occur upon lymphocyte stimulation, and for the early studies of signal transduction from the cell surface to its interior. The connection between lectins and cancer was emphasized by the discovery that some lectins, including wheat germ agglutinin, will preferentially agglutinate cancer cells [4]. This has become a productive field of research; several contributors to this volume describe the role of lectins, principally the family of animal lectins known as S-type lectins or galectins, in the development of cancer and inflammatory disease.

By 1975, enough was known about the importance of specific interactions between proteins and carbohydrates for Sharon to assert, in his acclaimed textbook, that "the specificity of many natural polymers is written in terms of sugar residues and not of amino acids or nucleotides" [12]. At that time, the vast majority of known lectins came from the plant kingdom; today, however, lectins are known to be ubiquitous. In May 2006 a simple search of the Uniprot protein database with the keyword "lectin" yielded over 2000 entries, for proteins from bacteria, archaea, viruses, and all eukaryotic kingdoms. And the truth of Sharon's statement has become clearer as more and more lectins and the carbohydrates they recognize have been elucidated.

The jack bean (*Canavalia ensiformis*) lectin concanavalin A features in several important "firsts" in the science of lectins. It was the first to be purified and the first to have both its amino acid sequence [13] and its three-dimensional structure [13,14] elucidated. The fold first observed in the concanavalin A structure, an elaborate arrangement of extended beta strands into two sheets, is now known as the jelly roll or lectin fold [15]. This fold is found in several different homologous families of lectins, including the legume lectins and the galectins. Other lectin families have different folds; ricin is among those that exhibit a different all-beta fold known as a beta-trefoil.

In 1988, Kurt Drickamer proposed that most animal lectins, including the galectins, bind carbohydrates via a fairly small, specific region of polypeptide chain termed the carbohydrate recognition domain (CRD) [16]. Animal lectins vary greatly in size; the galectins are small, soluble proteins whereas the large family of C-type lectins are large transmembrane glycoproteins which require calcium for activity and contain a single CRD. All three-dimensional lectin structures in the publicly available Protein Data Bank (PDB) are held in a database at http://www.cermav.cnrs.fr/lectines/, which currently (May 2006) has 528 entries. A new lectin database has also been set up at http://nscdb.bic.physics.iisc.ernet.in/.

Lectins are now known to be involved in molecular recognition events that have a variety of biological functions, including – and this is not a complete list – bacterial and viral infection, host defense, innate immunity, cell cycle regulation, and cell–cell interactions. Structural biology techniques, most often X-ray crystallography, have been used to elucidate these interactions in detail and describe the interatomic interactions that explain their specificity. This specificity is also now being exploited in biotechnology. Sharon has recently reviewed the potential of inhibitors of the interaction between bacterial lectins and carbohydrates on the surfaces of human cells as mild and non-toxic drugs for infectious diseases [17]. And, in this book, Rakefet Rosenfeld from Procognia in Israel and my co-editor Ofer Markman describe the use of lectins, arranged on protein arrays, in the analysis of the glycosylation patterns of glycoproteins.

3. LECTINS AND BEYOND

In this book we have attempted to put together contributions that reflect the wide range of Sharon's interests and some of the most interesting and important topics in glycoscience at the beginning of the twenty-first century. We are fortunate to lead with a short chapter by Raymond Dwek, Director of the Glycobiology Institute at the University of Oxford, setting out the prospects for drug intervention in the process of glycosylation, in diseases including cancer, infection, and inborn errors of metabolism. Two of the themes introduced here are described more fully later in the book. Tony Futerman, a colleague of Sharon at the Weizmann Institute, describes in detail some of the most important inborn errors in glycoconjugate metabolism, the glycosphingolipid (GSL) storage diseases, in which mutations in enzymes lead to GSL accumulation and, often, to severe and progressive neurological problems. Bryan Winchester of the Institute of Child Health in London reports on the development of iminosugars, metabolically inert molecules that mimic carbohydrate structures, as potential drugs for these diseases. One of these, N-butyl 1-deoxynojirimycin (N-butyl DNJ, Zavesca® or Miglustat) has been licensed to treat one type of Gaucher disease.

Synthesis is covered by contributions from Michel Monsigny (CNRS, Orléans, France) and Hansjörg Streicher, now at the University of Sussex, UK. We then move on to structural and functional analysis of glycoconjugates, with an emphasis on glycosaminoglycans; John Gallagher from Christie Hospital in Manchester follows this with a detailed review of the structure and functional properties of cell surface proteoglycans. Structural biology is represented by another of Sharon's long-time collaborators, Avadhesha Surolia, with an attractively illustrated review of plant lectins that adopt the beta-prism I fold, and developmental biology by a wide-ranging review by Pamela Stanley from the Albert Einstein College of Medicine, New York.

Sharon has always been interested in, and excited by, the potential application of his work in medicine; he is proud of the fact that some of his work on soybean agglutinin has been used as the basis of a successful technique for purging bone marrow for transplants to children with severe combined immune deficiency [18]. It is therefore particularly pleasing that we have been able to attract a number of excellent contributions on medical aspects of glycobiology. Glycoconjugate involvement in cancer and inflammatory disease, errors of glycan metabolism, and the role of lectin–sugar interactions in parasitic disease are all covered: the latter by Sharon's colleague, David Mirelman. The book ends with a section on applications of glycobiology in the biotechnology industry.

4. GLYCOSCIENCE COMES OF AGE

Even a book as comprehensive as we have tried to make this one cannot do justice to the immense advances that have been made in glycoscience during the last half century. It is even more impressive that this has been achieved with relatively little public funding. The total funding for glycoscience in the final decade of the last century, for instance, has been estimated at less than US$5 million – at the most conservative estimates, less than 10% of that allocated to the Human Genome Project [19]. It was perhaps easy to neglect sugars in the race to sequence genes and understand the proteins they encode. Now the sequence is complete, however, the molecular biology community is beginning to understand that the Central Dogma of Molecular Biology ("DNA makes RNA makes protein") is only a beginning, and that posttranslational modifications (of which glycosylation is one of the most important) play a crucial role in determining protein function. And there are encouraging signs that the science establishment is finally sitting up and taking notice. The last few years have seen the establishment of initiatives such as the Consortium for Functional Glycomics (http://www.functionalglycomics.org) in the US, and increased funding for glycoscience databases and Internet resources in Europe [20]. Glycoscientists in the UK recently set up the multidisciplinary UK Glycoscience Forum to promote their science and encourage collaborations via an online forum, conferences and advanced training.

Glycoscience has come an enormously long way since those nineteenth-century studies of hemagglutinins: particularly, of course, in the last half-century, during Nathan Sharon's career "at the carbohydrate–protein interface." We, as glycoscientists of the postgenome era, owe a great debt to him, and we trust that this book will be a worthy celebration of his achievements.

REFERENCES

1. Sharon N (2006) *Biochim. Biophys. Acta* **1760**, 527–537.
2. Lapid K & Sharon N (2006) *Glycobiology* **16**, 39R–45R.
3. Sharon N (2006) *The Biochemist* **28**, 13–17.
4. Sharon N & Lis H (2004) *Glycobiology* **14**, 53R–61R.
5. Sheridan C (2005) *Nat. Rev. Drug Discov.* **4**, 447.

6. Morgan WT & Watkins WM (2000) *Glycoconj. J.* **17**, 501–530.
7. Boyd WC & Shapleigh E (1954) *Science* **119**, 419.
8. Katchalsky A & Sharon N (1953) *Biochim. Biophys. Acta* **10**, 290–301.
9. Sharon N (1957) *Nature* **179**, 919.
10. Sharon N & Jeanloz RW (1960) *J. Biol. Chem.* **235**, 1–5.
11. Sharon N & Lis H (1972) *Science* **177**, 949–959.
12. Sharon N (1975) *Complex Carbohydrates*. Addison Wesley, Reading, MA, USA.
13. Edelman GM, Cunningham BA, Reeke GN Jr. *et al.* (1972) *Proc. Natl Acad. Sci. USA* **69**, 2580–2584.
14. Hardman KD & Ainsworth CF (1972) *Biochemistry* **11**, 4910–4919.
15. Srinivasan N, Rufino SD, Pepys MB, Wood S & Blundell TL (1996) *Chemtracts Biochem. Mol. Biol.* **6**, 149–164.
16. Drickamer K (1988) *J. Biol. Chem.* **263**, 9557–9560.
17. Sharon N (2005) *Cell Mol. Life Sci.* **62**, 1057–1062.
18. Schmidt K (2002) *New Scientist* **176**, 34–37.
19. Merry AH & Merry CLR (2005) *EMBO Rep.* **6**, 1–4.
20. Lütteke T, Bohne-Lang A, Loss A, Goetz T, Frank M & von der Leith C-W (2006) *Glycobiology* **16**, 71R–81R.

CHAPTER 2

Glycobiology – discovery, diagnostics and drugs

Timothy M. Block, Anand S. Mehta, Pauline M. Rudd, Mark R. Wormald, Dennis R. Burton, Ian A. Wilson, Stefana M. Petrescu, Frances M. Platt, Nicole Zitzmann, Terry D. Butters and Raymond A. Dwek

Five different approaches involving glycosylation illustrate strategies for providing therapy in different disease targets [1]. In the case of hepatitis C, an iminosugar (see Chapter 25) that inhibits an ion channel in the virus represents the conventional "target knockout" strategy. By contrast, in hepatitis B an iminosugar is used to create a long-lived misfolded surface antigen of the virus itself, which is retained in the cell and prevents virus formation. In the third example, the structural details of the clusters of oligosaccharides which are recognized by a neutralizing human immunodeficiency virus type 1 (HIV-1) antibody provide a way forward to creating an immunogen. The next example involves strongly activating the cellular immune response to destroy melanoma cells by educating the system to recognize tyrosinase peptides when presented (i.e. tyrosinase is the target antigen for autologous T-cell responses in melanoma patients). The final example involves partially inhibiting, rather than ablating, a carbohydrate-processing enzyme involved in glucosphingolipid biosynthesis so as to prevent storage disorders such as Gaucher's disease.

There are estimated to be 200 million people chronically infected with hepatitis C virus (HCV). The envelope glycoproteins E1 and E2 interact to form a dimer, which is thought to be the functional complex present on the surface of mature virions. Glucosidase inhibitors, such as N-butyl-deoxynojirimicin (NB-DNJ) (Fig. 1a), have been shown to be antiviral against HCV. The alpha-glucosidases (I and II) mediate the first steps in the glycan-processing pathway and allow for downstream glycan-processing events to occur. In addition, the actions of alpha-glucosidases allow for interaction with the endoplasmic reticulum (ER) chaperones calnexin and calreticulin (Fig. 1b), which can facilitate the productive folding of glycoproteins by binding to the monoglucosylated proteins. Preventing the early stages of glycan processing bypasses this chaperone system and can lead to protein misfolding. The antiviral effect is caused by a reduction in viral secretion from an impairment of viral morphogenesis [2]. A second antiviral effect arises from iminosugars carrying long alkyl chains, such as N7-oxanonyl-6 deoxy-DGJ [3]. These drugs work by inhibiting p7, an ion channel. Currently we are testing both classes of iminosugars to determine the extent of "viral rebound" following the withdrawal of the combination of interferon and ribavirin, the current standard-of-care treatment for HCV.

Worldwide, more than 350 million people are chronically infected with hepatitis B virus (HBV). Glucosidase inhibitors have been shown to be antiviral against HBV in tissue culture [4] and in the woodchuck model of chronic HBV infection [5]. The M surface antigen glycoprotein of HBV folds via the calnexin pathway. Glucosidase inhibitors that prevent this interaction prevent the formation and secretion of HBV. The misfolded M surface antigen is retained within the cell, has a lifetime of several days, and may itself act as the "drug"

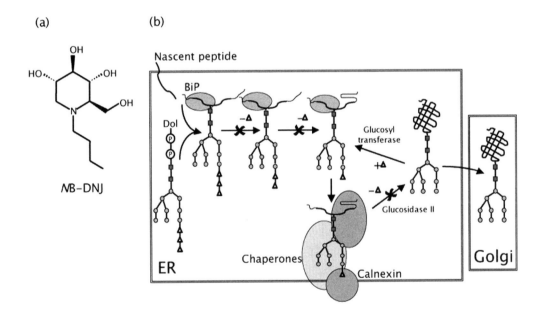

Figure 1. (a) Chemical structure of N-butyl-deoxynojirimicin (NB-DNJ). (b) The calnexin cycle: The Glc$_3$Man$_9$GlcNAc$_2$ oligosaccharide is transferred to the nascent peptide cotranslationally in the endoplasmic reticulum (ER) when interactions with glycan-independent chaperones such as BiP can occur. The glucose residues (triangles) are sequentially removed to give the Glc$_1$Man$_9$GlcNAc$_2$ species. This can bind to calnexin, recruiting the protein into the chaperone system. Glucosidase II removes the last glucose residue to release the peptide from calnexin, giving the protein a chance to fold. Incorrectly folded protein is reglucosylated by glucosyl transferase and rebinds to calnexin. Correctly folded protein exits to the Golgi. NB-DNJ blocks the removal of the glucose residues and so prevents interaction with calnexin.

which prevents new rounds of viral formation. As with HCV a second class of long-chain iminosugars which do not involve glycan processing are also potent antiviral agents and may act as an "oral interferon." Animal studies in woodchucks are underway as a prelude for clinical trials for the treatment of chronic HBV infection.

The humoral immune response to infection by HIV-1 is characterized by low levels of neutralizing antibodies, particularly those that have a broad specificity against many different isolates. One broadly neutralizing human monoclonal antibody is 2G12. This has a novel antibody structure [6] with three possible combining sites (Fig. 2) and recognizes a specific pattern of oligomannose glycans on the otherwise "immunologically silent" face of the HIV surface glycoprotein gp120 [7]. This recognition provides exciting challenges for immunogen design.

Activating the cellular immune response to destroy melanoma cells illustrates our approach to a tumor vaccine for a cancer that has an estimated risk of 1 in 70 people. The glycosylated enzyme tyrosinase, which is involved in melanin biosynthesis, is highly expressed in melanoma cells but only generates a weak cytotoxic T-lymphocyte (CTL) response. Antigen presentation is dependent on the processing of misfolded proteins so there is a direct relationship between the folding of tyrosinase and T-cell activation. Our strategy is to construct tyrosinase mutants that are retained in the ER [8] and retrotranslocated in the presence of chaperones for degradation and presentation, thus creating a strong CTL response.

The glucosphingolipid (GSL) storage diseases (see Chapter 23) are a family of progressive disorders in which GSL species are stored in the lysosome as a result of defects in the enzymes of the GSL-degradation pathway. Specific diseases include Gaucher, Tay-Sachs, Fabry, Sandhoff and GM1 gangliosidosis. GSL storage diseases

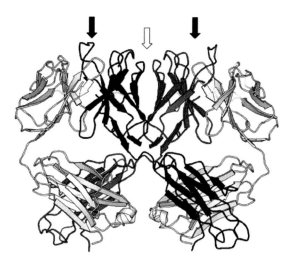

Figure 2. Crystal structure of the Fab dimer of 2G12 [6]. The structure is unique in that the V_H domains of the two Fabs are swapped, creating a rigid dimer. The two V_H/V_L binding sites (indicated by black arrows), each of which recognizes oligomannose glycans, are in a fixed geometry and a completely novel V_H/V_H groove (white arrow) is created which could act as a third antigen recognition site.

occur at a collective frequency of 1 in 18 000 live births and are one of the most common causes of neurodegenerative disease in infants and children. Our drug-based strategy for management of these diseases is to partially inhibit GSL synthesis using iminosugars. Slowing the rate of synthesis of GSLs will lead to fewer entering the lysosome for catabolism, reducing the rate of storage. This substrate reduction therapy (SRT) has led to an oral drug *N*B-DNJ (Zavesca) for Gaucher type-1 disease [9], which has been approved in the USA, Europe, and Israel. Since GSLs are abundantly expressed in the nervous system, the brain is frequently an organ affected by storage. Further experiments in animal models have shown that *N*B-DNJ crosses the blood/brain barrier and validate the SRT approach to treating CNS storage. A number of clinical trials are in progress, notably that for late onset Tay–Sachs, type 3 Gaucher, and Niemann–Pick type C.

REFERENCES

1. Dwek RA, Butters TD, Platt FM & Zitzmann N (2002) *Nat. Rev. Drug. Discov.* **1**, 65–75.
2. Zitzmann N, Mehta AS, Carrouee S *et al.* (1999) *Proc. Natl Acad. Sci. USA* **96**, 11878–11882.
3. Pavlovic D, Neville DC, Argaud O *et al.* (2003) *Proc. Natl Acad. Sci. USA* **100**, 6104–6108.
4. Block TM, Lu X, Platt FM *et al.* (1994) *Proc. Natl Acad. Sci. USA* **91**, 2235–2239.
5. Block TM, Lu X, Mehta AS *et al.* (1998) *Nat. Med.* **4**, 610–614.
6. Calarese DA, Scanlan CN, Zwick MB *et al.* (2003) *Science* **300**, 2065–2071.
7. Scanlan CN, Pantophlet R, Wormald MR *et al.* (2002) *J. Virol.* **76**, 7306–7321.
8. Popescu CI, Mares A, Zdrentu L, Zitzmann N, Dwek RA & Petrescu SM (2006) *J. Biol. Chem.* **281**, 21682–21689.
9. Cox T, Lachmann R, Hollak C *et al.* (2000) *Lancet* **355**, 1481–1485.
10. Patargias G, Zitzmann N, Dwek R & Fischer WB (2006) *J. Med. Chem.* **49**, 648–655.

SECTION 2
Synthesis of carbohydrate and glycoconjugates

CHAPTER 3
Sugar–lectin interactions: glycotargeting with synthetic glycoconjugates

Michel Monsigny, Éric Duverger, Oruganti Srinivas, Isabelle Fajac and Annie-Claude Roche

1. INTRODUCTION

In the early 1970s, while working on lectins [1], we presented our early results at the glycoconjugate meeting in Lille (France) in 1973, where we met Nathan Sharon for the first time. Since then, we have gone on to study the interactions between lectins and natural or synthetic glycoconjugates. We decided to develop neoglycoproteins [2] and later glycosylated cytochemical markers. The presence of both lectins and complex carbohydrates in biological fluids and on the surface of cells, as well as inside cells is now well documented (see [3] for a review). Numerous reviews and books dealing with the preparation and use of neoglycoproteins are available (see, for example, [4–8] and references therein). The recognition of an individual simple sugar by a lectin is usually in the low affinity (about 10^3L mol^{-1}) range. Conversely, complex oligosaccharides or glycoclusters, as well as neoglycoproteins, bind lectins in a high affinity range (up to 10^8L mol^{-1}). Multivalency is a strategy used by both binding partners to circumvent the intrinsic low affinity of carbohydrate–protein interactions. Multivalency leads to the possibility of establishing multiple separate connections, resulting in a strong attractive binding force.

Various methods are available for constructing multivalent structures. Some of them include conjugation of carbohydrate ligands with proteins or other polymers [6,7,9]: these glycoconjugates are called neoglycoproteins (glycosylated proteins) and glycopolymers, respectively. These approaches have been successful, although the products are ambiguous in composition and structure. Alternatively, high-affinity ligands of low molecular weight (complex oligosaccharides from natural or synthetic sources and small oligosaccharide clusters such as glycodendrimers or glycoclusters have been developed and can be used as tools to study the functions of endogenous lectins as well as devices to target molecules of interest (see [5–7] for reviews). Such oligosaccharides or glycoclusters may also be bound to a protein, conferring to the neoglycoprotein a very high apparent affinity.

2. SYNTHESIS OF NEOGLYCOPROTEINS AND GLYCOSYLATED POLYMERS

Neoglycoproteins have been prepared from various proteins, such as bovine and human serum albumin, ribonuclease, ferritin, diphtheria toxin, streptavidin, etc., as well as glycoproteins, such as ovalbumin, horse radish peroxidase, serum globulin, etc. Avery and Goebel [10], as early as 1929, prepared the first known neoglycoproteins by substituting horse serum globulin and crystalline egg albumin with diazonium derivatives of glucosides and galactosides. Those neoglycoproteins were used as antigens to induce the production of anti-sugar immune sera. Several years later, Iyer and Goldstein [11] prepared neoglycoproteins in order to study, in a quantitative approach, the interaction between glycoconjugates and concanavalin A, a plant lectin specific for mannosides and glucosides. Similarly, Privat et al. [2] prepared neoglycoproteins bearing chitin oligomers (GlcNAcβ-(4GlcNAcβ-)$_n$, where $n = 1$ to 3) and showed that those neoglycoproteins interact with another plant lectin, wheat germ agglutinin, inducing the precipitation of the complex in a concentration-dependent manner.

The synthesis of neoglycoproteins primarily involves random or defined coupling sites on the surface of the protein carrier and their covalent modification with oligosaccharides at their reducing end or functionalization of the oligosaccharides bearing a spacer arm. The coupling of saccharide residues to the surface of proteins has long been used as a straightforward strategy for the creation of high-valence neoglycoproteins. Serum albumins have often been used because such albumins (roughly M_r 67 000) are sugar-free and contain about 60 lysines dispersed throughout the molecule, allowing several tens of saccharides to be added per molecule. Various approaches have been described:

- A direct reductive amination: the conjugation of the reducing end of oligosaccharides with the ε-amino group of lysine, leading to the formation of a Schiff's base that is then reduced as a stable amine adduct [12] in the presence of sodium cyanoborohydride, or
- A formation of thiourea linkage: a large series of p-nitrophenyl glycosides have been synthesized from acetobromoglycosides and then converted to O-phenylisothiocyanate glycosides by using either thiophosgene or thiocarbonyldiimidazole. Alternatively, unprotected oligosaccharides, ending with a reducing sugar, were readily (within 6 h) converted to N-glycosyl-glutamyl-p-nitroanilide upon incubation of the oligosaccharide and α-glutamyl-p-nitroanilide in 1-methyl-2-pyrrolidone in the presence of imidazole [13]; the adduct was then converted within 30 min at room temperature into a stable N-glycosyl-pyroglutamyl derivative by the action of a carboxylate

activator. Oligosaccharyl pyroglutamyl-p-nitroanilides are easily converted to phenylisothiocyanate derivatives [13]. Such derivatives have been extensively used for conjugation of simple sugars or complex oligosaccharides onto lysines.
- A formation of amidine linkage: δ-alkylimidate glycosides (2-iminomethoxymethyl thioglycosides) were synthesized from acetobromoglycosides [14] and used to prepare neoglycoproteins upon reaction onto the lysyl side-chains of serum albumin. Such neoglycoproteins retain the number of positive charges of the lysines because the linkage is an amidine.
- A formation of amide linkage: other approaches to prepare neoglycoproteins via coupling to lysines involve condensation of oligosaccharides with spacer arms (carbohydrate or non-carbohydrate based) such as oligosaccharide-containing sugar lactones or synthetic oligosaccharides ending with a carboxylic group.

All these approaches are also suitable to prepare glycosylated polymers such as the partially glycosylated polylysines [15] as well as PEI (polyethylenimines) [16]. Such glycosylated polymers interact with DNA plasmids, leading to compact complexes called glycoplexes, based on ionic interactions between the basic glycopolymers and the acidic DNA; the glycoplexes are recognized by cell surface lectins and taken up inside the cell by an endocytosis process.

In order to develop more suitable tools for targeting purposes, several groups have developed small synthetic glycoclusters [17–20]. Small multivalent synthetic glycopeptides made of mannose residues covalently linked through a spacer arm to the α- and ε-amino groups of lysine, di- and trilysine [18] behave as competitive inhibitors (efficient in the micromolar concentration range) of rat alveolar macrophage uptake of mannosylated bovine serum albumin (BSA). Biessen and coworkers [17] synthesized a series of homologous lysine-based oligomannosides containing 2–6 terminal α-D-mannose groups linked to the α- and ε-amino groups of the linear oligolysine through a phenylthiocarbamyl bond. The affinity of these cluster mannosides for the mannose receptor continuously increased from 5×10^4 to $10^9 \, L \, mol^{-1}$ when the number of mannose units increased from two to six. With the aim of avoiding the use of a phenylthiocarbamyl bond which may lead to toxic derivatives *in vivo*, sugar clusters may be built from naturally occurring amino acids. We developed new glycoclusters [21,22] based on the substitution of oligolysines by glycosynthons (lactosyl-, Lewis oligosyl-, dimannosyl-pyroglutamyl derivatives). Using surface plasmon resonance [21], it was found that the apparent binding constant of clusters (containing five lactosyl units) binding *Ricinus communis* agglutinin was 2000 times larger than that of free lactose. A similar avidity effect was observed when Lewis oligosacchryl clusters were interacting with DC-SIGN, the dendritic cell-specific ICAM-3-grabbing non-integrin [22].

3. AVIDITY EFFECT

The success of neoglycoproteins, glycopolymers, and more recently of glycoclusters is linked to a phenomenon called "avidity effect" or "cluster effect."

Interactions between simple sugars and lectins are usually weak, in the 10^3 L mol^{-1} range; however, when simple sugars are gathered at the surface of a neoglycoprotein or in a glycocluster, the apparent binding constant is high, up to 10^8 L mol^{-1} [19,23]. For instance, the lectin from Solanum tuberosum agglutinates rabbit erythrocytes; this agglutination is inhibited by di-N-acetylchitobiose; a di-N-acetylchitobiose-substituted albumin was shown to be several hundred times more efficient inhibitor than the free disaccharide. An avidity effect was clearly demonstrated by Lee and coworkers [24,25] in the case of the human hepatic lectin. The affinity of a tri-antenna (or tetra-antenna derivative) ending with a galactosyl residue was found to be around 10^9 L mol^{-1}, which is 10^6 larger than the monovalent ligand. In the case of neoglycoproteins such as galactosylated BSA, the apparent affinity for the asialoglycoprotein receptor increased with the number of galactoses up to 25 residues per neoglycoprotein molecule.

4. ENDOGENOUS LECTINS AND DRUG TARGETING

Carrier-mediated delivery through sugar-specific recognition was first demonstrated [26] in 1971 and relies extensively on the presentation of clustered patches of carbohydrate moieties by membrane lectin receptors. Since then, carbohydrate-based macromolecular conjugates such as neoglycoconjugates, glycopolymers, polysaccharides, and liposomes have been widely used for drug delivery using glycotargeting approach [27,28].

The general aim in drug targeting is to increase the efficiency of a given drug by increasing the local drug concentration and/or decreasing its clearance rate. Many membrane lectins induce the internalization of their ligands and therefore, glycoconjugates specifically recognized by such lectins are suitable carriers of antiparasite, antiviral, and antitumor drugs.

The asialo-glycoprotein receptor, the first mammalian lectin known, was discovered by Ashwell and Morell and was then shown to internalize neoglycoproteins bearing galactosyl residues into hepatocytes [29]. Since then, many different membrane-bound receptors have been characterized in various cell types (see [5,30] for reviews). We found such an endocytosis process mediated by membrane lectins on Lewis lung carcinoma cells, on L1210 lymphoid cells, as well as on freshly isolated human blood monocytes and on immortalized normal and cystic fibrosis airway epithelial cells [31]. A mannosyl/fucosyl receptor, first evidenced on cultured primary macrophages, was also found on dendritic cells [32]. DC-SIGN, a membrane lectin recently identified on dendritic cells, was shown to mediate the internalization of mannosylated and fucosylated BSA as well as glycoclusters containing Lewis oligosaccharides [22].

Neoglycoproteins have been used successfully to carry and selectively deliver various types of drug to several types of cells. In order to be released inside a cell, the drug is associated through a linkage which is cleavable either in acidic medium as in the case of acido-labile heterobifunctional links, under reducing conditions as in the case of disulfide bridges or by endosome or lysosome hydrolase as in the case of peptidyl linkers.

4.1 Targeting antitumoral drugs

During the past 25 years, neoglycoproteins have been shown to be suitable to target toxic drugs, such as a pseudo-nucleoside Ara-A, gelonin, a plant toxin, methotrexate, or daunorubicine (see [27,33] for reviews). The chemotherapeutic index of Ara-AMP was improved when it was used as a lactosylated human serum albumin conjugate in the treatment of chronic hepatitis B [34].

4.2 Targeting macrophage immunomodulators

In vitro, muramyldipeptide (MDP) renders macrophages cytostatic and cytotoxic against tumorigenic target cells; its biological effect depends on internalization by pinocytosis; conversely, *in vivo*, free MDP does not induce any tumoricidal activity. In contrast, MDP linked to mannosylated serum albumin activate macrophage both *in vitro* and *in vivo*; in mice, upon systemic injections of this conjugate, established spontaneous metastases, originating from the primary tumor Lewis lung carcinoma cells, were shown to regress [35]. Similarly, the same conjugate was shown to be 50 times more efficient than free MDP in inhibiting the growth of *Leishmania donovani* inside peritoneal macrophages [36].

4.3 Targeting oligonucleotides

The use of antisense oligonucleotides as putative therapeutic agents is limited by their poor delivery into the cytosol and/or the nucleus because they are not able to efficiently cross lipid bilayers. Conversely, oligonucleotides, bound to either a neoglycoprotein or a glycosylated poly-L-lysine partially neutralized by gluconoylation, were shown to be more efficiently (up to 40 times) taken up by cells expressing a membrane lectin able to bind the glycoconjugate [37]. In the case of J774E cells (a macrophage cell line) both the uptake and the biological efficiency of an oligonucleotide specific for the sequence around the codon 12 of oncogene *ras* were enhanced when the oligonucleotide was linked to a neoglycoprotein containing mannose-6-phosphate [38].

On the basis of the "glycoside cluster effect", smaller and well-defined glycoclusters have been utilized instead of neoglycoproteins or glycopolymers [39] for targeting antisense oligonucleotides as selective gene expression inhibitors. The use of chemically and structurally homogeneous glycocluster–oligonucleotide conjugates, prepared by covalent attachment of oligonucleotides to multivalent synthetic glycoside clusters, hold great promise for significantly improving the cellular uptake and therefore the therapeutic efficacy of antisense oligonucleotides *in vivo* [40,41]. Multivalent carbohydrate-recognition motifs for the asialoglycoprotein receptor has been designed for tissue- and cell-specific delivery of antisense oligonucleotides to parenchymal liver cells [40,42].

5. GENE THERAPY

Gene therapy (i.e. the expression in cells of genetic material with therapeutic activity) can putatively be applied to the treatment of various human diseases. A delivery vehicle, a "vector" of either viral or non-viral origin, must be used to carry the foreign gene into a cell. While viral systems are by far the most effective means of DNA delivery, their use is hampered by their toxicity and immunogenicity. Non-viral systems are still less efficient than viral vectors but their low toxicity is highly attractive. These non-viral systems are synthetic molecules which should be able to compact the foreign DNA, protect it from degradation and promote its specific cellular uptake and hopefully efficient intracellular trafficking. Because all these functions cannot be performed with a single carrier molecule, synthetic multicomponents are being designed with a function assigned to each component. Thus bifunctional conjugates were constructed more than a decade ago, in which a cationic polymer allowed DNA binding and condensation and a ligand domain allowed receptor-mediated delivery of the nucleic acid. The cationic polymer was a poly-L-lysine able to form a complex with negatively charged DNA molecules. It was substituted with transferrin, a protein, in order to target the corresponding receptors expressed at the cell surface [43]. The drawbacks with this strategy of using a biologically active protein are that both the preparation and the purification of such (protein–polylysine) conjugates are difficult, their solubility may be very low and they may induce immune responses. One way of solving these problems may be to use cationic polymers substituted with small ligands such as sugar moieties (see [27,44] for reviews).

We developed a polylysine bearing lactosyl residues (glycofectin) and showed that plasmid DNA/lactosylated polylysine complexes (glycoplexes) efficiently transfected hepatocarcinoma (HepG2) cells and hepatocytes in primary culture, both of which express galactose-specific membrane lectins. Similarly, we have shown that plasmid DNA complexed with polylysine bearing mannosyl residues efficiently transfected human macrophages expressing a mannose/fucose-specific membrane lectin. We have used various glycosylated polylysine-based glycopolymers as vectors to study cystic fibrosis (CF) gene therapy. Some sugars enhanced the vector capability of polylysine and yielded high expression of the reporter gene luciferase in immortalized airway surface epithelial cells, airway gland serous cells, and differentiated airway epithelial cells in primary culture (see [45] for a review). Moreover, α-glucosylated polylysine allowed efficient *CFTR* gene transfer and led to the expression of a normal CFTR protein on the membrane of cystic fibrosis cells [46].

Polyethylenimine (PEI) is a cationic polymer with an ability to buffer the acidity of endosomes. An osmotic swelling occurs due to water entry, bursts the vacuole and releases the complex into the cytosol [47]. Lactosylated PEI was showed to efficiently transfect hepatocyte-derived cell lines [48] and to be nearly 10 times more efficient than lactosylated polylysine in transfecting airway epithelial cell lines [16].

The sugar moieties in glycoplex play a major role in both cell uptake and glycoplex intracellular trafficking (see [49] for a review). Hence, a high uptake of glycosylated complexes was not always related to an efficient gene expression. In the case of transfection of rabbit vascular smooth muscle cells, the uptake of glycoplexes prepared with polylysine substituted with lactose residues was maximal, whereas it was very poor with polylysine substituted with β-GalNAc or α-Gal residues. In contrast, complexes made with lactosylated polylysine were quite inefficient for gene transfer, whereas complexes made with polylysine bearing β-GalNAc or α-Gal residues led to a high luciferase gene expression. In human CF and non-CF airway epithelial cell lines, the cells efficiently bound and internalized neoglycoproteins as well as glycoplexes containing α-mannose residues but poorly took up neoglycoproteins and glycoplexes containing lactose. However, glycoplexes made with mannosylated polylysine were dramatically inefficient, whereas complexes made with lactosylated polylysine were very efficient [46]; furthermore, we showed that the main limiting steps for DNA transfer using mannosylated polylysine as compared with lactosylated polylysine, were a delayed exit from endosomes, high accumulation in lysosomes, and limited transcription of the complexed plasmid DNA. Similarly, lactosylated PEI was far more efficient for gene transfer than unsubstituted PEI in airway epithelial cells. This greater gene transfer efficiency observed with lactosylated complexes was attributed to a higher amount of lactosylated complexes incorporated by airway epithelial cells and a lower cytotoxicity that might be related to reduced endosomolytic properties and the size of the complexes [50].

Glycosylation of polycations has proved to be an important breakthrough in the study of gene transfer: it allows a massive entry of complexes into the cell of interest, a favorable intracellular trafficking providing the appropriate sugar moiety is used, and usually its toxicity is much lower than that observed with the unsubstituted polymer. However, one step that is still a barrier to an efficient gene transfer is the entry into the nucleus. Given the sizes of both the nuclear pore (lower than 25 nm) and that of complexes (70–500 nm), a poor crossing of the nuclear membrane by complexes is expected. The breakdown of the nuclear membrane during mitosis allows, in the case of dividing cells, the cytoplasmic complexes to access the nuclear compartment. In order to target therapeutic genes to the nucleus of non-dividing cells, bioconjugates with nuclear localization sequences (NLS) have been developed with little increase in gene expression so far [51]. Lectins have been documented in both the cytosol and the nucleus and glycosylated proteins lacking NLS were shown to enter the nucleus with the help of their oligosaccharides [52,53]. The sugar-dependent nuclear import is neither inhibited by an excess of NLS nor by known inhibitors of RanGTPase, demonstrating the existence of a new nuclear pathway independent of the classical peptide NLS nuclear import. A nuclear translocation of plasmid DNA/lactosylated polylysine complexes was reported that was not cell cycle dependent. It was attributed to the presence of the lactose residues on the polylysine [54]. However, such a nuclear translocation was not observed with lactosylated PEI [55].

Mannose, conjugated to a bifunctional spacer arm, was introduced on guanine residues of a plasmid by a diazotizing reaction (similar compounds containing lactose instead of mannose was used as a negative control). Mannosylated plasmid was shown to enter the nucleus in digitonin-permeabilized cells. The transfection efficiency of glycosylated plasmids was assayed upon lipofectin transfection of either HeLa cells or primary culture hepatocytes and was found to be enhanced up to 250 times with regards to sugar-free plasmid modified by diazotizing under similar conditions (unpublished results, Rondanino et al.). However, diazo-coupling of plasmid partially impaired gene expression. According to these results, more sophisticated devices allowing more efficient nuclear import and optimized gene expression must be developed.

The presence of lectins at the cell surface, in the cytosol, and in the nucleus of cells has been exploited with great success to enhance gene delivery. Sugar moieties substituting various polymers (a) increase cell uptake and selectivity, (b) are involved in intracellular trafficking in the cytoplasm, and (c) are putatively involved in nuclear import. Improvements in selecting more efficient ligands or in mildly glycosylating the plasmid may lead, in the future, to efficient and non-toxic gene delivery systems.

6. CONCLUSION

Neoglycoconjugates are not only useful for the basic understanding of protein–carbohydrate interactions but they have also many other practical applications. Obviously, some cautions should be mentioned with regards to therapeutic purposes. Due to their antigenic properties they are not suitable when their antigenicity is not the purpose of the therapy. With regard to bovine spongiform encephalopathy and related central nervous system diseases, neoglycoproteins made with animal proteins must be avoided; however, recombinant proteins do not suffer such a restriction. Similarly, the protein can be replaced by polymers such as polylysine or polyethyleneimine; glycosylated polymers have been used to obtain safe glycoconjugates usable as drug carriers as well as to target plasmids.

In conclusion, glycoconjugates (glycosylated polymers and glycoclusters) appear to be excellent tools for several major reasons: they are easily prepared from commercially available material, not only by using monosaccharides derivatives but also by using functionalized complex oligosaccharides of natural origin or glycoclusters obtained by synthesis, they are quite flexible (they can be made fluorescent, biotinylated, adsorbed on gold particles), they are highly soluble in usual buffers including physiological serum, they are quite stable (usually they can be freeze dried and quantitatively solubilized), they are powerful reagents suitable for many cell biology studies, they are excellent tools for targeting of drugs and delivering plasmids.

Acknowledgments

This work was partly supported by grants from the French foundations Agence Nationale de la Recherche sur le SIDA (AIDS), Cancéropole Grand Ouest, Association pour la Recherche sur le Cancer, Ligue Nationale contre le Cancer, Vaincre la Mucoviscidose, Fondation Recherche Médicale. M.M. is Emeritus

Professor at the University of Orléans, E.D. is Assistant Professor at the University of Orléans, O.S. is a post-doctoral fellow supported by Ligue Nationale contre le Cancer, I.F. is Assistant Professor at the University Paris-Descartes/Cochin Hospital, A.C.R. is Research Director at Inserm.

REFERENCES

1. Goldstein I, Hughes C, Monsigny M, Osawa T & Sharon N (1980) *Nature* **285**, 66.
2. Privat JP, Delmotte F & Monsigny M (1974) *FEBS Lett.* **46**, 224–228.
3. Lis H & Sharon N (1998) *Chem. Rev.* **98**, 637–674.
4. Schrével J, Gros D & Monsigny M (1981) *Prog. Histochem. Cytochem.* **14**, 1–269.
5. Monsigny M, Roche AC, Kieda C, Midoux P & Obrénovitch A (1988) *Biochimie* **70**, 1633–1649.
6. Lee YC & Lee RT (1994) *Neoglycoconjugates, Part A, Synthesis. Methods Enzymol.*, Vol. 242. Academic Press, San Diego.
7. Lee YC & Lee RT (1994) *Neoglycoconjugates, Part B, Biomedical Applications. Methods Enzymol.*, Vol. 247. Academic Press, San Diego.
8. Gabius HJ & Gabius S (2002) *Glycosciences: Status and Perspectives.* Chapman & Hall, London.
9. Monsigny M, Roche AC, Kieda C, Midoux P & Obrénovitch A (1988) *Biochimie* **70**, 1633–1649.
10. Avery OT & Goebel WF (1929) *J. Exp. Med.* **50**, 533–550.
11. Iyer RN & Goldstein IJ (1973) *Immunochemistry* **10**, 313–322.
12. Gray GR (1974) *Arch. Biochem. Biophys.* **163**, 426–428.
13. Quétard C, Bourgerie S, Normand-Sdiqui N *et al.* (1998) *Bioconjug. Chem.* **9**, 268–276.
14. Lee YC, Stowell CP & Krantz MJ (1976) *Biochemistry* **15**, 3956–3963.
15. Midoux P, Mendes C, Legrand A *et al.* (1993) *Nucleic Acids Res.* **21**, 871–878.
16. Fajac I, Thévenot G, Bédouet L *et al.* (2003) *J. Gene Med.* **5**, 38–48.
17. Biessen EA, Noorman F, van Teijlingen ME *et al.* (1996) *J. Biol. Chem.* **271**, 28024–28030.
18. Robbins JC, Lam MH, Tripp CS, Bugianesi RL, Ponpipom MM & Shen TY (1981) *Proc. Natl Acad. Sci. USA* **78**, 7294–7298.
19. Monsigny M, Mayer R & Roche AC (2000) *Carbohydr. Lett.* **4**, 35–52.
20. Roy R (1996) *Polymer News* **21**, 226–232.
21. Frison N, Marceau P, Roche AC, Monsigny M & Mayer R (2002) *Biochem. J.* **368**, 111–119.
22. Frison N, Taylor ME, Soilleux E *et al.* (2003) *J. Biol. Chem.*, **278**, 23922–23929.
23. Lee RT & Lee YC (2000) *Glycoconj. J.* **17**, 543–551.
24. Lee YC, Townsend RR, Hardy MR et al. (1983) *J. Biol. Chem.* **258**, 199–202.
25. Lee RT, Lin P & Lee YC (1984) *Biochemistry* **23**, 4255–4261.
26. Rogers JC & Kornfeld S (1971) *Biochem. Biophys. Res. Commun.* **45**, 622–629.
27. Monsigny M, Roche AC, Midoux P & Mayer R (1994) *Adv. Drug Delivery Rev.* **14**, 1–24.
28. Wadhwa MS & Rice KG (1995) *J. Drug Targeting* **3**, 111–127.
29. Connolly DT, Townsend RR, Kawaguchi K, Hobish MK, Bell WR & Lee YC (1983) *Biochem. J.* **214**, 421–431.
30. Sharon N & Lis H (1989) *Science* **246**, 227–234.
31. Fajac I, Briand P, Monsigny M & Midoux P (1999) *Hum. Gene Ther.* **10**, 395–406.
32. Avraméas A, McIlroy D, Hosmalin A *et al.* (1996) *Eur. J. Immunol.* **26**, 394–400.
33. Fiume L, Busi C & Mattioli A (1983) *FEBS Lett.* **153**, 6–10.
34. Fiume L, Busi C, Di Stefano G *et al.* (1995) *Ital. J. Gastroenterol.* **27**, 189–192.
35. Roche AC, Bailly P & Monsigny M (1985) *Invasion Metastasis* **5**, 218–232.
36. Sarkar K & Das PK (1997) *J. Immunol.* **158**, 5357–5365.
37. Bonfils E, Mendes C, Roche AC, Monsigny M & Midoux P (1992) *Bioconjug. Chem.* **3**, 277–284.
38. Sdiqui N, Arar K, Midoux P, Mayer R, Monsigny M & Roche A (1995) *Drug Delivery* **2**, 63–72.

39. Stewart AJ, Pichon C, Meunier L, Midoux P, Monsigny M & Roche AC (1996) *Mol. Pharmacol.* **50**, 1487–1494.
40. Maier MA, Yannopoulos CG, Mohamed N *et al.* (2003) *Bioconjug. Chem.* **14**, 18–29.
41. Hangeland JJ, Levis JT, Lee YC & Ts'o PO (1995) *Bioconjug. Chem.* **6**, 695–701.
42. Biessen EAL, Vietsch H, Rump ET *et al.* (1999) *Biochem. J.* **340**, 783–792.
43. Curiel DT, Agarwal S, Wagner E & Cotten M (1991) *Proc. Natl Acad. Sci. USA* **88**, 8850–8854.
44. Roche AC, Fajac I, Grosse S *et al.* (2003) *Cell Mol. Life Sci.* **60**, 288–297.
45. Fajac I, Grosse S, Roche AC & Monsigny M (2005) *Curr. Pediatr. Rev,* in press.
46. Allo JC, Midoux P, Merten M *et al.* (2000) *Respir. Cell Mol. Biol.* **22**, 166–175.
47. Sonawane ND, Szoka FC, Jr. & Verkman AS (2003) *J. Biol. Chem.* **278**, 44826–44831.
48. Zanta MA, Boussif O, Adib A & Behr JP (1997) *Bioconjug. Chem.* **8**, 839–844.
49. Monsigny M, Rondanino C, Duverger E, Fajac I & Roche AC (2004) *Biochim. Biophys. Acta* **1673**, 94–103.
50. Grosse S, Aron Y, Honore I *et al.* (2004) *J. Gene Med.* **6**, 345–356.
51. Cartier R & Reszka R (2002) *Gene Ther.* **9**, 157–167.
52. Duverger E, Carpentier V, Roche AC & Monsigny M (1993) *Exp. Cell Res.* **207**, 197–201.
53. Rondanino C, Bousser MT, Monsigny M & Roche AC (2003) *Glycobiology* **13**, 509–519.
54. Klink DT, Chao S, Glick MC & Scanlin TF (2001) *Mol. Ther.* **3**, 831–841.
55. Grosse S, Thévenot G, Monsigny M & Fajac I (2006) *J. Gene Med.* revised.

CHAPTER 4

Recent developments in the synthesis and application of sialylmimetics

Heike Busse and Hansjörg Streicher

1. INTRODUCTION

Sialic acids, a family of acidic nine-carbon monosaccharides (Fig. 1), are important and abundant components within the oligosaccharide parts of complex glycoconjugates in most living organisms (see Chapter 12) [1]. More than 50 members have been found to date, with the most prominent one being N-acetylneuraminic acid (5-acetamido-3.5-dideoxy-D-*glycero*-D-*galacto*-nonulosonic acid, Neu5Ac). Being located at exposed, typically terminal, positions on cell surfaces, they are thus key to a variety of physiological recognition events, reaching from endogenous cell–cell recognition to receptor functions for exogenous, often pathogenic, agents such as viruses, bacteria, and parasites [2].

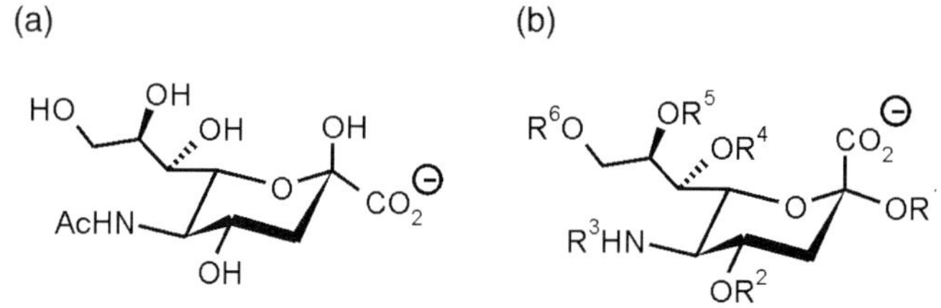

Figure 1. (a) *N*-Acetylneuraminic acid (NANA). (b) Naturally occurring sialic acids (R^1 = aglycon, R^2 = H, $COCH_3$, R^3 = $COCH_3$, $COCH_2OH$, $COCH_2COCH_3$, R^4 = H, $COCH_3$, R^5 = H, $COCH_3$, CH_3, SO_3H, R^6 = H, $COCH_3$, $COCHOHCH_3$, PO_3H_2).

Consequently, a number of proteins (lectins) involved in sialic acid-based recognition and the sialic acid-processing enzymes responsible for their turnover have become targets for low molecular weight inhibitor design. The most common approach so far has been the design and synthesis of compounds (either structure- or mechanism-based) that retain some structural features of a

Glycobiology (C. Sansom and O. Markman, eds.)
© Scion Publishing Limited, 2007

respective sialic acid known to be important for recognition and thus contribute to binding affinity towards the protein. At the same time, such mimics of sialic acid, or sialylmimetics, should remain chemically stable with respect to the given enzyme, test system, or physiological setting.

2. MAJOR APPLICATIONS OF SIALYLMIMETICS

There have been several examples of particularly intense efforts to find effective sialylmimetics in the last two or three decades.

First, (exo-)sialidases (neuraminidases) have been of interest. These cleave terminal sialic acids from cell surface glycoconjugates (Fig. 2), especially those from pathogenic microorganisms such as bacteria and viruses, where sialidase action has been shown or suspected to be essential at some stage of the infection or replication process. Influenza viruses require sialidase activity to allow viral progeny to be released from the infected host cell surface to ensure efficient viral spread. This is provided by the influenza virus' sialidase, which removes sialic acid receptors responsible for the initial viral hemagglutinin-mediated binding of the virus to the host cell. Hence the name "receptor-destroying enzyme." Based on the long-known transition-state analogous sialidase inhibitor DANA (2-deoxy-2.3-didehydro-N-acetylneuraminic acid, Fig. 2), highly potent inhibitors have been developed, two of which – the DANA-derivative zanamivir and the carbocycle oseltamivir (Fig. 2) – are now available as anti-influenza drugs [3,4]. The persistent influenza threat, especially in the light of avian flu, drives continuous efforts to find a second, improved generation of this type of antiviral.

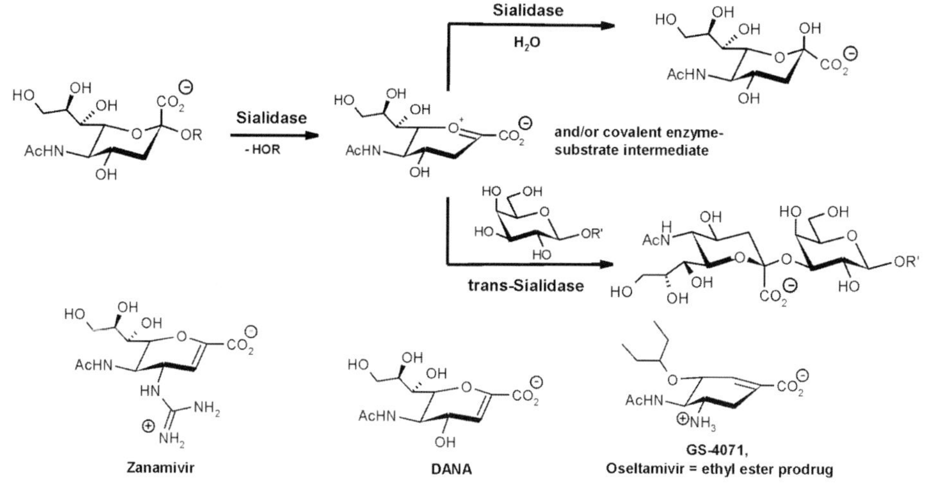

Figure 2. The sialidase and trans-sialidase reaction, respectively, classical sialidase inhibitor DANA and

More recently, the elucidation of the structures and functions of paramyxoviral hemagglutinin-neuraminidases has triggered hopes that anti-parainfluenza drugs (e.g. for the treatment of parainfluenza II-related respiratory disease in children) can be developed based on such inhibitors [5–7]. There is little doubt that research progress in the genetics and biochemistry of viral sialidases, not to mention other enzymes with a receptor-destroying function, will produce further demand for efficient sialylmimetics.

Attempts to target bacterial sialidases have thus far been less successful, in part because it has been difficult to find inhibitors, based on DANA or not, able to inhibit in the nanomolar range like the influenza drugs. In many cases, however, bacteria remain virulent in the absence of their sialidase, thus making the enzyme less attractive for inhibitor design. Prevention of a deleterious interplay between viruses and bacteria, if sialic acid based, such as in the case of secondary infection by *Streptococcus pneumoniae* (as a follow-up to extensive desialylation through parainfluenza or influenza virus sialidase), might very well emerge as a new interesting aspect of sialidase inhibition [8,9].

Trans-sialidases, which can be described as sialidases that transfer the sialyl moiety to an acceptor other than water (Fig. 2), have been shown to be essential for the survival of certain trypanosomes in their mammalian host and are thus interesting targets [10–12]. Finding lead structures has so far proven to be a challenge as DANA is ineffective.

A second area where sialylmimetics have been extensively applied is the inhibition of sialyltransferases, which transfer a sialyl moiety from the donor substrate (e.g. CMP-Neu5Ac) to acceptor sugars such as galactose in the final step of glycan biosynthesis (Fig. 3) [13,14].

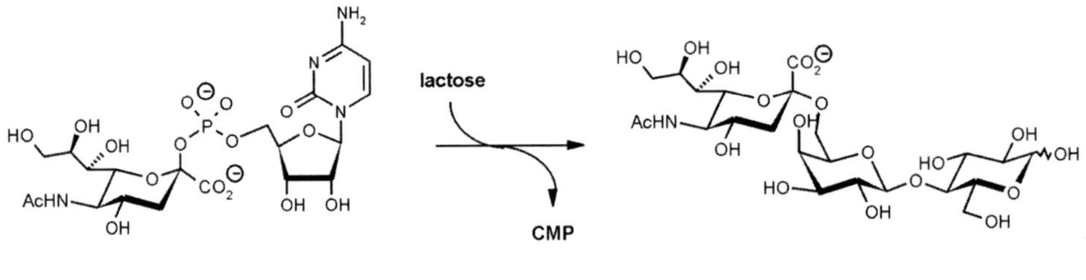

Figure 3. Transfer of a sialyl moiety from the natural sialyl donor CMP-Neu5Ac to the 6′-position of lactose by an α(2,6)-sialyltransferase.

This interest has mainly been due to the increased sialylation observed in various cancers, conferring a protective group function and enhanced mobility to migrating cells during metastasis. Donor analogous inhibition in the nanomolar range has been achieved (Fig. 3) and research is now focusing both on improved inhibition and on the introduction of less polar sialylmimetics to render the inhibitors more membrane-permeable.

Sialic acid-binding lectins of physiological significance are the third area of application for sialylmimetics. For instance, recruitment of leukocytes circulating in blood vessels to sites of inflammation is initiated by binding of sialyl-Lewis^x tetrasaccharides to selectins on the endothelial cell surface and vice versa, thus initiating adherence and, finally, extravasation of the leukocytes. In the most potent sialyl-Lewis^x mimetics developed to block this interaction, the sialic acid part could be reduced to a negative charge with an unpolar substituent, while other structural features of the tetrasaccharide (e.g. the L-fucose) had to be simulated more closely.

Another family of sialic acid-binding lectins are the siglecs, a family of mammalian sialic acid-binding proteins, members of which are involved in neurite growth and regulation of the immune system. The majority of studies so far involving sialylmimetics has aimed at exploring siglec specificity, but potent inhibitors could provide interesting insight into their function as well.

In this minireview, we aim to highlight some of the developments in the synthesis and application of sialylmimetics that have emerged during the past 2–3 years. For more comprehensive treatises, the reader is referred to recent reviews [13–17].

3. INHIBITION OF VIRAL SIALIDASES

The influenza virus neuraminidase remains the paradigm in sialidase inhibition, and much ongoing work is dedicated to the search for a second, more effective generation of anti-influenza drugs to replace the sialidase inhibitors Tamiflu (the carbocycle oseltamivir) and Relenza (the DANA derivative zanamivir) (Fig. 2) [17]. The current trend is towards 5-membered ring scaffolds like pyrrolidines and cyclopentanes but other structures are still pursued as well.

Quite recently, paramyxoviral sialidases, for instance the hemagglutinin-neuraminidases from human parainfluenza virus II, which causes respiratory infections in infants, or Newcastle disease virus have emerged as interesting applications for sialylmimetics as inhibitors. In contrast to the influenza virus, where the lectin (hemagglutinin) activity and the sialidase activity are provided by two distinct proteins on the cell surface, both activities reside in the same protein in paramyxoviruses. There is still dispute on whether there are one bifunctional or two distinct binding sites in the hemagglutinin-neuraminidases but in any case there will be room for intelligent inhibitor design [5–7].

3.1 Aromatic systems

The benzoic acid BCX-140 (Fig. 4) was reported almost a decade ago to inhibit the sialidase influenza virus A and B with an IC$_{50}$ of 2.5 µmol L^{-1}, which is an interesting value for a simple trisubstituted benzene as sialylmimetic. This value could not, however, be improved with aromatic compounds ever since and recent studies indicate that a plateau has been reached here, with higher substitution being counterproductive and heteroatom substitution to afford pyridines such as those in Fig. 4 (4 µmol L^{-1}) not improving efficiency [18,19]. Furan-based compounds have recently been suggested and synthesized but inhibition data have yet to come [20].

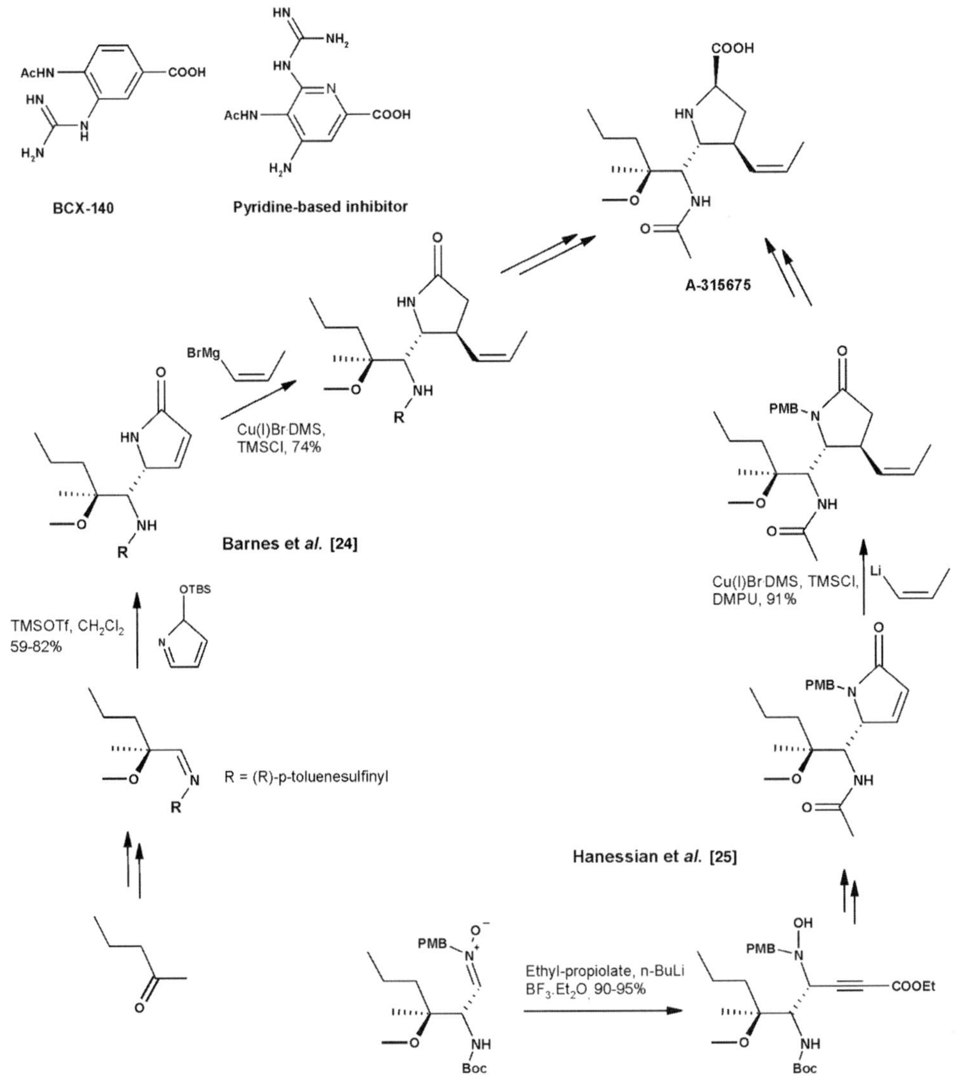

Figure 4. Syntheses of pyrrolidine-based sialidase inhibitor A-315675.

3.2 Five-membered rings

Abbott laboratories identified and optimized pyrrolidine-based inhibitors of influenza neuraminidase by a combination of structure-based and combinatorial techniques, eventually leading to the subnanomolar (IC_{50}) inhibitor A-315675, containing a Z-propenyl group and its carboxymethyl analog (Fig. 4) [21].

In both compounds, the largely electrostatic interactions of the amino- and guanidino- groups in oseltamivir and zanamivir, respectively, with the enzyme are replaced by novel hydrophobic interactions. This effect, which could lead to an inhibitor with better oral availability, was then systematically investigated by variation of the new hydrophobic substituent [22]. The fact that the initial synthesis of A-315675 was neither high-yielding nor diastereoselective prompted several interesting synthetic approaches. Researchers from the same laboratories introduced the preassembled ether side-chain into silyloxypyrrole by Mannich condensation followed by a Michael addition of propenyl cuprate as key steps [23]. Further improvement of diastereoselectivity then led to a synthesis in ~13% overall yield for the kilogram scale (Fig. 4) [24]. Hanessian and coworkers synthesized A-315675 in comparable overall yield from D-serine in a highly stereocontrolled fashion. Key steps are the addition of methylmagnesium bromide to a serine-derived ketone, the subsequent diastereoselective addition of an acetylide, and finally cyanide introduction [25].

A recent study on tetrahydrofuran-carboxylic acids as replacement for the respective pyrrolidine systems has revealed lower inhibitor potency than their pyrollidine counterparts [26].

Cyclopentane derivatives underwent a similar development, mainly pursued by BioCryst Pharmaceuticals, who developed BCX-1812 by structure-based drug design which is effective against many strains of influenza (Fig. 5). It reached phase III clinical trials but failed to show statistical efficacy, with low

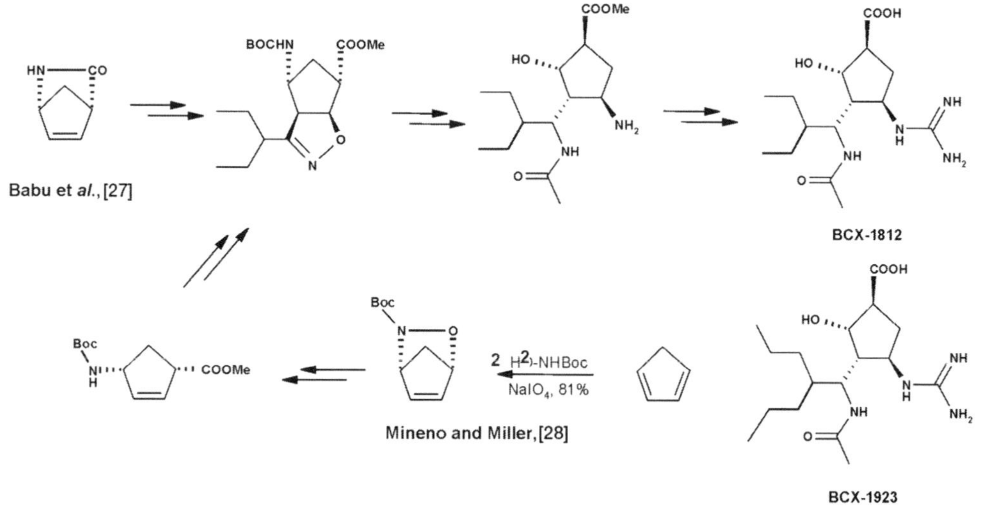

Figure 5. Syntheses and structures of potent cyclopentane-based influenza sialidase inhibitors.

bioavailability being a suspected cause. The original synthesis [27] started from commercially available chiral lactam (Fig. 5), which was converted into isoxazoline by 3+2 cycloaddition with 2-ethylbutyronitrile. Reductive opening and acetylation followed by BOC-cleavage gave the amine, which was converted into the guanidine and finally the ester was saponified to furnish BCX-1812. Mineno and Miller [28] developed a racemic approach to the methyl ester derived from the very same lactam which then essentially meets the original synthesis but allows introduction of diverse substituents for parallel or combinatorial approaches to improved inhibitors.

Recently, side-chain optimization based on BCX-1812 has led to inhibitors like BCX-1923 surpassing oseltamivir and zanamivir *in vivo* when alternate methods of administration such as the intranasal route are chosen [29,30]. Given that the compounds still contain carboxylates, the speculation is justified that the respective esters or other prodrugs have potential to become drugs.

3.3 DANA-based

Zanamivir, the most active DANA-derived sialidase inhibitor and the basis of Relenza, has recently been taken one step further when di- and oligomeric zanamivir conjugates were synthesized [31–34]. In all cases, the 7-OH was chosen as the anchor point of the tether as the X-ray structure indicates that this hydroxyl group does not interact with the enzyme and hexanediaminecarbamate linkers were employed to link zanamivir to tri- and tetravalent aromatic and aliphatic scaffolds (Fig. 6). Long-lasting inhibition of infectivity was achieved when administered intranasally, even though the linkers were too short to effectively span two binding sites within the same enzyme

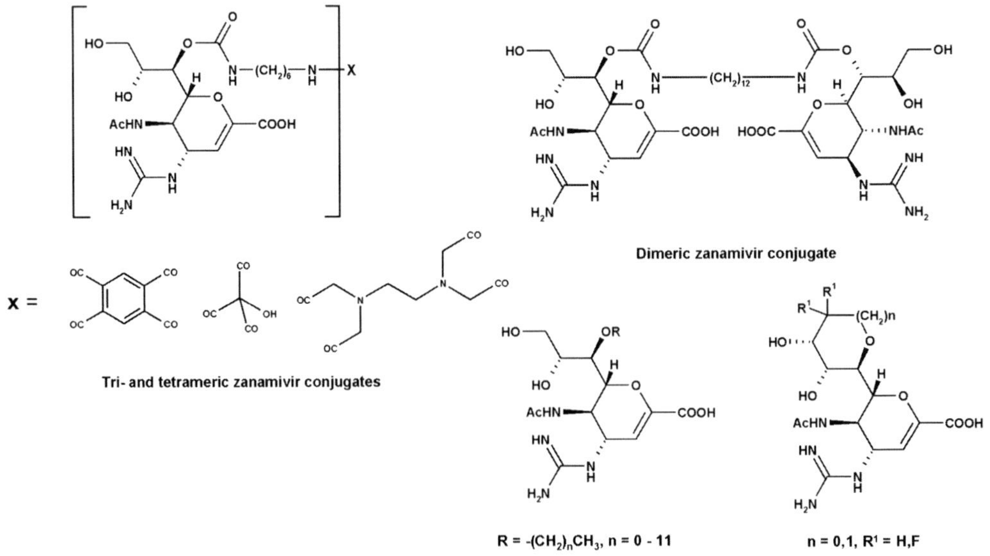

Figure 6. Novel DANA-derived influenza sialidase inhibitors.

tetramer. A subsequent study investigating linkers revealed that effectivity *in vivo* can be enhanced by an appropriate length even when the enzyme inhibition itself is not improved [31].

McKimm-Breschkin et al. [35] applied zanamivir–biotin conjugates in an ELISA-type assay to capture influenza virions (Fig. 6) with the distinct advantage over antibody-based assays that strain-independent detection can be expected. Visualization by transmission-electron microscopy of the virion–zanamivir interaction was achieved by the same researchers by applying zanamivir-coated microspheres. For both cases, specific binding was demonstrated.

In an attempt to overcome the inferior oral efficacy of zanamivir as compared to oseltamivir, Honda and colleagues introduced alkyl chains at C-7 and cyclic mimetics of the glycerol side-chain into zanamivir via ring-closing metathesis and significant improvements were reported (Fig. 6) [34,35].

3.4 Oseltamivir-derived compounds

The novel hydrophobic interactions detected of pyrrolidine A-315675 with the enzyme prompted a study to test this shift of paradigm with oseltamivir [36]. Replacement of the amino group by hydrophobic substituents resulted in a K_i of 45 nmol L^{-1} in the case of a vinyl group while retaining the stereochemistry of the carbocycle, but failed to reach the potency of oseltamivir (Fig. 7).

Another approach towards oseltamivir analogs which continues to be applied is the generation and evaluation of dynamic combinatorial libraries. Amine derivatives of oseltamivir are condensed with ketone libraries and the resulting

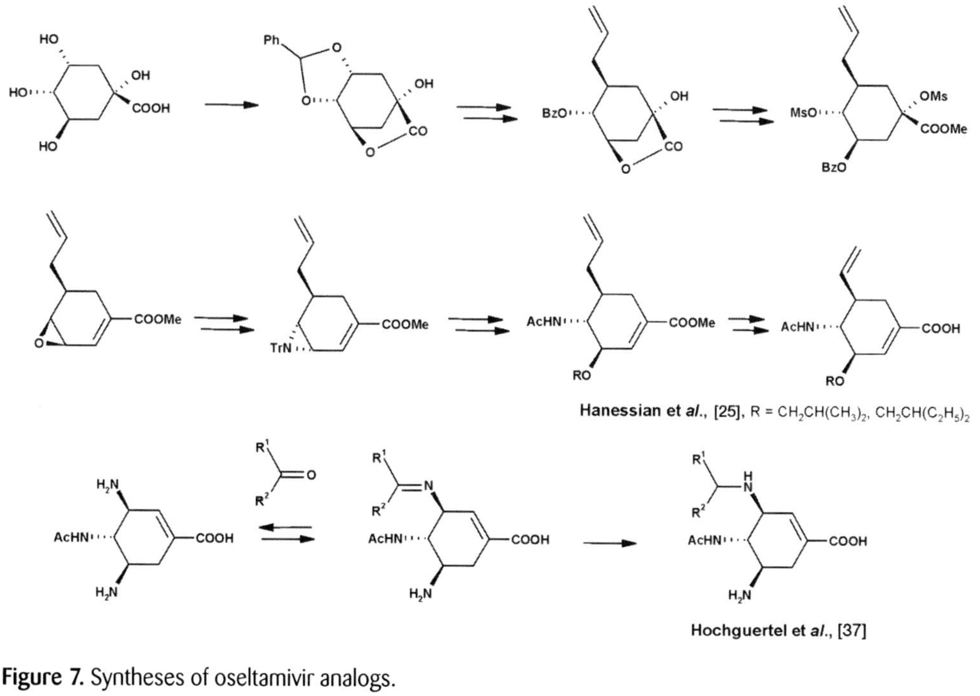

Figure 7. Syntheses of oseltamivir analogs.

most active imines are trapped as amines by reduction. No compound surpassing oseltamivir could be found, but interestingly a 3-pentyl amine resembling the drug's 3-pentyl oxy substituent was found to be most active (Fig. 7) [37].

4. OTHER SIALYLMIMETICS AS INHIBITORS OF VIRAL AND BACTERIAL SIALDASES AND PARASITIC TRANS-SIALIDASES

Several research groups have exploited the stereochemical relationship between DANA and N-acetyl-D-glucosamine (D-GlcNAc) for the synthesis of sialyl-transferase inhibitors [38] and inhibitors tested against *Vibrio cholerae* sialidase [39–41] and the influenza sialidase [42]. In β-glycosides of GlcNAc, the absolute configuration at C-1, C-2, and C-3 corresponds to the one at C-6, C-5, and C-4 in Neu5Ac. Consequently, oxidation at C-6 in GlcNAc combined with β-elimination directly gives DANA analogs, with the glycerol side-chain being replaced by the anomeric alkyloxy substituent (Fig. 8).

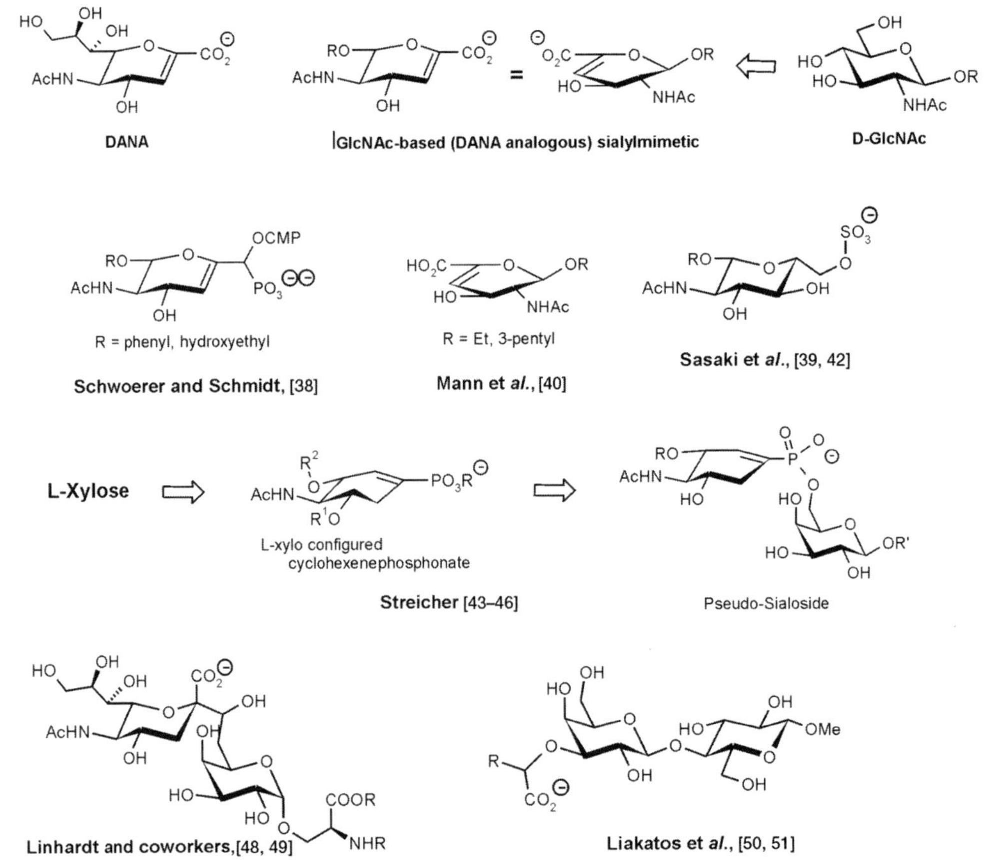

Figure 8. Sialylmimetics derived from GlcNAc and xylose; T-antigen and sialyl lactose mimetics.

Kobayashi and coworkers realized carboxylate mimicry without oxidation by introducing a sulfate group at C-6 of GlcNAc and neglected elimination (Fig. 8) [39,42]. The resulting $^4C_1$-chairs still contain a hydroxy group at the position corresponding to C-3 in Neu5Ac, so weak inhibition of Vc sialidase (IC_{50} = 2 mmol L^{-1}) is not surprising. However, this type of compound has shown considerable activity when presented as oligomers to selectins.

In our laboratory, we have developed a chiral pool-based approach to *xylo*-configured cyclohexenphosphonates (Fig. 8). Whilst in the L-xylo series they can be seen as phospha analogs of oseltamivir, the D-series resembles carba-DANA with a phosphonate instead of a carboxylate and an alkyloxy side-chain mimetic [43–45]. The phosphonate allows attachment of substituents as monoesters while retaining a negative charge under physiological conditions known to be important for binding by sialic acid-processing enzymes. We were indeed able to synthesize pseudo-sialosides like the one depicted in Fig. 8 and found promising inhibitory properties of our scaffolds with selected bacterial sialidases (e.g. *Salmonella typhimurium*) and *Trypanosoma cruzi* trans-sialidase. In fact, we thus introduced the cyclohexene motif, successful in the cases of oseltamivir and the glycosidase inhibitor acarbose with its active component valienamine, into sialoside mimicry (Fig. 8) [46].

Despite their sometimes tedious synthesis, C-glycosides of sialic acid have for some time now been attractive mimetics of sialosides and have been successfully applied in sialidase inhibition previously by Linhardt and coworkers (e.g. *Clostridium perfringens* sialidase). The same laboratory recently synthesized various C-sialosides, among them a C-glycoside analog of the tumor-associated Tn-antigen (Fig. 8) by chemical and chemo-enzymatic methods [47–49].

Another recent approach to sialoside mimicry is the simulation of the important and abundant α-(2,3)-galactose linkage of sialic acid by alkylating a 3'-thiolactose and 5-thiofructofuranoside derivatives with various α-halo esters [50–52]. The resulting conjugates should find interesting applications where only few structure elements of sialic acid are required, for instance selectins or sialyltransferases.

Mechanism-based inhibitors of a trypanosomal sialidase and a trans-sialidase as well as potentially of a sialyltransferase were synthesized and applied by Withers and collaborators [11,53]. While the 2-fluoro-3-hydroxy analog of Neu5Ac was obtained chemically from DANA, the 2,3-difluoro derivative and the CMP conjugate were synthesized chemoenzymatically with the aldolase-catalyzed synthesis of the C9-chain from *N*-acetylmannosamine and fluoropyruvate being the first step (not shown). Chemical synthesis of 2,3-difluoro sialic acids further derivatized at C-4 and targeted against parainfluenza sialidases was achieved from DANA by Ikeda and coworkers [54].

5. SIALYLTRANSFERASES

Inhibition of α(2,6)-sialyltransferases based on analogs of the natural sialyl donor CMP-Neu5Ac containing a flattened (DANA analogous) transition-state mimetic and two negative charges (e.g. Fig. 9) reached the submicromolar level several years ago [55]. Recently, synthetic efforts have focused on both facilitating synthesis and rendering the molecules more hydrophobic to achieve membrane permeability. It has been known for some time that the CMP part must not be changed but replacement of the sialic acid by more hydrophobic, mainly aromatic, moieties proved very successful [56,57]. This way, relatively simple to synthesize and more hydrophobic inhibitors, which are nevertheless even more effective, such as those in Fig. 9, have been obtained.

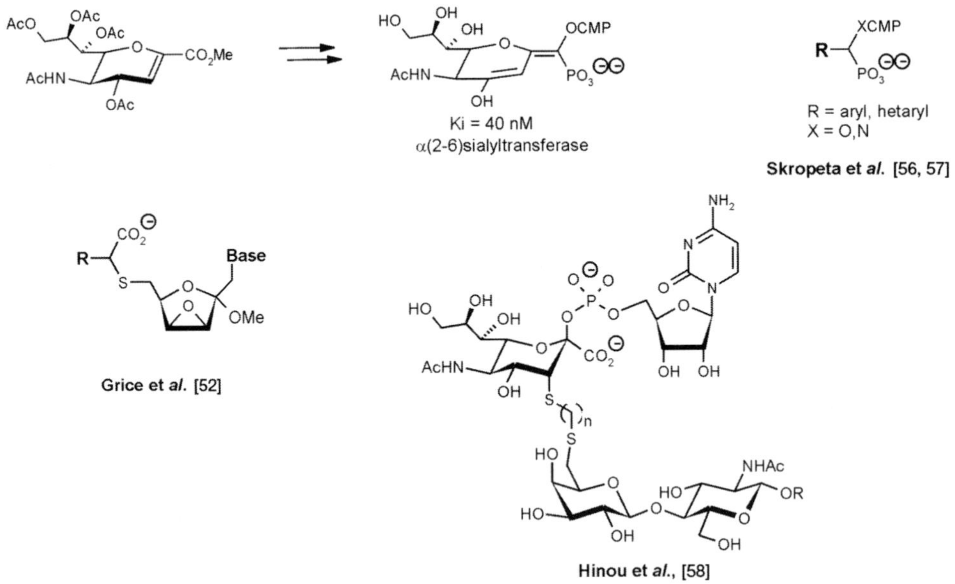

Figure 9. Recently synthesized sialylmimetics aimed at inhibiting sialyltransferases.

A different approach is the generation of bisubstrate-type inhibitors which seek to exploit the fact that the sialyltransferase interacts both with the donor substrate and the acceptor substrate more or less simultaneously.

Ito and coworkers [58] realized this concept for α(2,3)-sialyltransferase by connecting NeuAc containing an axial bromine in the 3 position to the natural acceptor N-acetyl lactosamine via an alkyldithiol linker. Retention of configuration at C-3 was most likely achieved by double displacement involving the 2.3-epoxide. This way, an acceptor position "below" the sialic acid ring was ensured, a situation which is thought to transiently exist. Coupling with activated CMP and deprotection afforded a series of bisubstrate mimetics, the inhibitory potency of which displayed an interesting dependency of the spacer length (Fig. 9).

6. OUTLOOK

In conclusion, the large number of designs of and synthetic approaches towards sialylmimetics published in recent years continues to reflect the variety of important protein–sialic acid interactions occurring in nature. Inducing high-affinity binding between a distinct sialic acid-binding protein and a corresponding sialylmimetic by rational design remains a challenge but the progress in sialidase and sialyltransferase inhibition demonstrates that it can be achieved. More and more lectins and enzymes involved in physiologically relevant sialic acid recognition and metabolism events are emerging and newly gained insight into their structure, mechanism, and sialic acid counterparts will undoubtedly further increase the demand for design and synthesis of new sialylmimetics.

REFERENCES

1. Schauer R & Kamerling, JP (1997) In: *Glycoproteins II*, pp. 241–400. Edited by J Montreuil, JFG Vliegenthardt and H Schachter. Elsevier, Amsterdam.
2. Angata T & Varki A (2002) *Chem. Rev.* **102**, 439–469.
3. von Itzstein M, Wu W, Kok GB *et al.* (1993) *Nature* **363**, 418–423.
4. Kim CU, Lew W, Williams MA *et al.* (1997) *J. Am. Chem. Soc.* **119**, 681–690.
5. Crennell S, Takimoto T, Portner A & Taylor G (2000) *Nat. Struct. Biol.* **7**, 1068–1074
6. Yuan P, Thompson TB, Wurzburg BA, Paterson RG, Lamb RA & Jardetzky TS (2005) *Structure* **143**, 803–815.
7. Alymova IV, Taylor G, Takimoto T *et al.* (2004) *Antimicrob. Agents Chemother.* **48**, 1495–1502.
8. Alymova IV, Portner A, Takimoto T, Boyd KL, Babu YS & McCullers JA (2005) *Antimicrob. Agents Chemother.* **49**, 398–405.
9. McCullers JA & Bartmess KC (2003) *J. Infect. Dis.* **187**, 1000–1009.
10. Buschiazzo A, Amaya MF, Cremona ML, Frasch AC & Alzari PM (2002) *Mol. Cell* **10**, 757–768.
11. Watts AG, Damager I, Amaya ML *et al.* (2003) *J. Am. Chem. Soc.* **125**, 7532–7533.
12. Amaya MF, Watts AG, Damager I *et al.* (2004) *Structure* **12**, 775–784.
13. Wang X, Zhang L-H & Ye X-S (2003) *Med. Res. Rev.* **23**, 32–47.
14. Jung K-H, Schwörer R & Schmidt RR (2003) *Trends Glycosci. Glycotechnol.* **15**, 275–189.
15. Kiefel MJ & von Itzstein M (2002) *Chem. Rev.* **102**, 471–490.
16. Streicher H (2004) *Curr. Med. Chem. Anti Infect. Agents* **3**, 149–161.
17. Wilson JC & von Itzstein M (2003) *Curr. Drug. Targets* **4**, 389–408.
18. Chand P, Kotian PL, Morris PE, Bantia S, Walsh DA & Babu YS (2005) *Bioorg. Med. Chem.* **13**, 2665–2678.
19. Brouillette WJ, Bajpaj SN, Ali SM *et al.* (2003) *Bioorg. Med. Chem.* **11**, 2739–2749.
20. Bianco A, Brufani M, Dri DA, Melchioni C & Filocamo L (2005) *Lett. Org. Chem.* **2**, 83–88.
21. Stoll V, Stewart KD, Maring CJ *et al.* (2003) *Biochemistry* **42**, 718–727.
22. Stoll V, Stewart KD, Maring CJ *et al.* (2005) *J. Med. Chem.* **48**, 3980–3990.
23. DeGoey DA, Chen H-J, Flosi WJ *et al.* (2002). *J. Org. Chem.* **67**, 5445–5453.
24. Barnes DM, Bhagavatula L, DeMattei J *et al.* (2003) *Tetrahedron Asymmetry* **14**, 3541–3551.
25. Hanessian S, Bayrakdarian M & Luo X (2002) *J. Am. Chem. Soc.* **124**, 4716–4721.
26. Wang GT, Wang S, Gentles R *et al.* (2005) *Bioorg. Med. Chem. Lett.* **15**, 125–128.
27. Babu YS, Chand P, Bantia S *et al.* (2000) *J. Med. Chem.* **43**, 3482–3486.
28. Mineno T & Miller MJ (2003) *J. Org. Chem.* **68**, 6591–6596.
29. Chand P, Babu YS, Bantia S *et al.* (2004) *J. Med. Chem.* **47**, 1919–1929.

30. Chand P, Bantia S, Kotian PL, El-Kattan Y, Lin T-H & Babu YS (2005) *Bioorg. Med. Chem.* **13**, 4071–4077.
31. Macdonald SJF, Cameron R, Demaine DA *et al.* (2005) *J. Med. Chem.* **48**, 2964–2971.
32. Watson KG, Cameron R, Fenton RJ *et al.* (2004) *Bioorg. Med. Chem. Lett.* **14**, 1589–1592.
33. Masuda T, Yoshida S, Arai M, Kaneko S, Yamashita M & Honda T. (2003) *Chem. Pharm. Bull.* **51**, 1386–1398.
34. Masuda T, Shibuya S, Arai M *et al.* (2003) *Bioorg. Med. Chem. Lett.* **13**, 669–673.
35. McKimm-Breschkin JL, Colman PM, Jin B *et al.* (2003) *Angew. Chem. Int. Ed. Engl.* **42**, 3118–3121.
36. Hanessian S, Wang J, Montgomery D *et al.* (2002) *Bioorg. Med. Chem. Lett.* **12**, 3425–3429.
37. Hochgürtel M, Biesinger R, Kroth H *et al.* (2003) *J. Med. Chem.* **46**, 356–358.
38. Schwörer R & Schmidt RR *J. Am. Chem. Soc.* **124**, 1632–1637.
39. Sasaki K, Nishida Y, Uzawa H & Kobayashi K. (2003) *Bioorg. Med. Chem. Lett.* **13**, 2821–2823.
40. Mann MC, Thomson RJ & von Itzstein M (2004) *Bioorg. Med. Chem. Lett.* **14**, 5555–5558.
41. Mann MC, Thompson RJ *et al.* (2006) *Bioorg. Med. Chem.* **14**, 1518–1537.
42. Sasaki K, Nishida Y, Kambara M *et al.* (2004) *J. Med. Chem.* **12**, 1367–1375.
43. Streicher H (2004) *Bioorg. Med. Chem. Lett.* **14**, 361–364.
44. Streicher H & Bohner C (2002) *Tetrahedron* **58**, 7573–7581.
45. Streicher H, Meisch J & Bohner C (2001) *Tetrahedron* **57**, 8851–8859.
46. Streicher H & Busse H (2006) *Bioorg. Med. Chem.*, **14**, 1047–57.
47. Ress DK, Baytas SN, Wang Q *et al.* (2005) *J. Org. Chem.* **70**, 8197–8200.
48. Baytas SN, Wang Q, Karst NA, Dordick JS & Linhardt RJ (2004) *J. Org. Chem.* **69**, 6900–6903.
49. Kuberan B, Sikkander SA, Tomiyama H & Linhardt RJ (2003) *Angew. Chem. Int. Ed. Engl.* **42**, 2073–2075.
50. Liakatos A, Kiefel MJ & von Itzstein M (2003) *Org. Lett.* **5**, 4365–4368.
51. Liakatos A, Kiefel MJ, Fleming F, Coulson B & von Itzstein M (2006) *Bioorg. Med. Chem.* **14**, 739–757.
52. Grice ID, Whelan C, Tredwell GD & von Itzstein M *Tetrahedron Asymmetry* **16**, 1425–1434.
53. Watts AG & Withers SG (2004) *Can. J. Chem.* **82**, 1581–1588.
54. Ikeda K, Kitani S *et al.* (2004) *Carb. Res.* **339**, 1367–1372.
55. Mueller B, Schaub C & Schmidt RR (1998) *Angew. Chem.* **110**, 3021–3024, *Angew. Chem. Int. Ed. Engl.* **37**, 2893–2897.
56. Skropeta D, Schwörer R & Schmidt RR (2003) *Bioorg. Med. Chem. Lett.* **13**, 3351–3354.
57. Skropeta D, Schwörer R, Haag T & Schmidt RR (2004) *Glycoconjugate J.* **21**, 205–219.
58. Hinou H, Sun X-L & Ito Y (2003) *J. Org. Chem.* **68**, 5602–5613.

SECTION 3
Analysis and structure

CHAPTER 5
Structure and function of the chondroitin/dermatan sulfates

Robert M. Lauder

1. INTRODUCTION

There is considerable interest in the detailed molecular structure and function of glycosaminoglycans (GAGs), including chondroitin and dermatan sulfate (CS/DS) [1–3]. These structurally diverse polymers are abundant components of extracellular matrices and cell surfaces in humans [4], other mammals, invertebrates [5], *Drosophila melanogaster* [6], and *Caenorhabditis elegans* [7].

Structural roles have been ascribed to CS/DS, but data now show roles as binding partners and receptors or coreceptors in processes including development [8], neurite outgrowth [9], disease [10], growth factor binding [11], and as therapeutics [5].

The potential for CS/DS chain diversity structure is expanding as new structures are identified, however, our understanding of the actual diversity found *in vivo* is also being developed by studies of chain structure and sequence and of the specificities of the enzymes involved in CS/DS biosynthesis [12]. It is clear that not all possible structures are found *in vivo*, and that structures may be clustered, highlighting the non-random nature of CS/DS chains. In addition we are gaining an understanding of the way the structure of these chains changes with, for example development, age, and pathology [2,13].

2. CHONDROITIN/DERMATAN SULFATE CHAIN STRUCTURE

Chondroitin/dermatan sulfate chains are linear carbohydrate polymers synthesized attached, via a linkage region, to a protein core forming a proteoglycan. The linkage region leads to the central repeat region which is terminated by a chain cap (Fig. 1) [1,2].

Glycobiology (C. Sansom and O. Markman, eds.)
© Scion Publishing Limited, 2007

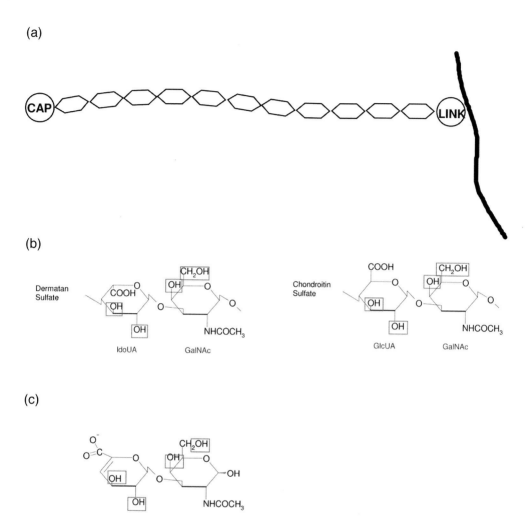

Figure 1. Structure of chondroitin/dermatan sulfates. (a) The intact chain incorporating the linkage region, repeat region and chain cap. (b) The repeating region disaccharide motifs with potential sulfation positions boxed. Note, however, that not all the potential combinations have been observed *in vivo*. (c) The disaccharides generated by chondroitin lyase digestion incorporating an unsaturated uronic acid and losing the distinction between IdoA and GlcA.

2.1 The repeat region

The repeat region of CS is a repeating disaccharide of glucuronic acid (GlcA) and N-acetylgalactosamine (GalNAc) [-4)GlcA(β1–3)GalNAc(β1-]$_n$ which may be O-sulfated on any of the four free OH groups (i.e. C-4 and C-6 of GalNAc) [1] and C-2 and C-3 of GlcA [10,14] (Fig. 1). These sulfate modifications are non-random and show an association with species, tissues, and chain position [13,15].

Other modifications have been identified; glucose has been found attached to C-3 of GalNAc in squid CS [14] and fucose has been found at C-3 of GlcA in king crab [16] and echinoderm CS [5]. Fucose modification at C-3 of galactose is well known from keratan sulfate [4,17], however, the fucose found in echinoderm CS is sulfated, mainly at C-4 but forms disulfated at C-2 and C-4 and at C-3 and C-4 were found.

The GlcA residues of CS may be epimerized at C5 [18,19] to iduronic acid (IdoA) (Fig. 1), forming the repeating disaccharide [-4]IdoA(α1-3)GalNAc(β1-]$_n$ of DS. Sulfation at GalNAc C6 is rare in human DS although it is abundant in DS from *Ascidia nigra* [20] and to date no IdoA C3-O-sulfation has been identified.

Thus, CS/DS may be found as pure polymers or as a mixed co-polymer, in which IdoA residues, and sulfation isoforms may be located together in clusters or distributed throughout the chain. These will have very different impacts upon molecular interactions and biological function.

The potential diversity of CS/DS is vast (Fig. 1); there are 16 possible sulfation isoforms for disaccharides containing a GlcA and 16 for disaccharides containing IdoA, each can form 256 tetrasaccharides or 1024 for copolymeric CS/DS, where either uronic acid may be GlcA or IdoA. However, as we understand more about the specificity of the enzymes involved in CS/DS biosynthesis [12], and explore the actual CS/DS diversity found *in vivo*, it is clear that while they are complex, they do not display the full spectrum of random structures possible.

Indeed, it is their non-random nature that allows information to be contained in the chain and interactions to be controlled.

2.2 The linkage region

The linkage region, common to CS, DS, heparan sulfate (HS), and heparin, comprises UA(β1-3)Gal(β1-3)Gal(β1-4)Xyl(β1-O)-Ser and shows C-4 and C-6 sulfation of both Gal residues [2,3,15,21,22] along with Xyl C-2 phosphorylation [23]. In CS chains the uronic acid has been found to be GlcA, while in DS both GlcA and IdoA residues have been identified.

Xylose phosphorylation occurs dynamically and may represent a sorting or chain elongation signal, most Xyl residues being dephosphorylated when the chain reaches tetrasaccharide size [23]. The sulfation of Gal residues (a feature absent from HS and heparin) may also be reversible and a signal for specific modification of the growing chain; enzymes involved in CS chain synthesis interact with the protein core [24] and so may also interact with Gal(S). In aggrecan a wide variety of GalNAc sulfation forms have been observed without any Gal sulfation [15] and so the biological importance of these modifications remains unclear.

The repeat region close to the linkage region differs in structure from the average chain composition. In articular cartilage aggrecan the two GalNAc residues closest to the linkage region show a lower level of 6-O-sulfation and higher level of 4-O-sulfated and unsulfated residues than the chain average [2,15]. The abundance of 6-O-sulfation increases with age, the most dramatic being during the first 20 years of life in humans [2], but is always lower than the chain average. Sulfation in this region of the chain appears to be controlled, but again the biological importance of these modifications remains unclear.

2.3 The chain caps

The chain caps of CS/DS are either GlcA or GalNAc residues, marking a difference from keratin sulfate (KS) in which chain caps are specialized residues absent from the main repeat region [17,25].

However, although the chain cap is not made up of specialized residues, they show highly sulfated motifs which are absent from the repeating region [13,19]. Importantly Plaas et al. [13] have shown that the chain cap structure changes with age and pathology.

2.4 Nomenclature

The original names for CS were CSA for the 4-O-sulfated material, CSB for another 4-O-sulfated form, dermatan sulfate, and CSC for a 6-O-sulfated form. These letters are widely used for CS/DS disaccharide unit identification (Table 1). Thus, the A-unit is -4)GlcA(β1-3)GalNAc4S(β1- when it is still part of the polymer, Δdi-A, or Δdi-4S for the disaccharide generated by chondroitin lyase digestion and di-A or di-4S for the saturated disaccharide generated by hyaluronidase digestion. Oligosaccharides containing IdoA are indicated by the prefix i (see Table 1). However, describing chains in this way obscures much detail as they are rarely exclusively sulfated by a single form. The ABC nomenclature has also been retained in the names given to the enzymes which depolymerize these CS/DS chains (see Section 2.7).

Table 1 Disaccharide IDs of chondroitin/dermatan sulfates

	ID	Structure	Reference
0	Di-0S	-4)GlcA(β1-3)GalNAc(β1-	
i0	Di-i0S	-4)IdoA(α1-3)GalNAc(β1-	1
A	Di-A	-4)GlcA(β1-3)GalNAc4S(β1-	26
B/iA	Di-B/Di-iA	-4)IdoA(α1-3)GalNAc4S(β1-	26
C	Di-C	-4)GlcA(β1-3)GalNAc6S(β1-	26
iC	Di-iC	-4)IdoA(α1-3)GalNAc6S(β1-	
		-4)IdoA2S(α1-3)GalNAc4S(β1-	20
D	Di-diS$_D$	-4)GlcA2S(β1-3)GalNAc6S(β1-	26
iD	Di-diS$_{iD}$	-4)IdoA2S(α1-3)GalNAc6S(β1-	20
E	Di-diS$_E$	-4)GlcA(β1-3)GalNAc46S(β1-	53
iE	Di-diS$_{iE}$	-4)IdoA(α1-3)GalNAc46S(β1-	53
	Di-triS	-4)GlcA2S(β1-3)GalNAc46S(β1-	34
		-4)GlcA3S(β1-3)GalNAc46S(β1-	14
		-4)GlcA3S(β1-3)GalNAc4S(β1-	14
		-4)GlcA2S(β1-3)GalNAc6S(β1-	14
		-4)[Fuc(α1-3)]GlcA(β1-3)GalNAc(β1-[a]	5

Disaccharide structures are shown with the letters used in this review and widely in the literature to describe them. The presence of a IdoA in the unit is indicated by the prefix i. Following depolymerization by chondroitin lyase enzymes the non-reducing terminal uronic acid is unsaturated between C-4 and C-5 and the distinction between IdoA and GlcA is lost. Disaccharides generated are described using the prefix Δdi-. Disaccharides generated using other strategies may retain the structure of the non-reducing uronic acid and are described using di-.

[a]The fucosylated structure reported by [5] is found as multiple sulfation forms but is shown unsulfated for clarity.

2.5 Variable sulfation

The sulfation of CS/DS shows variation amongst the same tissue from differing species, different tissues from the same organism [1,26], with age and location within a single tissue [27], different GAG attachment positions along a proteoglycan core and also within a single chain [28].

For example, in adult human articular cartilage, extensive 6-O-sulfation of GalNAc is observed (ca. 95%) [1] while in the adjacent meniscus the 6-O-sulfation level is lower (ca. 70%) [29]. Porcine tracheal cartilage has ca. 20% 6-O-sulfation while in bovine this is higher (ca. 30%). In shark cartilage 6-O-sulfation levels are ca. 70% with 4-O-sulfation making up the balance but ca. 25% 2-O-sulfation of the uronic acid residues [1] is seen mainly between a 4-O-sulfate on the non-reducing side and a 6-O-sulfate on the reducing [1]. In *Drosophila melanogaster* 4-O-sulfation but not 6-O-sulfation is observed, while in *Caenorhabditis elegans* the chondroitin is unsulfated [6,7]. In contrast, CS/DS chains with very high levels of GalNAc 4,6-O-sulfation, have been found in squid cartilage and hagfish notochord [9]. DS is an abundant component of skin, where it shows mainly GalNAc 4-O-sulfation and only modest levels of IdoA 2-O-sulfation. However, these levels are much higher in *Styela plicata*, and *Ascidia nigra* DS comprises GalNAc 6-O-sulfation and abundant IdoA 2-O-sulfation [20].

Within a tissue the sulfation profile and levels of epimerization of GlcA to IdoA change with age. For example, the level of GalNAc 6-O-sulfation reported above for human articular cartilage applies only to the adult; at birth this level is close to zero but rises significantly during the first 20 years of life [2,27]. The levels of IdoA fall with age in meniscal tissues (Lauder, unpublished) as do levels of 4-O-sulfation, while 6-O-sulfation rises.

Chains show a variation in sulfation along their length; a 4,6-disulfated GalNAc residue, rare in the repeat region of human articular cartilage CS, represents over 50% of the chain caps for a normal adult. Changes in pathology are also evident, with this level falling to only ca. 30% in osteoarthritic cartilage [13,30]. High levels of this GalNAc sulfation form are also found adjacent to a 2-sulfated uronic acid as the cap in whale CS [19].

While the CS chain caps may be highly sulfated, the linkage regions show low levels of sulfation relative to residues within the repeat region [2,3,15] with preferential localization of unsulfated and 4-O-sulfated GalNAc residues close to linkage regions [2,3,15].

Although data are available on the structure of the two chain ends, and we know about repeat region composition, we cannot yet say if there is a relationship between the structures of the linkage region, chain cap, and the sequence of sulfate modifications in the repeat region.

2.6 Chain size

The size of CS/DS chains is challenging to establish, and because electrophoretic techniques are inappropriate, size exclusion chromatography (SEC) is widely used. However, because CS/DS chains adopt a rigid rod-like structure in solution they cannot be examined without appropriate standards. In addition, the impact of sulfation forms upon apparent hydrodynamic volume, by influencing the rigidity of the chain [31], may result in chains with the same number of disaccharides but different sulfation profiles having different behaviors upon SEC.

Data show that tracheal cartilage CS is around 22–25 kDa with porcine chains appearing slightly smaller than bovine. Plaas et al. found that normal corneal CS/DS is ca. 17.5 kDa while those from people with macular corneal dystrophy are only ca. 6.7–16.2 kDa.

In addition to CS/DS chain size being modified by pathology, in vitro newly synthesized CS chains are longer in loaded articular tissues than unloaded [30]. CS chains isolated from adult human articular cartilage were found to be ca. 10 kDa. DS from porcine skin and intestinal mucosa are reported as 19 kDa and 21 kDa, respectively, while the CS/DS chains of hagfish notochord were found to be ca. 18 kDa [9].

However, CS from shark cartilage and CS/DS from shark skin are significantly larger, with sizes up to 70 kDa reported [32].

2.7 Study of CS/DS structure

Glycosaminoglycans are not primary gene products and therefore their analysis cannot rely upon genomic approaches. Our work, and that of other laboratories, has used the paradigm of isolation and depolymerization of GAG chains to generate oligosaccharides which are isolated and characterized [3,26] before being integrated into fingerprinting methods for the analysis of biological samples [1].

Chondroitin lyase enzymes are eliminases that cleave the -3)GalNAc(β1-4)GlcA(β1-/IdoA(α1- bond in CS/DS in the case of chondroitin ABC lyase, while chondroitin AC and C lyases act on GlcA-containing bonds alone and chondroitin B lyase acts upon IdoA-containing bonds only. Chondroitin AC and ABC lyases can generate di- and tetrasaccharides [1] and are widely used for CS/DS composition analysis. These studies allow an understanding of species-, tissue-, age-, and pathology-related differences and changes in CS/DS abundance and composition. However, the reduction of the polymer to small units, while making analysis easier, removes all sequence data. In addition, the action of chondroitin lyase enzymes generates a 4,5-unsaturated hexuronic acid (ΔUA) from the uronic acid of the cleaved bond (Fig. 1). Thus, the distinction between IdoA and GlcA, epimerization at C-5, is lost and it is impossible to distinguish between disaccharides derived from an A unit and those from an iA unit.

While chondroitin B and C lyase also generate ΔUA non-reducing termini, their restricted substrate specificity results in large oligosaccharides of resistant structures being derived limit digests. Thus, information on the size of blocks of specific structures can be obtained.

These enzymes are unable to act upon highly sulfated bonds; for example -3)GalNAc4,6S(β1–4)GlcA2S(β1- is resistant to cleavage by all chondroitin lyase enzymes while -3)GalNAc4S(β1–4)GlcA2S(β1- can be cleaved by chondroitin ABC lyase but not by chondroitin C lyase (Lauder, unpublished). It is important to note that many of the CS/DS functions discussed later are associated with highly sulfated motifs.

For each of these depolymerization strategies a heterogeneous population of oligosaccharides is generated which may be examined by chromatographic methods [1], although electrophoretic methods can also be used [33]. Detection can rely upon the ΔUA residue which absorbs strongly at 232 nm [34] or the oligosaccharide may be labeled by conjugation of a probe to the reducing terminus. These strategies allow the rapid quantitative analysis of CS/DS from small biological samples.

Mass spectrometry (MS) techniques are being developed for CS/DS analysis with the advantage of rapid and sensitive quantitative analysis of an unmodified digest. A major challenge is that the sulfation forms of CS and DS, and indeed the distinction between CS/DS, represent structural isomers with the same mass (Fig. 1). However, from our own work and that of others [35] it is clear that MS_2 and MS_3 techniques permit the distinction between these isomers by their distinct fragmentation patterns.

Oligosaccharides may also be examined by nuclear magnetic resonance (NMR) spectroscopy; we have reported ^1H and ^{13}C NMR data for di-, tetra-, hexa-, and octa-saccharides from CS/DS [26] and now have data from oligosaccharides up to 20-mer (Lauder, unpublished). NMR spectroscopy has the advantages of being non-destructive and, unlike other methods, data can be acquired from intact polymers. Work in the author's laboratory is directed at gaining sulfation data from intact CS/DS chains, as we have already achieved for KS [36,37].

2.8 Getting CS/DS for interaction studies

Studies of CS/DS structure/function often use compositional data to explore these relationships, looking for structures associated with a function, or specifically removing features to abrogate an interaction [10]. However, fuller studies are required to establish the binding detailed motifs and the size of binding partners, for these defined oligosaccharides are required.

These have been obtained from tissue sources following enzyme digestion [1,26,38], although both reducing terminal and non-reducing terminal residues are modified by chondroitin lyase enzyme action (see Fig. 1). The use of hyaluronidase avoids the eliminase action which generates the non-reducing terminal ΔUA, alternatively the ΔUA residue may be removed chemically [39].

Other strategies include the chemical modification of CS/DS by addition or removal of specific sulfate esters by enzymatic or chemical approaches [10]. In addition, synthetic approaches, although less well developed than those for HS, have successfully generated hyaluronan (HA) and an unmodified CS [24]. It is possible that the addition of sulfotransferase enzymes to such approaches will permit the generation of sulfated CS, while C-5 epimerase addition could generate DS. However, generation of the complex structures implicated in interactions important for cellular function will not be simple.

3. CHONDROITIN/DERMATAN SULFATE CHAIN FUNCTION

The best-described function of CS is its structural role within articular cartilage – the osmotic swelling generated by the large charge associated with the CS supporting weight bearing. However, the well established age-related changes in sulfation profile point to other more fundamental impacts upon biochemical processes [2,27].

3.1 Development

The nematode *C. elegans* synthesizes chondroitin, unmodified by sulfates, which is essential for embryonic development [8]. However, the mechanism for this is not yet established and may rely upon the osmotic swelling generated by chondroitin [40]. In addition, while *C. elegans* with disrupted chondroitin synthesis had failures of cell division, Chinese hamster ovary (CHO) cells unable to synthesize CS showed no such defect [41].

3.2 Central nervous system

CS/DS proteoglycans are abundant in the CNS and studies are finding roles for these GAGs in CNS development. The receptor-type protein tyrosine phosphatase zeta (RPTP beta/PTPζ) is a CS proteoglycan, the extracellular region of which is found as phosphacan (6B4 PG/DSD-1-PG).

Interactions involving PTPζ and pleiotropin, which regulate cerebellar Purkinje cell morphogenesis, requires the highly sulfated CS D-unit (Table 1) on CS chains of the receptor [42]. Hippocampal neurite outgrowth is also impacted by highly sulfated CS/DS, with the D or iD-units promoting multiple dendrite-like processes while the E- or iE-units promoted the development of long axon like neurites [43].

The E-unit may impact neurite outgrowth and neuronal cell adhesion by interaction with midkine [44], a process which requires the disulfated E-unit but not the GlcA 3-*O*-sulfation common in tissues rich in the E-unit [45]. In addition to midkine the E-unit has also been shown to interact very strongly with pleiotropin, FGF-16 and 18 [46], and more weakly with FGF-2 and 10.

Chondroitin sulfate with E- and D-units has been found in the brain and, while it is likely that, in general, CS/DS is inhibitory to neurite outgrowth, differences in structure and abundance promote controlled development.

The possibility of using our understanding of the roles of defined CS/DS structures to repair CNS damage is of enormous importance. Following injury there appears to be an increase in the RNA levels for 6-sulfotransferase, suggesting that postinjury CS may be synthesized with elevated levels of 6-*O*-sulfate [47]. However, CS/DS removal by chondroitin ABC lyase can disrupt controlled axon growth, or permit movement through damage/scar tissue, which represents a barrier to regrowth [48].

3.3 Turnover

A role for CS in extracellular matrix turnover has been highlighted by work showing that the activation of ADAMTS is dependent upon cell surface CS on syndecan [49]. This enzyme is active in the degradative processes associated with aggrecan turnover, but is implicated in the pathological processes of osteoarthritis.

3.4 Pathology

CS has a central role in the sequestration of *Plasmodium falciparum*-infected red blood cells to the human placenta. Initial studies implicated 4-*O*-sulfated CS [50,51] and these data were confirmed by chemical removal of 6-*O*-sulfates which enhanced binding, suggesting that not only was 4-*O*-sulfate required but that the presence of 6-*O*-sulfates inhibited the interaction. Indeed, highly sulfated CS was shown not to bind [10]. Viral infection has also been shown to utilize CS and herpes simplex virus infection was inhibited by CS rich in the E-unit (Table 1) [52].

3.5 Therapeutics

A widely studied GAG/protein interaction is the regulation, by heparin, of thrombin activity. DS also shows anticoagulant activity by activation of heparin cofactor II (HCII) through a series of three adjacent -4)IdoA2S(α1–3) GalNAc4S(β1- units. Most DS from mammalian tissues comprises the iA-unit (see Table 1) and contain less than 5% of this active motif. In mammalian tissues these units represent less than 5% of the total chain and so for these GAGs to have any effect these units must be clustered, clearly indicating a non-random chain structure. However, this motif is abundant in *Styeca plicata* DS, which shows significant anticoagulant activity. In contrast *Ascidia nigra* DS, rich in the iD-unit, which has the same sulfation level but differs in position, shows no significant anticoagulant activity [20].

The fucosylated CS of echinoderms has been shown to have an anticoagulant activity [5] that depends mainly upon the presence of fucose residues. While DS acts through HCII, fucosylated CS can act through HCII or antithrombin. However, the latter route does not achieve full inhibition, suggesting action via a different mechanism to heparin.

The use of heparin has undesirable hemorrhagic side-effects and data point to a reduction in these effects with DS.

4. THE FUTURE

Although we are acquiring a detailed understanding of the structure/function relationships of CS/DS, challenges still remain. Techniques are needed to study the sequence of modifications on CS/DS chains and for the isolation of known structures in sufficient quantity for use as inhibitors or binding partners in interaction analysis. Others challenges are core questions, such as the processes controlling the sulfate modification of CS/DS chains and their length.

We must also increase our understanding of the structures and mechanisms central to biologically important interactions and how what we know about age- and pathology-related changes in sulfate composition translate into changes in the abundance of structural motifs and how this impacts function.

Acknowledgments

Studies of CS/DS structure and function in the author's laboratory are funded by Laboratoires Genévrier, The Joy Welch Trust and the Bingham Trust.

REFERENCES

1. Lauder RM, Huckerby TN & Nieduszynski IA (2000) *Glycobiology* 10, 393–401.
2. Lauder RM, Huckerby TN, Brown GM, Bayliss MT & Nieduszynski IA (2001) *Biochem. J.* **358**, 523–528.
3. Huckerby TN, Lauder RM & Nieduszynski IA (1998) *Eur. J. Biochem.* 258, 669-76.
4. Lauder RM, Huckerby TN & Nieduszynski IA (1997) *Glycoconj. J.* 14, 651–660.
5. Mourao PA, Pereira MS, Pavao MS *et al.* (1996) *J. Biol. Chem.* **271**, 23973–23984.
6. Toyoda H, Kinoshita-Toyoda A & Selleck SB (2000) *J. Biol. Chem.* **275**, 2269–2275.
7. Yamada S, Van Die I, Van den Eijnden DH, Yokota A, Kitagawa H & Sugahara K (1999) *FEBS Lett.* **459**, 327–331.
8. Mizuguchi S, Uyama T, Kitagawa H *et al.* (2003) *Nature* **423**, 443–448.
9. Nandini CD, Mikami T, Ohta M, Itoh N, Akiyama-Nambu F & Sugahara K (2004) *J. Biol. Chem.* **279**, 50799–50809.
10. Fried M, Lauder RM & Duffy PE (2000) *Exp. Parasitol.* **95**, 75–78.
11. Bao X, Nishimura S, Mikami T, Yamada S, Itoh N & Sugahara K (2004) *J. Biol. Chem.* **279**, 9765–9776.
12. Kusche-Gullberg M & Kjellen L (2003) *Curr. Opin. Struct. Biol.* **13**, 605–611.
13. Plaas AH, West LA, Wong-Palms S & Nelson FR (1998) *J. Biol. Chem.* **273**, 12642–12649.
14. Kinoshita-Toyoda A, Yamada S, Haslam SM *et al.* (2004) *Biochemistry* **43**, 11063–11074.
15. Lauder RM, Huckerby TN & Nieduszynski IA (2000) *Biochem. J.* **347**, 339–348.
16. Sugahara K, Tanaka Y, Yamada S *et al.* (1996) *J. Biol. Chem.* **271**, 26745–26754.
17. Nieduszynski IA, Huckerby TN, Dickenson JM *et al.* (1990) *Biochem. J.* **271**, 243–245.
18. Kobayashi M, Sugumaran G, Liu J, Shworak NW, Silbert JE & Rosenberg RD (1999) *J. Biol. Chem.* **274**, 10474–10480.
19. Ohtake S, Kimata K & Habuchi O (2005) *J. Biol. Chem.* **280**, 39115–39123.
20. Pavao MS, Aiello KR, Werneck CC *et al.* (1998) *J. Biol. Chem.* **273**, 27848–27857.
21. Sugahara K, Ohi Y, Harada T, de Waard P & Vliegenthart JF (1992) *J. Biol. Chem.* **267**, 6027–6035.
22. de Waard P, Vliegenthart JF, Harada T & Sugahara K (1992) *J. Biol. Chem.* **267**, 6036–6043.
23. Moses J, Oldberg A, Cheng F & Fransson LA (1997) *Eur. J. Biochem.* **248**, 521–526.
24. Kitagawa H, Izumikawa T, Uyama T & Sugahara K (2003) *J. Biol. Chem.* **278**, 23666–23671.
25. Tai GH, Nieduszynski IA, Fullwood NJ & Huckerby TN (1997) *J. Biol. Chem.* **272**, 28227–28231.
26. Huckerby TN, Lauder RM, Brown GM *et al.* (2001) *Eur. J. Biochem.* **268**, 1181–1189.
27. Bayliss MT, Osborne D, Woodhouse S & Davidson C (1999) *J. Biol. Chem.* **274**, 15892–15900.
28. Lauder RM, Huckerby TN & Nieduszynski IA (1999) *Carbohydr. Lett.* **3**, 381–388.
29. McNicol D & Roughley PJ (1980) *Biochem. J.* **185**, 705–713.
30. Sauerland K, Plaas AH, Raiss RX & Steinmeyer J (2003) *Biochim. Biophys. Acta* **1638**, 241–248.
31. Bathe M, Rutledge GC, Grodzinsky AJ & Tidor B (2005) *Biophys. J.* **88**, 3870–3887.
32. Nandini CD, Itoh N & Sugahara K (2005) *J. Biol. Chem.* **280**, 4058–4069.
33. Plaas AH, West L, Midura RJ & Hascall VC (2001) *Methods Mol. Biol.* **171**, 117–128.
34. Lauder RM (2000) In: *Encyclopaedia of Analytical Chemistry*, pp. 860–895. Edited by RA Myers. John Wiley & Sons, Chichester.
35. Zaia J, McClellan JE & Costello CE (2001) *Anal. Chem.* **73**, 6030–6039.

36. Huckerby TN, Nieduszynski IA, Bayliss MT & Brown GM (1999) *Eur. J. Biochem.* **266**, 1174–1183.
37. Huckerby TN & Lauder RM (2000) *Eur. J. Biochem.* **267**, 3360–3369.
38. Huckerby TN, Nieduszynski IA, Giannopoulos M, Weeks SD, Sadler IH & Lauder RM (2005) *FEBS J.* **272**, 6276–6286.
39. Ludwigs U, Elgavish A, Esko JD, Meezan E & Roden L (1987) *Biochem. J.* **245**, 795–804.
40. Hwang HY, Olson SK, Esko JD & Horvitz HR (2003) *Nature* **423**, 439–443.
41. Bai X, Wei G, Sinha A & Esko JD (1999) *J. Biol. Chem.* **274**, 13017–13024.
42. Tanaka M, Maeda N, Noda M & Marunouchi T (2003) *J. Neurosci.* **23**, 2804–2814.

CHAPTER 6

Revealing the hidden dimensions of glycosaminoglycans in solution and protein complexes with saccharide libraries, FTIR, and SRCD

E.A. Yates, T.R. Rudd, L. Duchesne, M.A. Skidmore, R.J. Nichols, D.T. Clarke and D.G. Fernig

1. INTRODUCTION

The glycosaminoglycans (GAGs) and heparan sulfate (HS) in particular have a notably high degree of structural diversity. Consequently, they form an important part of the glycome, the complete repertoire of expressed glycan structures (see Chapter 8). The appearance of these polysaccharides coincides with the evolution of metazoans and a massive expansion of the glycome, which reflects the need for a vastly increased chemical information space to accommodate intercellular communication and coordination, without a commensurate expansion of the genome [1].

Heparan sulfate is a negatively charged, linear polysaccharide of varying length that is made of 1–4 linked disaccharide repeating units consisting of a uronic acid and an amino sugar. The uronic acid is either α-L-iduronate or β-D-glucuronate and the amino sugar is α-D-glucosamine, either as D-glucosamine or N-acetylglucosamine. The uronate residues may be O-sulfated at position 2 while the glucosamine can be N-sulfated and/or 6-O-sulfated (Fig. 1a). In addition, some glucosamine residues carry O-sulfation at position 3. Modifications to the basic backbone structure, which is initially synthesized as alternating units of GlcA-GlcNAc are the result of the action of a series of enzymes, but these do not act to completion to produce homogeneously substituted polymers. Rather, the chains are modified to a substoichiometric extent, resulting in considerable structural heterogeneity both within single chains and between chains. Analysis of HS chains from cell lines has indicated that there is an additional structural characteristic present in HS that is absent in the related polysaccharide heparin. Long, unsulfated or sparsely sulfated regions termed NA domains are interspersed with more highly sulfated domains termed S-domains. At the junction of these two types of domains, a region of more mixed structures is found, termed NA/NS domains (Fig. 1b). The expression of HS chains with varying sequences and domain structures at the cell surface is both spatially and developmentally (temporally) regulated.

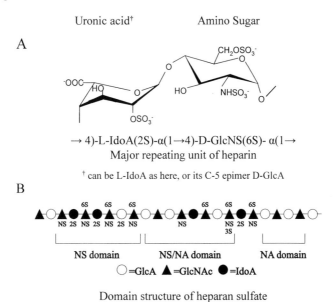

Figure 1. (a) The major repeating unit of heparin. (b) A schematic of the domain structure of heparan sulfate.

The interactions of proteins with GAGs represent one of the major axes of cellular regulation in metazoans. Hitherto a major focus has been the identification of short sequences of sugars that interact with protein targets and the biological and biophysical characterization of these interactions. Examples of such interactions, which have been studied in depth and that provide many of the paradigms for the field include the interactions of antithrombin III with a heparin pentasaccharide [2], of fibroblast growth factor (FGF)-1, -2, and -7 with diverse sizes and structures of oligosaccharides derived from heparin and HS itself [3–8] and of hepatocyte growth factor/scatter factor with HS and dermatan sulfate [9,10]. One recurring functional theme of the structures within HS that interact with effectors such as growth factors is that of coreceptor, whereby the polysaccharide enables the effector protein to transduce efficiently intracellular signals via its signaling receptor.

1.1 Oligosaccharides: just part of the GAG chain

Whilst these and similar experiments provide a large amount of information on the molecular basis of the interaction between HS and proteins, the structure/function relationships that are responsible for mediating GAG–protein interactions arise from more than simply the atomic groups responsible for molecular binding. One of the first indications that this was the case was a survey of the ability of different HS proteoglycans to function as coreceptors for FGF-2. In this work, perlecan acted as a coreceptor, whereas other HS proteoglycans failed to do so [11]. Later work showed that HS chains isolated from a panel of mammary cell lines representative of some of the different cell types found in the parenchymal and stromal compartments of this tissue possessed markedly different FGF-binding activities and coreceptor activities [4]. This study provided clear evidence for the existence of HS chains capable of regulating differently FGF-1 and FGF-2 and, importantly, for the existence of FGF-2-binding motifs that were inactive due to the context of the remainder of the GAG chain in which they were embedded. Thus, liberation of the FGF-2-binding structures by digestion with heparinase III resulted in their acquisition of FGF-2 coreceptor activity.

In a mouse model of skin wound healing the proteoglycan syndencan-1 bound FGF-2, but these binding sites were, again, without activity. The partial degradation of the HS chains of syndecan-1 by endogenous heparanase, an early event in wound healing, resulted in the release of complexes of FGF-2 and oligosaccharides; the released oligosaccharides were relieved of the inhibition caused by the context of the HS chain in which they were embedded and were fully active as coreceptors [12]. In a different set of experiments, HS chains from a malignant mammary cell line, MDA-MB-231, were observed not to possess FGF-2 coreceptor activity, though they were shown qualitatively to bind the growth factor. Reduction of the sulfation of these chains, achieved by including in the cell culture medium different amounts of the competitive inhibitor of PAPS synthesis, chlorate, resulted in the chains acquiring coreceptor activity [13]. Subsequent work in this system showed that whereas FGF-2 coreceptor activity depended on the level of sulfation of the entire GAG chains, the equivalent activity for FGF-1 was not affected [14]. These results suggest that the richness of GAG structure arises not only from the diversity of saccharide sequences that bind proteins and which, in isolation, may have specifically identifiable activities in terms of regulating protein activity, but also from the context of these structures in terms of the GAG chain in which they are embedded.

1.2 Higher order structure in GAG chains

Clearly, an entirely reductionist approach, which focuses solely on the identification of protein-binding motifs, is unlikely to resolve the structure/function relationships represented by protein–GAG interactions. There is considerable indirect evidence for structural information outside of the protein binding sites in the HS chain regulating the function of the binding site itself. A key question, therefore, is how such regulation might occur. Most of the above work on GAG–protein interactions ignores a key structural feature of the sugar, the multiplicity of conformations, which arise from the complexity of linkage and ring geometry. If this level of structural information is used by biology to modulate the activity of protein-binding saccharide sequences, it would provide ample scope for such regulation. There is, however, little direct evidence that this is the case, other than two static structures derived from X-ray crystallography of co-complexes of HS-derived oligosaccharides with FGF-2 and FGF receptors, in which different conformations of sugars are evident within a single short sequence [8]. Therefore, a major challenge with respect to understanding the glycome must be to establish the extent to which polysaccharide conformation contributes to the regulatory activity of protein-binding sites.

1.3 The information content of heparan sulfate

The structural complexity of carbohydrates gives them a potential information content several orders of magnitude greater than proteins or nucleic acids. If we consider the number of theoretically possible disaccharide units for HS, it is clear that the total number of possible combinations is vast; for a dodecasaccharide it is in excess of 10^6. The variety of structures isolated from HS so far, however, suggests that most of these are not found naturally, although this still gives HS very substantial heterogeneity and, therefore, significant potential information content. It is not clear that each sequence should be considered a unique entity, in the sense that exquisite specificity exists, because the same activities are often observed for different structures. The challenge is to understand which features are responsible for bestowing a particular activity and to achieve this, it will be necessary to gain a detailed understanding not only of particular sequences, but also their resulting conformation, dynamics and other factors including the effects of cation binding and the specificity of that binding.

The structural complexity of HS, hence the subtly different nature of particular sequences, is brought to bear through distinct conformational properties in solution. This facet of carbohydrates is considerably less well understood than for proteins and nucleic acids, not only because of this complexity, but also because crucial structural features, such as linkage and ring geometry, have been more difficult to access experimentally, particularly in aqueous solution. Spectroscopic techniques have been at the heart of studies in this area, but relatively little systematic data of the more complex carbohydrates has been amassed that can be directly related to their solution conformation.

We have approached the issue using a library of systematically modified carbohydrates and are currently correlating complementary spectroscopic data (synchrotron radiation circular dichroism (SRCD), Fourier transform infrared (FTIR) and nuclear magnetic resonance (NMR)) with molecular modeling (MM) results to elucidate carbohydrate conformation in solution and so gain a fundamental understanding of their complex solution conformation.

2. SPECTROSCOPIC APPROACHES TO UNDERSTANDING HIGHER ORDER GAG STRUCTURE

Most work concerning the physicochemical characteristics of these molecules has used heparin, usually as a proxy for the sulfated domains of HS. It is now well established that the 2-*O*-sulfated IdoA residue exists in an equilibrium between chair ($^1C_4$) and skew-boat ($^2S_0$) forms [15]. Unsubstituted IdoA and GlcA lack this property, the former resides mainly in the $^1C_4$ chair form and the latter in $^4C_1$, irrespective of substitution [15–17]. It has also been suggested that substitution pattern may influence the conformation around the glycosidic linkages [18]. The dynamics of heparin have also been studied using NMR by several groups and, depending on the mathematical model used to interpret the data, give somewhat contradictory results, ranging from relatively rigid [19] to semi-flexible [20,21], although the latter group report a distinct difference in flexibility between relatively highly sulfated heparin derivatives and the unsulfated backbone structure (GlcA-GlcNAc repeating units). It seems most likely that HS, with its domain structure, probably contains regions with different degrees of flexibility.

We have shown by NMR that the glycosidic linkage geometry of these macromolecules varies widely with substitution pattern [17,18] and have identified probable hydrogen bonding patterns (e.g. between N-sulfates and 3-hydroxyl groups of adjacent iduronate residues in the sequence IdoA2S-GINS,6S), suggesting the presence of distinct secondary structure. Analysis of glycosidic and ring geometries depends on interpreting spectral data using a suitable model [19–21]. Despite these efforts, the conformation of the polysaccharides in solution remains largely inaccessible, particularly as the substitution pattern varies. This is partly due to the difficulty of observing changes around the glycosidic bonds of these molecules. Obtaining a sound physical description of model compounds will be crucial if a detailed understanding of complex polysaccharides is to be achieved.

Ultimately, the aim will be to study the contribution of HS chains to the formation of the large number of HS–protein complexes that are important for a host of signaling, coordination, and regulation events in biological systems. Some considerations relevant to the study of HS–protein complexes will also be covered in the latter sections of this review. In the following sections, the basic principles of how each technique works, what information each technique can provide and their complementarity are described.

2.1 Circular dichroism: The advent of synchrotron radiation circular dichroism

Circular dichroism (CD) arises when a chromophore in an asymmetric environment absorbs right and left circularly polarized light, through the excitation of electrons in molecular orbitals, to a different extent. CD's best-known application has been to quantify protein secondary structure; α helices, β sheets, turns, and unstructured regions all exhibit characteristic spectra [22, 23] and changes in tertiary structure can also be detected through chromophores in aromatic amino acid side-chains. Several well-documented algorithms have been compiled [24–26], which allow quantification of protein secondary structure from CD spectra. A recent technological advance has been the introduction of synchrotron radiation CD (SRCD). Facilities such as the CD12 beamline at Daresbury Laboratory (UK) have rejuvenated the subject because lower wavelengths (in practice, with conventional cell windows, down to 168 nm) than were conventionally available (around 190 nm) are accessible and possess much higher light intensity.

We have recently undertaken a systematic study of heparin and heparan sulfate in solution using SRCD. Signals arise principally from two chromophores; the amide bonds present in *N*-acetyl groups of glucosamine residues and the carboxylate groups of uronic acids. The technique has also highlighted a number of distinct modes of cation binding to heparin (Fig. 2). The ability of SRCD to operate at low wavelength with unprecedented accuracy, has recently made it possible to detect changes in conformation through electronic transitions originating from oxygen atoms of the glycosidic linkages (that were hitherto undetectable), as well as through changes in uronic acid conformation originating in the carboxylic acid chromophore. Furthermore, it is possible to follow conformational changes (e.g. of individual compounds or, interestingly, complexes of several molecular species) over the submillisecond timescale relevant to conformational change [27] that may encompass binding events and can be combined with other spectroscopic techniques [22], including FTIR, NMR as well as molecular modeling.

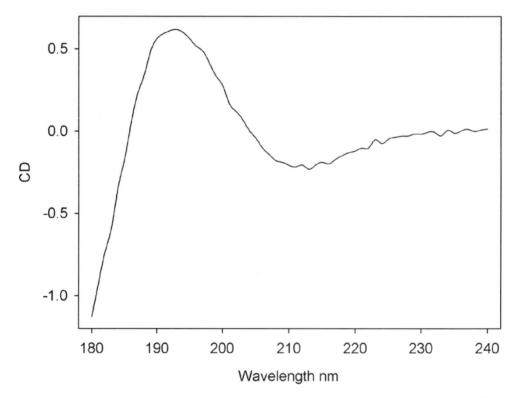

Figure 2. Synchrotron radiation circular dichroism spectrum of a heparin oligosaccharide (10 mg mL^{-1}) in the presence of Ca^{2+} cations.

For carbohydrates, CD has been restricted principally to structurally regular helical gelling polysaccharides [28] and others containing suitable chromophores, e.g., amide bonds (I), and carboxylic groups in uronic acids [29,30]. This latter work established the principles which govern CD of uronic acid derivatives between 190 and 260 nm (π to π^* transitions of the carboxylic acid) and showed they were sensitive to configuration at C-4 and C-5. The dependence of CD spectra on the counterion in these acidic derivatives was also demonstrated. At lower wavelengths, higher energy transitions arising from OH, ring oxygen, and glycosidic oxygens are observed (π to π^* and σ to σ^* from OH and ring oxygens at 175 nm, π to π^* in the carboxylic acid at 185 nm [28], $2p_x$ to σ^* and 3s at 183 nm and $2p_x$ to $3p_z$ at 171 nm of the glycosidic oxygen [31]. Hence, glycosidic linkage geometry, which is intimately related to conformation, can now start to be analyzed. The high flux beamline (CD-12) at Daresbury was purpose built for UV and vacuum UV applications and is many orders more intense than conventional CD, while exhibiting lower stray light characteristics (<0.01%), which is particularly important for low wavelength CD investigations. This facility has several advantages: (a) lower wavelengths (to 165 nm in water) are available than for conventional CD (ca. 190 nm in water) accessing this information rich region for the first time; (b) the intensity of the light source is up to 10^5 times higher, giving much greater sensitivity and precision. Preliminary SRCD spectra on some model compounds, heparin with different counterions and chemical derivatives of heparin (Fig. 2) demonstrate the feasibility of this approach.

2.2 Fourier transform infrared spectroscopy

Fourier transform infrared (FTIR) spectroscopy is a complementary technique [32] arising from the absorption of infrared radiation through resonance of non-centrosymmetric (infrared active) modes of vibration and, like CD, is also a useful tool for quantifying secondary structure in proteins. FTIR can detect changes in the conformational environment of sulfate/acetyl groups and we are also investigating whether altered iduronate residue conformation and putative hydrogen bonding patterns in HS derivatives can be detected (Fig. 3). NMR combined with molecular modeling (MM, using density functional theory) [33] will ultimately back these methods up to provide the most detailed conformational information in solution, particularly sugar-ring and glycosidic linkage geometry.

The position of the amide I band (between 1600 and 1700 cm^{-1} arising from C=O stretching) of the peptide backbone is sensitive to the short-range order imposed by distinct hydrogen bonding arrangements in protein secondary structure and can be quantified by deconvolution techniques [34,35]. In the presence of HS, this region can coincide with signals from amide bands originating in the N-acetyl groups of glucosamine residues, but it may be possible to subtract these from the spectrum of the complex because they are likely to be relatively insensitive to conformational changes in the saccharide. Deuteration of the protein and protein/HS complex following exchange with D_2O can aid the process of assignment (e.g. amide II' bands due to the difference between N-H and N-D bending move from ca. 1546 cm^{-1} to 1430/55 cm^{-1}), although the exchange is not

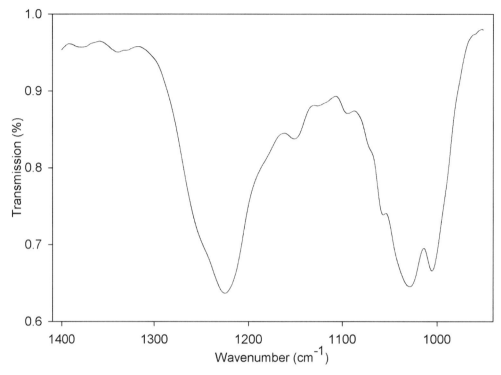

Figure 3. Fourier transform infrared spectrum of a heparin oligosaccharide (10 mg mL^{-1}) in the presence of Mg^{2+} cations. The spectrum is made up of 1000 scans at a resolution of 4 cm^{-1}.

normally complete. The bands of interest are near the middle of the spectral range, which is relatively uncluttered, but attempts to quantify the different secondary structural types are hampered by uncertainty over the extent of deuteration. Amide III bands (1350–1200 cm^{-1}) [36], which arise from several modes of vibration in the amide chromophore of the protein backbone, are also sensitive to secondary structural type; α helix (1328–1289 cm^{-1}), β sheet (1255–1224 cm^{-1}), and random coil (1288–1256 cm^{-1}). These can also be quantified but, in the presence of HS/heparin, are less useful because they coincide with strong sulfate stretches (centered between 1230 and 1240 cm^{-1} and around 1360 cm^{-1}). There is also some overlap between the positions of some spectral bands arising from particular secondary structural features, particularly in the case of amide I and II' bands. Subtraction of HS FTIR spectra (particularly in the amide I region) from that of the complex is probably currently the best approach, if analysis of the protein secondary structure is required in the presence of HS ligands, although the limitations and assumptions inherent in such an approach should be borne in mind. We discuss some possible alternatives in Section 2.5.

FTIR results from the absorption of infrared radiation through resonance of non-centro-symmetric (infrared active) modes of vibration and can monitor conformational changes involving sulfate and acetyl groups, which NMR studies have indicated, may be involved in hydrogen bonding (i.e. secondary structural elements) (Fig. 3). Sulfate stretches are observed (S=O) at 1230–1240 cm^{-1} [37], O=S=O stretches around 1360 cm^{-1}, and those arising from the N-acetyl groups (amide I; C=O) between 1600 and 1500 cm^{-1}, amide II (N-H) at 1546–1455/30 cm^{-1} and amide III between 1350 and 1200 cm^{-1}. These principal bands of interest lie in the middle of the IR frequency range at which the sensors operate at close to optimal sensitivity. Variation between 800 and 950 cm^{-1} in sulfated algal polysaccharides [38,39] has been attributed to sulfates in axial or equatorial orientation (i.e. are conformationally dependent) and this is of particular relevance to the flexible iduronate-2-sulfate group.

The approaches being developed will also be applicable to future studies of these and structurally more complex molecules alone and when interacting with proteins. This will be important to understand the properties of carbohydrates from chemical and biological perspectives and for the rational design of drugs and applications in bioengineering.

2.3 Summary of our recent findings highlighting the complementary nature of FTIR and SRCD

The relationship between substitution pattern and spectral features was recently investigated by SRCD, FTIR, and NMR, using a series of model compounds and polysaccharides related to HS [40]. In selected cases, this was also related to biological activity. FTIR and SRCD spectra indicated that subtle structural changes occurred following substitution of N-sulfate with N-acetyl groups in heparin (Fig. 3). The SRCD spectra were found to be particularly useful indicators of conformational change in the flexible iduronate-2-O-sulfate residues and these were shown to be sensitive to the substitution pattern surrounding them in adjacent glucosamine residues. This helped indicate which structures were of interest for more in-depth NMR investigations. NMR demonstrated that N-acetylated heparin exhibits altered glycosidic linkage and iduronate 2-O-sulfate ring geometries compared to heparin. The nature of interactions between heparin and a series of common cations (Na$^+$, K$^+$, NH$_4^+$, Mg^{2+}, Ca^{2+}, Cu^{2+}, Mn^{2+}, Zn^{2+}, and Fe^{3+}) were also studied by SRCD and two principal modes of interaction (one involving Na$^+$, K$^+$, Mn^{2+}, and Fe^{3+} and the other Mg^{2+}, Ca^{2+}, Zn^{2+}, NH$_4^+$ ions) with the carboxylic acid were indicated, together with a third, unique to Cu^{2+} ions. Again, more detailed studies by NMR revealed that Cu^{2+} ions bind HS initially through the carboxylic acid of iduronic acid 2-O-sulfate residues, but also interact with the glycosidic oxygen atom between iduronate and glucosamine residues, as well as ring oxygen atoms in IdoA2S residues (in the sequence IdoA2S-GlcNS6S). Binding of Cu^{2+} by heparin dodecasaccharides was also shown in an optical biosensor binding assay to increase the binding of FGF-2.

2.4 Future developments in the study of HS saccharides and their protein complexes by FTIR and SRCD

SRCD and FTIR, therefore, represent two methods of detecting conformational changes in these molecules, but their origins are distinct. They also serve to indicate which molecules could usefully be studied by more revealing, but much more laborious, NMR techniques. Another aim of HS structural studies is to be able to unravel the contributions of the individual components in HS–protein complexes. The observation, characterization, and measurement of secondary structural changes in protein–ligand interactions in solution also relies principally on spectroscopic methods, particularly CD and FTIR spectroscopy. These typically work well for relatively small ligands, such as drug molecules and hormones, in which significant interference of the protein spectrum by the ligand is not an issue. However, our recent work has involved the use of state-of-the-art FTIR and SRCD spectroscopy. These involve the study of protein interactions with a library of analogs [17,41] of HS. We found that the established CD and FTIR methods cannot cope well with the presence of HS ligands because they also exhibit spectral features, which interfere with those of the proteins.

Hundreds of proteins have been identified as heparin (a close analog of HS) binding, both by biochemical (e.g. binding) or genetic experiments (e.g. sulfotransferase knockouts) and genome analysis [42]. Many are already known to bind HS *in vivo* and many more will, no doubt, be confirmed. Some have been shown to be of critical importance to biological systems including the fibroblast growth factor/receptor family (FGF/FGFR), vascular endothelial growth factor (VEGF), glial derived neurotrophic factor (GDNF) as well as others that are involved in normal as well as aberrant processes such as lipoprotein metabolism and the regulation of extracellular superoxide dismutase. Notably, β-secretase, which acts on amyloid precursor protein (APP) to form plaques, binds and is inhibited by HS *in vivo* [43]. Most protein–HS interactions have yet to be studied quantitatively, but understanding the detailed structural changes that occur during all of these interactions with current techniques is likely to be difficult for the reasons outlined below.

The chromophores which give rise to signals in CD spectra, *N*-acetyl and carboxylic acid groups contribute in the same spectral regions (240–180 nm) as those arising from the protein backbone, the analysis of which (using a number of algorithms [24–26]) provides an estimate of the secondary structural composition of the protein. Recent work to find a suitable material for the manufacture of cell windows for SRCD, which would allow measurements to lower wavelengths than currently possible will allow these regions rich in information, particularly concerning the glycosidic linkages, to be accessed [24–26,44,45]. However, it is clear that significant levels of extraneous signals from the HS ligand undermine the sensitivity and ultimately the validity of these structural analyses. Unfortunately, there is currently no way of deconvoluting those signals arising from protein and those originating from HS except by making a simple subtraction of the spectrum of the HS ligand alone from that of the complex, a procedure that ignores the CD signal arising from any change in ligand conformation and the effects of ligand–protein interactions, to which CD is sensitive. The result will be reduced sensitivity to the structural changes that occur, which could ultimately lead to doubtful estimates of secondary structural content and, in extreme cases, will lead to incorrect structural assignments. However, perhaps the most likely outcome is that the spectrum will be indecipherable and the structural changes will remain unknown.

FTIR possesses three spectral regions that can be employed for the analysis of protein secondary structure; the amide I, II′, and III regions [46,47]. These are the result of a number of modes of vibration arising from the amide backbone of the protein and, like CD, their spectral positions are sensitive to the geometric arrangement of the amide chromophores that gave rise to them. Unfortunately, these same modes also exist within the *N*-acetyl groups in many HS samples and, furthermore, sulfate-stretching modes coincide with the amide III protein region, likewise interfering in the protein spectrum. The same problem of loss of sensitivity and erroneous structural analysis of protein–HS complexes that occurs with CD therefore also applies to transmission FTIR.

There are two related approaches that might be applied to resolve this problem and allow accurate analysis of the structural changes in protein–HS complexes to be realized. The first is the selection of active HS ligands that are free of amide groups from libraries of HS analog structures. The oligosaccharides will need to be separated, screened by conventional FTIR and selected on the basis of their activities. These will be essentially "transparent" in the FTIR amide I region, permitting transmission FTIR to be tested with active ligands lacking interfering amide groups, as well as providing an independent external verification of the second method, which is described below. The second involves the use of vibrational circular dichroism (VCD), which relies on the fact that right and left circularly polarized light (IR frequency) interact to a different extent, and are therefore differentially absorbed, with protein secondary structural elements (in an analogous way to CD, but in the IR frequency range where the physical basis of the absorption is due to modes of vibration, not electronic transitions). This phenomenon arises from the short-range, regular geometric arrangement of the amide backbone present in protein secondary structure. VCD has been applied to protein structural analysis before [48–53], but here the fundamental difference between the amide bonds of the protein backbone and those present in HS might be exploitable. The difference is that in the case of HS, the amide groups are not connected to each other and so lack the short-range, regular geometric relationship that those of the protein backbone possess. This regular geometric relationship gives rise to the VCD spectrum of protein secondary structure. It is, therefore, highly likely that the VCD spectrum of HS will lack signals in the amide regions and VCD will selectively report the protein secondary structure in the presence of HS ligands. Furthermore, the sulfate groups are non-chiral and will not present a VCD signal. This approach is still in its infancy. One obvious drawback is the large quantity of materials that will be required because VCD signals are very weak. However, the establishment of VCD on a suitable synchrotron source beamline should alleviate some of these problems.

3. PERSPECTIVES

Whilst the application of SRCD and FTIR to the elucidation of the conformation of GAGs in solution is still very much in development, we believe that together they form a pair of high-powered, complementary tools that permit fundamentally new insights into structure/function relationships in GAG–protein interactions. Thus, these techniques are enabling a transition in the field from a reductionist focus on the structure and activities of oligosaccharides that represent minimal binding fragments of the polysaccharide, as exemplified by combined biophysical and cell signaling approaches [3,54], to a holistic approach that establishes structure/function relationships at the level of entire GAG chains or proteoglycans. Particularly exciting is that these techniques are from the outset providing evidence that biology accesses both the sequence and conformational information of GAG chains for the purposes of regulating protein–GAG interactions and so cellular activities. Given the clear evidence for differential regulation of protein–GAG interactions by whole polysaccharide chains and for the association of differences in chain structure with tissue development, pathological changes and wound repair, we can now look forward to designing experiments that yield data that are commensurate with the actual biological complexity of GAGs and their regulatory functions. This is an essential step towards integrating knowledge on molecular function to gain insights into the integration of cell behavior.

Acknowledgments

The authors thank the Biotechnology and Biological Sciences Research Council, the Cancer and Polio Research Fund, the Human Frontiers Science Program and the North West Cancer Research Fund for financial support.

REFERENCES

1. Delehedde M, Lyon M, Sergeant N, Rahmoune H & Fernig DG (2001) *J. Mammary Gland Biol.* **6**, 253–273.
2. Kjellen L & Lindahl U (1991) *Annu. Rev. Biochem.* **60**, 443–475.
3. Delehedde M, Lyon M, Gallagher JT, Rudland PS & Fernig DG (2002) *Biochem. J.* **366**, 235–244.
4. Rahmoune H, Gallagher JT, Rudland PS & Fernig DG (1998) *J. Biol. Chem.* **273**, 7303–7310.
5. Turnbull JE, Fernig DG, Ke YQ, Wilkinson MC & Gallagher JT (1992) *J. Biol. Chem.* **267**, 10337–10341.
6. Ostrovsky O, Berman B, Gallagher J, et al. (2002) *J. Biol. Chem.* **277**, 2444–2453.
7. Powell AK, Fernig DG & Turnbull JE (2002) *J. Biol. Chem.* **277**, 28554–28563.
8. Faham S, Hileman RE, Fromm JR, Linhardt RJ & Rees DC (1996) *Science* **271**, 1116–1120.
9. Lyon M, Deakin JA, Rahmoune H, Fernig DG, Nakamura T & Gallagher JT (1998) *J. Biol. Chem.* **273**, 271–278.
10. Rahmoune H, Gallagher JT, Rudland PS & Fernig DG (1998) *Biochemistry* **37**, 6003–6008.
11. Aviezer D, Hecht D, Safran M, Eisinger M, David G & Yayon A (1994) *Cell* **79**, 1005–1013.
12. Kato M, Wang H, Kainulainen V et al. (1998) *Nat. Med.* **4**, 691–697.

13. Delehedde M, Deudon E, Boilly B & Hondermarck H (1996) *Exp. Cell Res.* **229**, 398-406.
14. Fernig DG, Chen HL, Rahmoune H, Descamps S, Boilly B & Hondermarck H (2000) *Biochem. Biophys. Res. Commun.* **267**, 770-776.
15. Ferro DR, Provasoli A, Ragazzi M *et al.* (1986) *J. Am. Chem. Soc.* **108**, 6773-6778.
16. Desai UR, Wang HM, Kelly TR & Linhardt RJ (1993) *Carbohydr. Res.* **241**, 249-259.
17. Yates EA, Santini F, Guerrini M, Naggi A, Torri G & Casu B (1996) *Carbohydr. Res.* **294**, 15-27.
18. Yates EA, Santini F, De Cristofano B *et al.* (2000) *Carbohydr. Res.* **329**, 239-247.
19. Mulloy B, Forster MJ, Jones C & Davies DB (1993) *Biochem. J.* **293**, 849-858.
20. Hricovini M, Guerrini M, Torri G & Casu B (1997) *Carbohydr. Res.* **300**, 69-76.
21. Hricovini M, Guerrini M, Torri G, Piani S & Ungarelli F (1995) *Carbohydr. Res.* **277**, 11-23.
22. Jones GR & Clarke DT (2004) *Faraday Disc.* **126**, 223-236.
23. Wallace BA (2000) *Nat. Struct. Biol.* **7**, 708-709.
24. Greenfield NJ (1996) *Anal. Biochem.* **235**, 1-10.
25. Greenfield NJ (1999) *Trends Anal. Chem.* **18**, 236-244.
26. Wallace BA (2006) http://www.cryst.bbk.ac.uk/cdweb/html/.
27. Clarke DT, Doig AJ, Stapley BJ & Jones GR (1999) *Proc. Natl Acad. Sci. USA* **96**, 7232-7237.
28. Arndt ER & Stevens ES (1996) *Carbohydr. Res.* **280**, 15-26.
29. Buffington LA, Pysh ES, Chakrabarti B & Balazs EA (1977) *J. Am. Chem. Soc.* **99**, 1730-1734.
30. Morris ER, Rees DA, Sanderson GR & Thom D (1975) *J. Chem. Soc. Perkin Trans.* **2**, 1418-1425.
31. Arndt ER & Stevens ES (1993) *J. Am. Chem. Soc.* **115**, 7849-7853.
32. Oberg KA, Ruysschaert JM & Goormaghtigh E (2004) *Eur. J. Biochem.* **271**, 2937-2948.
33. Hricovini M, Nieto OP & Torri G (2002) In: *NMR Spectroscopy of Glycoconjugates*, pp. 189-229. Edited by J Jiménez-Barbero and T Peters. Wiley-VCH Verlag, Weinheim, Germany.
34. Manning MC (2005) *Expert Rev. Proteomics* **2**, 731-743.
35. Surewicz WK, Mantsch HH & Chapman D (1993) *Biochemistry* **32**, 389-394.
36. Cai SW & Singh BR (1999) *Biophys. Chem.* **80**, 7-20.
37. Longas MO & Breitweiser KO (1991) *Anal. Biochem.* **192**, 193-196.
38. Chiovitti A, Kraft GT, Bacic A, Craik DJ & Liao ML (2001) *J. Phycol.* **37**, 1127-1137.
39. Villanueva RD & Montano MNE (2003) *J. Phycol.* **39**, 513-518.
40. Guimond SE, Rudd TR, Skidmore MA *et al. J. Biol. Chem.* submitted.
41. Yates EA, Jones MO, Clarke CE *et al.* (2003) *J. Mater. Chem.* **13**, 2061-2063.
42. Selleck SB (2000) *Trends Genet.* **16**, 206-212.
43. Scholefield Z, Yates EA, Wayne G, Amour A, McDowell W & Turnbull JE (2003) *J. Cell Biol.* **163**, 97-107.
44. Matsuo K & K.Gekko K (2004) *Carbohydr. Res.* **339**, 591-595.
45. Matsuo K, Sakai K, Matsushima Y, Fukuyama T & Gekko K (2003) *Anal. Sci.* **19**, 129-132.
46. Byler DM & Susi H (1986) *Biopolymers* **25**, 469-487.
47. Prestrelski SJ, Arakawa T, Kenney WC & Byler DM (1991) *Arch. Biochem. Biophys.* **285**, 111-115.
48. Baello BI, Pancoska P & Keiderling TA (1997) *Anal. Biochem.* **250**, 212-221.
49. Dukor RK & Keiderling TA (1991) *Biopolymers* **31**, 1747-1761.
50. Pancoska P, Bitto E, Janota V & Keiderling TA (1994) *Faraday Disc.* 287-310.
51. Pancoska P, Janota V & Keiderling TA (1996) *Appl. Spectrosc.* **50**, 658-668.
52. Pancoska P, Wang LJ & Keiderling TA (1993) *Prot. Sci.* **2**, 411-419.
53. Vass E, Hollosi M, Besson F & Buchet R (2003) *Chem. Rev.* **103**, 1917-1954.
54. Delehedde M, Lyon M, Vidyasagar R, McDonnell TJ & Fernig DG (2002) *J. Biol. Chem.* **277**, 12456-12462.

CHAPTER 7
Structural determination of *N*-linked glycans by matrix-assisted laser desorption/ionization and electrospray ionization mass spectrometry

David J. Harvey

1. INTRODUCTION

This chapter reviews methods for the analysis of *N*-linked glycans by mass spectrometry and covers topics such as the examination of intact glycoproteins, the release of glycans by chemical and enzymatic approaches and their analysis by matrix-assisted laser desorption/ionization (MALDI) and electrospray ionization (ESI) mass spectrometry. MALDI mass spectrometry provides a rapid method for profiling neutral *N*-linked glycans as their $[M + Na]^+$ ions but is less satisfactory for sialylated glycans because of laser-induced sialic acid loss. Stabilization can be achieved by methylation. ESI mass spectrometry is a softer technique that preserves the sialic acid but is less useful for quantitative glycan profiling because of its tendency to produce several ions (e.g. $[M + Na]^+$, $[M + 2H]^{2+}$) from each glycan. Details of glycan structure can be acquired by techniques such as sequential exoglycosidase digestion with product monitoring by high-performance liquid chromatography (HPLC) or mass spectrometry, or by mass spectrometric fragmentation. Fragmentation of positive ions ($[M + H]^+$ and $[M + Na]^+$) gives sequence and linkage information but some spectra are ambiguous because of multiple pathways to single fragment ions. Negative ions such as $[M - H]^-$ or the more stable anion adducts formed by addition of, for example, chloride, nitrate, or phosphate, are stable and, furthermore, fragment to give much more informative spectra than their positive ion counterparts. Negative ion fragmentation enables structural details of the glycans to be determined that were difficult to obtain by classical methods.

The function and physicochemical properties of many proteins are influenced by posttranslational events of which glycosylation is one of the most important [1,2]. Attached carbohydrates are classified into two main groups depending on their general structure and site of attachment. N-Linked glycans are attached with an amide bond to asparagine in an Asp-Xxx-Ser (or Thr or, occasionally Cys) motif, where Xxx is any amino acid except proline. Their structures are generally well defined, unlike the O-linked glycans attached to serine or threonine which, although usually smaller, have less well-defined structures and are, consequently, more difficult to analyze.

N-Linked glycans are biosynthesized by attachment of the preformed tetradecasaccharide $Glc_3Man_9GlcNAc_2$-dolichol (**I**) (Scheme 1) to the nascent protein, followed by cleavage of the three glucose residues (**II**), the latter process being involved with protein folding. Once the protein has folded correctly, mannosidase I trims two mannose residues from each of the two antennae to give a glycan of structure $Man_5GlcNAc_2$ (**III**), commonly referred to as Man-5, a member of the high-mannose group of N-linked glycans. GlcNAc-transferase I can then add a GlcNAc residue to the mannose of the 3-antenna which produces a glycan (**IV**) that can be further elaborated along two main pathways. Either the 3-antenna can be extended by addition of galactose (**V**) and N-acetylneuraminic acid (sialic acid) to give a series of glycans known as hybrid glycans or two further mannose residues can be removed from the 6-antenna (**VI**) opening the way for GlcNAc (**VII**) and galactose addition to the remaining mannose to form biantennary complex glycans (**VIII**, both antenna processed). Further addition of N-acetyl-lactosamine (Gal-β-(1–4)-GlcNAc) to one or both of the 4-position of the 3-linked mannose or the 6-position of the 6-linked mannose produces tri- (**XI**) and tetra-antennary (**XII**) complex glycans respectively. Antennae usually terminate in sialic acid but can also contain substituents such as fucose, sulfate, additional N-acetyl-lactosamine groups or α-linked galactose. Fucose is commonly added to the reducing-terminal GlcNAc residue at the 6-position in mammals (**X**) and the 3-position in plants and insects. Fucose can also be added to ether the galactose or GlcNAc residues of the antennae. Another common addition is a GlcNAc residue at the 4-position of the core mannose (**IX**) to give what is commonly known as a bisecting GlcNAc residue.

It can be seen, therefore, that N-linked glycans have well-defined structures, making their analysis comparatively straightforward. However, because of the branched nature of these compounds, their structural characterization has not yet achieved the sophistication of that of the protein moiety of glycoproteins whose primary structure consists only of a linear chain of amino acids. All N-linked glycans have a common trimannosylchitobiose ($Man_3GlcNAc_2$) core and their structures can be determined by identification of factors such as: the type of glycan (high-mannose, hybrid or complex), the number and composition of each antenna, number and location of fucose residues, presence or absence of a bisecting GlcNAc residue and the presence of any further substitution.

CHAPTER 7: STRUCTURAL DETERMINATION OF N-LINKED GLYCANS

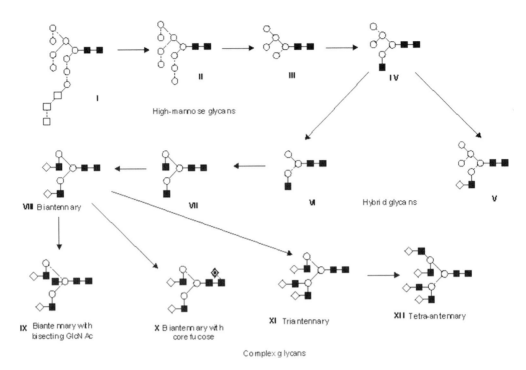

Scheme 1. Biosynthesis of N-linked glycans. Arrows are illustrative and do not necessarily depict intact biochemical pathways. Structures are only a few of those possible. Key to symbols used for the constituent monosaccharides in this scheme and subsequent figures: ◇ = galactose, ■ = GlcNAc, ○ = mannose, ◇ = fucose, ★ = N-acetylneuraminic (sialic) acid.

2. MASS SPECTROMETRY OF INTACT GLYCOPROTEINS AND GLYCOPEPTIDES

The number of N-linked glycans attached to glycoproteins can vary from one to several tens but most compounds contain only a few N-linked sites, typically from one to five. Each N-linked site can be variously occupied from zero to 100% and contain anything from one to over 100 different structures. Glycoproteins with several occupied sites can, therefore, consist of a large number of glycoforms, making their resolution by techniques such as mass spectrometry difficult or impossible (Fig. 1b). Nevertheless, glycoforms of relatively small glycoproteins with one site occupied by a limited set of glycans, such as ribonuclease B (15 kDa) (Fig. 1a) or soluble CD59 (10 kDa) (Fig. 1c) can be resolved reasonably well by both MALDI time-of-flight (TOF) or ESI mass spectrometry.

If the mass of the protein is known from its sequence, and the protein contains a single occupied N-linked site and no additional posttranslational modifications,

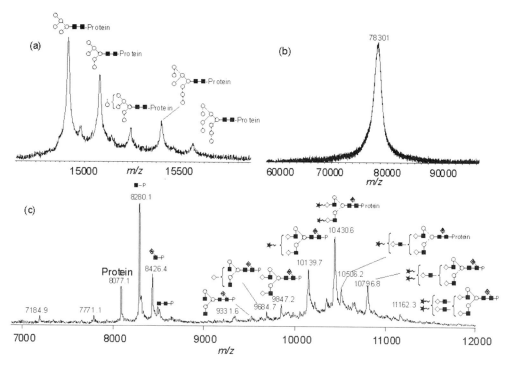

Figure 1. Positive ion MALDI mass spectrum of glycoforms of (a) ribonuclease B and (b) transferrin, and (c) soluble CD59.

the glycan masses can be obtained by subtraction of the protein mass from that measured experimentally. Glycan compositions, in terms of the constituent isobaric monosaccharides, can then be calculated because N-linked glycans contain only a limited number of monosaccharide types (mainly hexose (mannose, glucose, galactose) N-acetylaminohexose, deoxyhexose (fucose), and sialic acid). Combinations of the masses of these monosaccharides give unique molecular weights for most combinations in the mass range of the common N-linked glycans. Table 1 lists the masses of the common monosaccharide constituents. The glycan composition provides a good prediction of the type of glycan (high-mannose etc.) that is present and provides a guide for the next stage of the analysis.

Most glycoproteins do not behave in this ideal way, however, and the mass spectrum only indicates the presence of glycosylation by, for example, containing a broad unresolved peak with a mass higher that that predicted from the amino acid sequence (Fig. 1b). Proteolysis, commonly by trypsin, is thus used to split the protein into smaller units that can be resolved. If cleavage isolates each N-linked site to a single glycopeptide, it is possible to determine the glycosylation at each site both with respect to occupancy and to the composition of the glycans present. However, glycopeptides often give very weak mass spectra in the presence

Table 1 Residue masses of common monosaccharides

Monosaccharide	Residue formula	Residue mass[a]
Pentose	$C_5H_8O_4$	132.042
		132.116
Deoxyhexose	$C_6H_{10}O_4$	146.078
		146.143
Hexose	$C_6H_{10}O_5$	162.053
		162.142
Hexosamine	$C_6H_{11}NO_4$	161.069
		161.157
HexNAc	$C_8H_{13}NO_5$	203.079
		203.179
Hexuronic acid	$C_6H_8O_6$	176.032
		176.126
N-Acetylneuraminic acid	$C_{11}H_{17}NO_8$	291.095
		291.258
N-Glycoylneuraminic acid	$C_{11}H_{17}NO_9$	307.090
		307.257

[a]Top figure is the monoisotopic mass (based on C = 12.000000, H = 1.007825, N = 14.003074, O = 15.994915); lower figure is the average mass (based on C = 12.011, H = 1.00794, N = 14.0067, O = 15.9994).
The masses of the intact glycans can be obtained by addition of the residue masses given above, plus the mass of the terminal group (H_2O for an unmodified glycan) 18.011 (monoisotopic) and 18.152 (average) and the mass of any reducing terminal or other derivative.

of peptides and some form of fractionation, such as by HPLC, is usually necessary. Several methods for identifying glycopeptides in peptide/glycopeptide mixtures separated by HPLC have been published [3]. The glycopeptides can be located by high-energy fragmentation, under which conditions the glycans fragment to give diagnostic fragments such as m/z 163 (hexose), 204 (HexNAc), and 366 (Hex-HexNAc). Ion chromatograms of these ions can be compared with the total ion chromatogram to identify which of the peaks corresponds to a glycopeptide.

3. RELEASE OF N-GLYCANS FROM GLYCOPROTEINS

A more common method of glycan analysis is to release them from the glycoprotein before any structural work is attempted. This approach overcomes any resolution problems and ambiguities resulting from determination of protein or peptide masses but limits the available information on the attachment site to the protein. Glycan release can be by either chemical or enzymatic methods. Hydrazine release has been used expensively in the past [4] but is now falling out of favor. It cleaves amide bonds and has the advantage of being capable of releasing all of the glycans as intact reducing sugars. Its disadvantages are that the protein is totally destroyed and that the acetyl groups are removed from *N*-acetylaminosugars, necessitating a re-*N*-acetylation step with concomitant *O*-acetylation often occurring as a side-reaction. Enzymatic release is usually performed with protein *N*-glycosidase F (PNGase F) which releases most glycans except those containing a 3-linked fucose attached to the reducing terminal GlcNAc residue. In these cases, PNGase A is effective. Both enzymes cleave the glycans as glycosylamines, as a consequence of the amide link to the protein, but these compounds hydrolyze to the intact reducing sugars. The conversion can be accelerated by lowering the pH, usually by the addition of acetic acid, but care must be taken to avoid concomitant sialic acid loss. Another common endoglycosidase, endoglycosidase H (endo H) cleaves between the GlcNAc residues of the chitobiose core but has the disadvantage that information on the reducing terminal GlcNAc residue, such as the presence or absence of fucose, is lost.

Glycoproteins are frequently isolated and purified by sodium dodecylsulfate polyacrylamide gel electrophoresis (SDS PAGE). Several methods are available for releasing the glycans within the separating gel. The method that we use was developed by Küster et al. [5] and involves the diffusion of PNGase F into a low-density gel after protein separation, followed by extraction of the glycans with water and acetonitrile. Clean-up of neutral glycans with a three-bed microcolumn containing AG50, C18, and AG3 to remove cations, organics, and anions respectively has largely been replaced by allowing an aqueous solution of the glycans to sit on a Nafion 117 membrane [6] for about 15 min. In any case, the three-bed column procedure is unsuitable for sialylated glycans because these are removed by the AG3 resin. Clean-up of samples is particularly important before analysis by mass spectrometry because the presence of contaminants frequently causes suppression of the intensity of ions from the analyte.

Several other methods of glycan release have been developed. Of particular interest are those that involve the release of the glycan from two-dimensional gels [7] and those that involve release from glycoproteins directly on the MALDI target [8].

4. MASS SPECTROMETRIC PROFILING OF *N*-GLYCANS BY MASS SPECTROMETRY

4.1 MALDI

The use of MALDI for the analysis of carbohydrates was first reported by Karas and Hillenkamp in one of their early papers on the technique [9] and it was applied to *N*-linked glycans by Mock et al. in 1991 [10]. The technique, as it applies to carbohydrate analysis was reviewed in 1999 [11] and this review is being updated on a two-yearly basis. MALDI ionization produces predominantly $[M + Na]^+$ ions from neutral glycans although other cations can be introduced by doping the matrix with an appropriate salt. Only singly charged ions are produced and the technique is ideal for obtaining quantitative profiles of neutral glycans (Fig. 2a). Sialylated glycans produce a mixture of $[M - H]^-$, $[M + Na]^+$, and $[M - nH + (n+1)Na)]^+$ ions, the latter being due to salt formation but there is also a pronounced tendency for loss of sialic acid (Fig. 2b). Early studies on the quantitative nature of MALDI mass spectrometry showed that there was a linear relationship between the signal produced by the $[M + Na]^+$ ion and the amount applied to the target [12] and that reasonably accurate quantitative measurements could be made with and without an internal standard if precautions were taken to allow for the inhomogeneous nature of the target surface. Although accuracy was reasonably good, precision was lower, a feature common to quantification of other compounds by MALDI. N-Linked glycans with masses above about 1 kDa were found to give comparable signals, irrespective of structure [13], probably as the result of a similar ionization mechanism in the formation of the $[M + Na]^+$ ion. In contrast, peptides give signals of very different abundance depending on the proton affinity of their constituent amino acids.

Many matrices have been used to ionize carbohydrates, but one of the earliest, 2,5-dihydrobenzoic acid (DHB) [14] is still the most popular. Although the dried

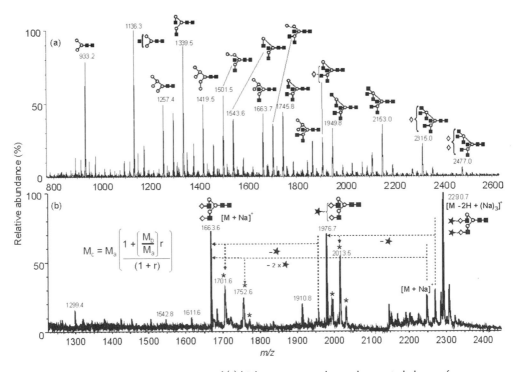

Figure 2. Positive ion MALDI mass spectra of (a) high-mannose and complex neutral glycans from chicken egg white glycoproteins (mainly ovalbumin) and (b) a sialylated biantennary glycan showing extensive metastable ion formation (broad peaks marked with an asterisk). The formula enables the mass of metastable ions to be calculated.

droplet method of sample preparation works reasonably well, more homogeneous targets and, consequently, better performance can be achieved by techniques such as mixing the DHB with various additives such as 2-hydroxy-5-methoxy-benzoic acid [15], 1-amino-*iso*-quinoline (HIQ) [16] or spermine [17], or by recrystallization of the initially dried sample spot from ethanol [12]; the technique that we prefer. 2,4,6-Trihydroxyacetophenone (THAP) is more appropriate for sialylated glycans, particularly when mixed with ammonium citrate [18] because it minimizes laser-induced loss of sialic acid.

The fragmentation of sialylated carbohydrates produces both in-source and post-source (PSD) fragmentation. When recorded with reflectron-TOF instruments, the PSD fragments produce prominent broad metastable peaks (Fig. 2b) whose apparent mass can be related to the parent and fragment ion masses by means of the equation shown in the figure [19]. In order to minimize the effects of fragmentation many investigators examine sialylated glycans with TOF instruments operated in linear mode. However, this only removes the post-source component of the fragmentation.

The loss of sialic acid can be overcome by formation of methyl esters [20] because this technique removes the acidic carboxylic proton, the loss of which catalyzes the loss of sialic acid. Methylation not only stabilizes the sialic acids but also inhibits negative ion formation and the production of positive ions produced by the sodium salts, thus allowing quantitative glycan profiling to be made in the positive ion mode. The method cited above for methyl ester formation employed a relatively lengthy procedure involving conversion of the sialic acids into their sodium salts and their subsequent reaction with methyl iodide. A simpler, recently developed procedure, involves simply heating the glycans in methanolic solution with 4-(4,6-dimethoxy-1,3,5-triazin-2-yl)-4-methyl-morpholinium chloride, a method that has the additional advantage of converting all $\alpha2\rightarrow6$-linked sialic acids to methyl esters but forming lactones from the $\alpha2\rightarrow3$-linked acids. The 32 mass unit difference is readily detectable by mass spectrometry.

4.2 Electrospray ionization

Electrospray ionization produces several types of ion, depending on the ion source conditions and additives to the solvent. In positive ion mode, [M + Na]$^+$ ions can be obtained, as in MALDI, at high cone voltages [21] but these are usually accompanied by a considerable amount of in-source fragmentation. Sensitivity, however, generally falls as a function of increasing mass such that it is difficult to record [M + Na]$^+$ ions from the larger glycans. Both [M + H]$^+$ and doubly charged ions produced by protonation are formed under milder ion-source conditions. An advantage of electrospray over MALDI is that the milder ionization conditions cause little, if any, sialic acid loss but, on the other hand, the production of several types of ion and some cross-ring in-source fragmentation limits the technique for glycan profiling.

Prominent [M – H]$^-$ ions are formed in negative ion mode but these ions frequently produce in-source fragments. However, stable negative ions of the type [M + A]$^-$, where A is an anion such as SO_4^{2-} [22], Cl$^-$ [23], Br$^-$ or NO_3^- [24], can be produced by adding the appropriate salt to the electrospray solvent. Addition of ammonium nitrate to the ESI solvent has been found to be an efficient way to produce these ions; the adducts are stable and give an increase in sensitivity of about an order of magnitude over formation of [M – H]⁻ ions. Addition of metal salts, such as silver nitrate, can produce a high background as the result of cluster formation, thus limiting the sensitivity. Consequently, ammonium salts are better. Although nitrate adducts form readily, residual phosphate and chloride anions in glycan samples recovered from SDS gels frequently result in these anions forming the major species in ESI spectra. However, as described below, all of these adducts fragment in a similar way and, thus, the type of adduct has little effect on the information that can subsequently be acquired from the fragmentation spectra.

5. STRUCTURAL INFORMATION FROM RELEASED GLYCANS

5.1 Exoglycosidase digestion

A traditional method for structural determination of N-glycans is to make use of the many specific exoglycosidases that are currently available. A profile of the native compounds is first obtained by gel filtration or normal-phase HPLC [25] of the glycans labeled with a fluorescent dye such as 2-aminobenzamide (2-AB) [26]. The glycans are then incubated with an exoglycosidase that removes monosaccharide residues from the non-reducing termini of the antennae (see Dwek et al. [27] for suitable enzymes). Reprofiling the remaining glycans reveals the number of residues removed and their composition and linkage is determined by the specificity of the exoglycosidase. The technique was adapted for MALDI mass spectrometry in 1994 [28] and has the advantage that resolution, in terms of the number of glycans seen is higher than with the chromatographic techniques and linking parent and product glycans is easier. This approach also does not require fluorescence labeling and, furthermore, labeling can be a distinct disadvantage because of sample loss during clean-up and by the production of less informative spectra. On the other hand, resolution of isomers is not possible by direct mass measurement although this information can be recovered by negative ion fragmentation as described below.

5.2 Fragmentation

Although exoglycosidase sequencing is a robust and convenient technique, mass spectral fragmentation is becoming more generally accepted as a method for obtaining structural information because there are currently problems in obtaining exoglycosidases in a sufficiently pure state. Although FAB mass spectrometry has been used in this context for many years [29] it is now being replaced by the more sensitive techniques of electrospray or MALDI ionization.

Carbohydrates show two main types of fragmentation, cleavages of the bonds between the sugar rings, known as glycosidic fragmentation, and cleavages of the rings themselves, a process known as cross-ring cleavages. Glycosidic cleavage ions provide information mainly on carbohydrate sequence with some information on branching, whereas the cross-ring cleavage ions provide the most informative information on linkage and branching. Glycosidic cleavages from even-electron ions, such as $[M + Na]^+$, result in the loss of neutral molecules and are, thus, always accompanied by hydrogen migrations.

The nomenclature usually used for describing these ions is that introduced by Domon and Costello in 1988 [30] (Scheme 2). In this scheme, ions retaining the charge on the non-reducing terminus are named A (cross-ring), B and C (glycosidic), whereas those ions retaining charge on the reducing terminus are X (cross-ring), Y and Z as outlined in Scheme 2. Numbering (as subscripts) is from the reducing terminus for the X, Y, and Z ions and from the non-reducing end for the others. Cross-ring cleavage ions additionally have a superscript number defining the bonds that are broken. In order to avoid the problem of the subscript number changing for similar fragmentations as the result of different chain lengths, we use a modification of this system in which B-ions produced by cleavage of the reducing-terminal GlcNAc are termed B_R (R = reducing) ions and Y-ions from the non-reducing terminus are Y_{NR} ions. By extension, cleavages at the next glycosidic bonds are B_{R-1} and Y_{NR-1} respectively [31].

Until recently, recording of the fragment ions from MALDI-generated ions with TOF instruments was restricted mainly to PSD technology [32] with its inherent problems of low resolution and the necessity to record spectra in segments with the majority of instruments. More recently, however, instruments with curved-field reflectrons or specific acquisition methods have enabled PSD spectra to be acquired in single scans. Higher quality fragmentation spectra can be obtained using collision-induced decomposition (CID) with, for example quadrupole-TOF (Q-TOF)-type instruments using either MALDI [33,34] or ESI sources. With these instruments, MS/MS spectra recorded following each ionization method are virtually identical [34].

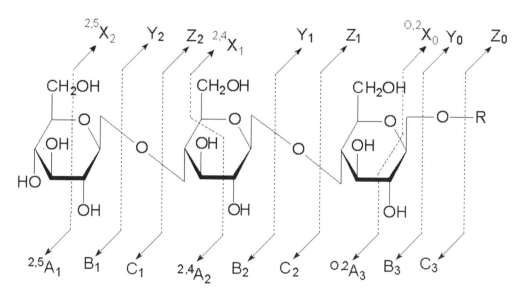

Scheme 2. Nomenclature introduced by Domon and Costello (30) for describing the fragmentation of carbohydrates.

72 ■ CHAPTER 7: STRUCTURAL DETERMINATION OF N-LINKED GLYCANS

Fragmentation of [M + H]$^+$ ions

CID spectra of [M + H]$^+$ ions from underivatized N-linked glycans are relatively simple and not particularly informative as they contain mainly B- and Y-type glycosidic cleavage ions produced by one or more cleavages (Fig. 3a) [21]. Cross-ring cleavage ions are generally absent. A complicating factor is the appearance of

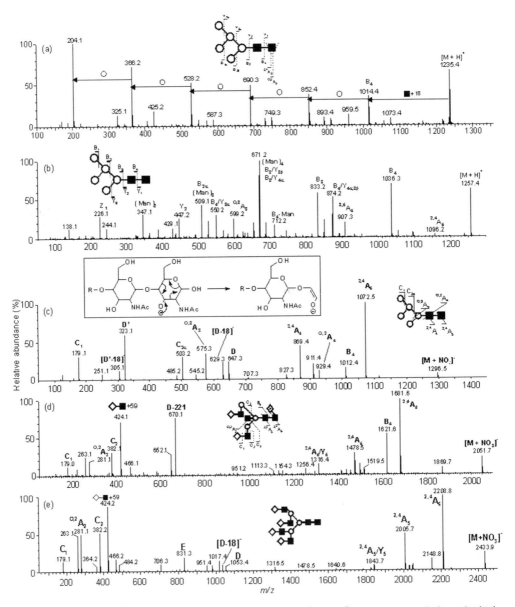

Figure 3. (a and b) Positive ion CID spectra of the [M + H]$^+$ and [M + Na]$^+$ ions respectively from the high-mannose glycan, Man$_5$GlcNAc$_2$. (c–e) Negative ion CID spectra of the nitrate adducts of Man$_5$GlcNAc$_2$, a bisected, fucosylated biantennary glycan and a tetra-antennary glycan respectively. The insert to (c) shows the proposed mechanism for the formation of $^{2,4}A_R$ ions.

rearrangement ions [35], particularly from glycans derivatized at the reducing terminus and containing fucose [36,37]. The presence of such ions can considerably complicate the interpretation of the spectra of unknown glycans. Glycans that are derivatized at the reducing terminus by reductive amination such that the derivative contains a quaternary ammonium group or an amine that can be protonated, retain the charge on the derivative and, consequently, their fragmentation spectra are dominated by Y-type cleavages that provide predominantly sequence information. Glycans derivatized with neutral fluorophores such as 2-AB and ionized as $[M + H]^+$ ions, fragment in a similar manner to the underivatized glycans. When examined by liquid chromatography/mass spectrometry (LC/MS) using normal-phase (NP)-HPLC columns and an ammonium formate buffer, a range of $[M + 2X]^{2+}$ ions where X = H^+, NH_4^+, or Na^+ can be produced. If one of the X atoms is hydrogen, fragmentation is normally similar to that of the $[M + H]^+$ ion with the production of abundant singly charged Y ions providing sequence information.

Fragmentation of [M + alkali metal]$^+$ ions

In contrast to the fragmentation of $[M + H]^+$ ions, fragmentation of $[M + Li]^+$ or $[M + Na]^+$ (Fig. 3b) ions from N-linked glycans are much more complicated. Internal fragments (losses from two or more terminal sites) are more common and, in addition, A-type cross-ring fragments are present [21,38] although generally in fairly low abundance from the low-energy spectra that are typical of spectra produced by the collision cells of Q-TOF type instruments [21,34,39]. The most significant of these ions are the $^{0,2}A_R$ and $^{2,4}A_R$ ions from the reducing terminus and $^{0,3}A$, $^{0,4}A$, and $^{3,5}A$ ions from the branching mannose residue; the latter two ions can be used to define the composition of the individual antennae attached to the core mannose. An additional advantage of the fragmentation of the [M + metal]$^+$ ions is that the rearrangement ions reported from the $[M + H]^+$ ions are absent [40]. It is increasingly difficult to fragment adducts of the higher alkali metals to the extent that $[M + Cs]^+$ ions yield practically no fragments other than Cs^+ [41]. Under high-energy conditions, such as those produced by magnetic sector [42] or TOF/TOF instruments [43,44], additional abundant X-type cross-ring fragment ions are produced. These high-energy spectra are useful in other respects because of the production of other structure-specific ions. Thus, a prominent diagnostic ion is produced by loss of a bisecting GlcNAc residue as 221 mass units (C-cleavage) from the internal cleavage ion, designated "D ion" that is formed by loss of the chitobiose core and 3-antenna.

Further information on glycan structure and the fragmentation mechanisms leading to the diagnostic ions in their MS/MS spectra can be obtained by performing additional successive stages of fragmentation with instruments such as the ion trap, the so-called "MS^n" technique. Thus, for example, we have recently used the Shimadzu–Kratos MALDI-quadrupole ion trap (MALDI-QIT) tandem ion trap/TOF instrument to define fragmentation pathways in the positive ion MS/MS spectra of N-linked glycans and have confirmed that many ions are formed by several pathways [31] and, consequently, positive ion fragmentation spectra can sometimes give ambiguous results. However, for structural studies, fragment ions derived from specific parts of a molecule, such as one antenna of an N-linked glycan, can be selected and fragmented further for detailed structural information. Several research groups have extended this concept to include database matching and have developed strategies for the automatic sequencing of these compounds [45,46].

Fragmentation of negative ions

Until recently, fragmentation of negative ions has received less attention than that of positive ions even though the spectra are more informative and contain fewer of the ambiguities present in the positive ion spectra. Although both glycosidic and cross-ring fragments are produced, as in positive ion spectra, the glycosidic ions tend to be of the C-type. These and A-type cross-ring fragments often dominate [47–50] the spectra (Fig. 3c–e). Acidic carbohydrates, such as those containing sialic acid or carboxyl-containing derivatives such as 2-aminobenzoic acid (2-AA) [51], ionize by loss of one or more protons from acidic groups to give $[M - nH]^{n-}$ ions with localized charges and, consequently, restricted fragmentation, whereas neutral carbohydrates ionize by proton loss from one of the many hydroxyl groups and produce very specific and structurally diagnostic fragment ions (Fig. 3c–e) by mechanisms similar to that shown in Fig. 3c (insert) [52,53]. The $[M - H]^-$ ions formed from neutral carbohydrates tend to be rather unstable and can show extensive fragmentation in the ion source of the mass spectrometer. However, ionization also occurs by anion adduction with anions such as Cl^-, Br^-, NO_3^-, or $H_2PO_4^-$ to give stable species that, nevertheless, produce similar fragmentation spectra to the $[M - H]^-$ ions after the initial loss of the corresponding acids (HCl etc.) by hydrogen abstraction [24].

Many of the diagnostic ions produced from neutral N-linked glycans provide very specific information on structure and often information that is not available directly from techniques such as exoglycosidase digestions. One of the most prominent fragments is a $^{2,4}A_R$ cleavage of the reducing-terminal GlcNAc residue (Fig. 3c–e) which confirms the β1→4-linkage between the GlcNAc residues and indicates the presence or absence of 6-linked fucose on the core GlcNAc (fucose is eliminated in the neutral fragment, Fig. 3d). Although this $^{2,4}A_R$ ion is not present in the spectra of glycans derivatized at the reducing terminus by reductive amination, because of the open nature of the GlcNAc ring, B_R and $^{2,4}A_{R-1}$ (cleavage of the penultimate GlcNAc residue) carry similar information with respect to core fucosylation. Another very diagnostic ion contains the 6-antenna and the branching mannose residue from the core (named D ion) and, consequently, allows the composition of both the 6- and 3- (by difference) antennae to be determined. The mass of this ion allows isomers of, for example, the high-mannose glycans to be identified [52]. The D ion is accompanied by another ion 18 mass units lower formed, presumably, by loss of H_2O. When a bisecting GlcNAc residue is present, the D ion, which will contain the bisecting GlcNAc residue, eliminates this GlcNAc as a GlcNAc molecule (221 u) to give what is usually one of the most abundant ions in the spectrum (Fig. 3d). Identification of bisecting GlcNAc residues is difficult by exoglycosidase digestion or NP-HPLC. Antenna composition is revealed by a cross-ring cleavage of the mannose residues to give an ion consisting of the antenna plus H and -O-CH=CH-O⁻ (59 u), e.g. Gal-GlcNAc-O-CH=CH-O- (m/z 424) and, consequently, the presence of substituents such as fucose or α-galactose can easily be spotted. An ^{0,4}A cleavage of the mannose residue from each antenna also gives a prominent fragment allowing the two possible triantennary complex glycans to be differentiated because this ion contains the substituent at the 4- but not the 6-position. These and other diagnostic fragments are summarized in Table 2.

Table 2 Ions defining structural features in the negative ion spectra of N-linked glycans

Structural feature	Ion
Composition	Molecular ion (data from Table 1)
Antenna sequence	C Ions
Fucose at 6-position of reducing terminus	$[M - X - 307]^-$ ($^{2,4}A_R$ ion)[a]
No fucose at 6-position of reducing terminus	$[M - X - 161]^-$ ($^{2,4}A_R$ ion)
Composition of 6-antenna	D and $[D - 18]^-$ ions (e.g. m/z 688, 670 = Gal-GlcNAc-Man; 647, 629 = Man₃). Tetra- and tri-antennary glycans with a branched 6-antenna give a prominent ion at $[D - (2\times18)]^-$ (m/z 1017) $^{0,3}A_{R-2}$ $^{0,4}A_{R-2}$
Presence of bisect	Absent or weak D ion, strong D - 221
Composition of 3-antenna	Substituents plus 101 u from mannose (E ion). E ions can also arise from the 6-antenna but those from the 3-antenna can be identified by taking note of the presence of D and $[D - 18]^-$ ions. Glycans with Gal-GlcNAc₂ in the 3-antenna (tri- or tetra-antennary) give a prominent E ion at m/z 831).
Presence of sialic acid	m/z 290.1
Presence of α2→6-linked sialic acid	Additional m/z 306.1
Gal-GlcNAc in antenna	m/z 424.1 (Gal-GlcNAc + 59)
Presence of α-galactose on antenna	m/z 586 (Gal-Gal-GlcNAc + 59)
Presence of fucose on antenna	m/z 570 (Gal-(Fuc)GlcNAc + 59)

[a] X = Anionic adduct or H.

6. CONCLUSIONS

MALDI/mass spectrometry provides a rapid method for obtaining qualitative and quantitative information on neutral and derivatized sialylated N-linked glycans. Measured mass leads directly to the constituent monosaccharide composition and additional structural information can be obtained by fragmentation. ESI mass spectrometry, although more sensitive to the presence of contaminants, minimizes the problem of sialic acid instability but tends to produce several types of ion from many compounds. However, negative ions of neutral compounds fragment by specific pathways to give definitive structural information from a single spectrum, some of which is difficult to obtain by classical methods and offer considerable advantages for the identification of the glycans attached to glycoproteins in proteomic and glycoproteomic studies.

Acknowledgments

I thank Professor R.A. Dwek, Director of the Glycobiology Institute for his help and encouragement and the Wellcome Trust and the Biotechnology and Biological Sciences Research Council for equipment grants to purchase the Q-TOF and TOFSpec mass spectrometers that were used in the work discussed in this chapter.

REFERENCES

1. Varki A (1993) *Glycobiology* **3**, 97–130.
2. Dwek RA (1996) *Chem. Rev.* **96**, 683–720.
3. Carr SA, Barr JR, Roberts GD, Anumula KR & Taylor PB (1990) *Methods Enzymol.* **193**, 501–518.
4. Patel T, Bruce J, Merry A et al. (1993) *Biochemistry* **32**, 679–693.
5. Küster B, Wheeler SF, Hunter AP, Dwek RA & Harvey DJ (1997) *Anal. Biochem.* **250**, 82–101.
6. Börnsen KO, Mohr MD & Widmer HM (1995) *Rapid Commun. Mass Spectrom.* **9**, 1031–1034.
7. Charlwood J, Skehel JM & Camilleri P (2000) *Anal. Biochem.* **284**, 49–59.
8. Mechref Y & Novotny MV (1998) *Anal. Chem.* **70**, 455–463.
9. Karas M, Bachmann D, Bahr U & Hillenkamp F (1987) *Int. J. Mass Spectrom. Ion Processes* **78**, 53–68.
10. Mock KK, Davy M & Cottrell JS (1991) *Biochem. Biophys. Res. Commun.* **177**, 644–651.
11. Harvey DJ (1999) *Mass Spectrom. Rev.* **18**, 349–451.
12. Harvey DJ (1993) *Rapid Commun. Mass Spectrom.* **7**, 614–619.
13. Naven TJP & Harvey DJ (1996) *Rapid Commun. Mass Spectrom.* **10**, 1361–1366.
14. Strupat K, Karas M & Hillenkamp F (1991) *Int. J. Mass Spectrom. Ion Processes* **111**, 89–102.
15. Karas M, Ehring H, Nordhoff E et al. (1993) *Org. Mass Spectrom.* **28**, 1476–1481.
16. Mohr MD, Börnsen KO & Widmer HM (1995) *Rapid Commun. Mass Spectrom.* **9**, 809–814.
17. Mechref Y & Novotny MV (1998) *J. Am. Soc. Mass Spectrom.* **9**, 1292–1302.
18. Papac DI, Wong A & Jones AJS (1996) *Anal. Chem.* **68**, 3215–3223.
19. Harvey DJ, Hunter AP, Bateman RH, Brown J & Critchley G (1999) *Int. J. Mass Spectrom. Ion Processes* **188**, 131–146.
20. Powell AK & Harvey DJ (1996) *Rapid Commun. Mass Spectrom.* **10**, 1027–1032.
21. Harvey DJ (2000) *J. Mass Spectrom.* **35**, 1178–1190.
22. Wong AW, Cancilla MT, Voss LR & Lebrilla CB (1999) *Anal. Chem.* **71**, 205–211.
23. Cole RB & Zhu J (1999) *Rapid Commun. Mass Spectrom.* **13**, 607–611.
24. Harvey DJ (2005) *J. Am. Soc. Mass Spectrom.* **16**, 622–630.
25. Guile GR, Rudd PM, Wing DR, Prime SB & Dwek RA (1996) *Anal. Biochem.* **240**, 210–226.
26. Bigge JC, Patel TP, Bruce JA, Goulding PN, Charles SM & Parekh RB (1995) *Anal. Biochem.* **230**, 229–238.
27. Dwek RA, Edge CJ, Harvey DJ, Wormald MR & Parekh RB (1993) *Annu. Rev. Biochem.* **62**, 65–100.
28. Sutton CW, O'Neill JA & Cottrell JS (1994) *Anal. Biochem.* **218**, 34–46.
29. Dell A & Thomas-Oates JE (1989) In: *Analysis of Carbohydrates by GLC and MS*, pp. 217–235. Edited by CJ Biermann and GD McGinniss. CRC Press, Boca Raton.
30. Domon B & Costello CE (1988) *Glycoconj. J.* **5**, 397–409.
31. Harvey DJ, Martin RL, Jackson KA & Sutton CW (2004) *Rapid Commun. Mass Spectrom.* **18**, 2997–3007.
32. Spengler B, Kirsch D, Kaufmann R & Lemoine J (1995) *J. Mass Spectrom.* **30**, 782–787.
33. Shevchenko A, Loboda A, Shevchenko A, Ens W & Standing KG (2000) *Anal. Chem.* **72**, 2132–2141.

34. Harvey DJ, Bateman RH, Bordoli RS & Tyldesley R (2000) *Rapid Commun. Mass Spectrom.* **14**, 2135–2142.
35. Kovácik V, Hirsch J, Kovác P, Heerma W, Thomas-Oates J & Haverkamp J (1995) *J. Mass Spectrom.* **30**, 949–958.
36. Harvey DJ, Mattu TS, Wormald MR, Royle L, Dwek RA & Rudd PM (2002) *Anal. Chem.* **74**, 734–740.
37. Franz AH & Lebrilla CB (2002) *J. Am. Soc. Mass Spectrom.* **13**, 325–337.
38. Orlando R, Bush CA & Fenselau C (1990) *Biomed. Environ. Mass Spectrom.* **19**, 747–754.
39. Reinhold VN, Reinhold BB & Costello CE (1995) *Anal. Chem.* **67**, 1772–1784.
40. Brüll LP, Kovácik V, Thomas-Oates JE, Heerma W & Haverkamp J (1998) *Rapid Commun. Mass Spectrom.* **12**, 1520–1532.
41. Ngoka LC, Gal J-F & Lebrilla CB (1994) *Anal. Chem.* **66**, 692–698.
42. Harvey DJ, Bateman RH & Green MR (1997) *J. Mass Spectrom.* **32**, 167–187.
43. Mechref Y, Novotny MV & Krishnan C (2003) *Anal. Chem.* **75**, 4895–4903.
44. Stephens E, Maslen SL, Green LG & Williams DH (2004) *Anal. Chem.* **76**, 2343–2354.
45. Ashline D, Singh S, Hanneman A & Reinhold V (2005) *Anal. Chem.* **77**, 6250–6262.
46. Kameyama A, Kikuchi N, Nakaya S *et al.* (2005) *Anal. Chem.* **77**, 4719–4725.
47. Chai W, Piskarev V & Lawson AM (2001) *Anal. Chem.* **73**, 651–657.
48. Chai W, Piskarev V & Lawson AM (2002) *J. Am. Soc. Mass Spectrom.* **13**, 670–679.
49. Sagi D, Peter-Katalinic J, Conradt HS & Nimtz M (2002) *J. Am. Soc. Mass Spectrom.* **13**, 1138–1148.
50. Wheeler SF & Harvey DJ (2000) *Anal. Chem.* **72**, 5027–5039.
51. Harvey DJ (2005) *J. Mass Spectrom.* **40**, 642–653.
52. Harvey DJ (2005) *J. Am. Soc. Mass Spectrom.* **16**, 631–646.
53. Harvey DJ (2005) *J. Am. Soc. Mass Spectrom.* **16**, 647–659.

CHAPTER 8

Integrated approach to glycan structure-function relationships

Jonathan R. Behr and Ram Sasisekharan

1. INTRODUCTION

In glycobiology it is becoming extremely important to integrate data sets into a systems biology framework in order to determine the biologically relevant structure-function relationships for glycans. This integration is necessitated by the complexity of glycan mixtures and the multitude of methods currently applied to sequence these molecules. After sequencing, the structure-function relationship for these glycans needs to be determined, and there needs to be a system for describing and presenting these structures in terms of the properties that lead to their functions. In addition, the structure-function relationships for glycans are determined by more factors than just the glycan fine structure, such as the expression and functions of relevant genes and proteins. Understanding the complete picture of glycan structure-function relationships will help to develop the glycan component of glycoprotein biologicals, therapeutics that target glycans, and standalone saccharide drugs (see Chapters 2 and 25).

The functional changes imparted by posttranslational modifications seem to explain how human and other mammalian cells can have such diverse phenotypes even with fewer genes than other organisms. These posttranslational modifications include phosphorylation, ubiquitination, glycosylation, and others. Glycosylated proteins dominate the extracellular space, both as secreted and as cell surface proteins. At the interface between cells and their environment, glycoproteins have been shown to interact with many important extracellular proteins, including signaling molecules, adhesion molecules, and immune receptors. It is through interactions with proteins such as these that glycans impinge upon biological processes from development to viral pathogenesis to cancer progression [1-3]. Therefore, glycoproteins are potentially important therapeutic targets.

There are two broad categories of glycans: linear and branched [4,5]. Glycosaminoglycans (GAGs) are the major linear polysaccharides, and the four types of GAGs in this class are all formed of disaccharide repeat units of alternating uronic acids linked to hexosamines. These chains may be variably sulfated, and they are O-linked to a core protein, forming a proteoglycan aggregate (Fig. 1a) [6]. Branched glycans are composed of a multitude of different monosaccharide building blocks that may be O-linked or N-linked to glycoproteins or glycolipids (Fig. 1b) [7]. Due to the various building blocks, linkages, and patterns of sulfation, glycans are information dense [8]. Determining the structure–function relationships for these sugar structures is a key challenge in glycobiology.

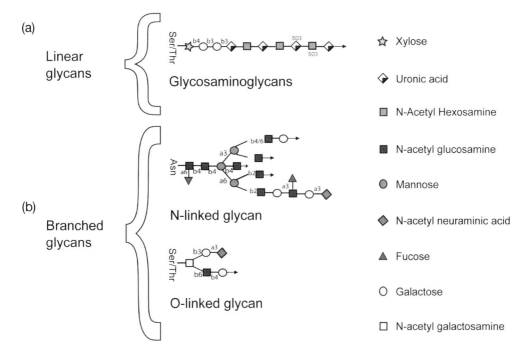

Figure 1. Structures of glycans. Examples of linear and branched sugars. (a) Linear glycosaminoglycans are polymers of repeated disaccharide units of uronic acids linked to hexosamines, attached to a core tetrasaccharide. Heparin/heparan sulfate (HS) is formed of glucuronic acid 1–4 linked to N-acetylglucosamine, while chondroitin/dermatan sulfate (CS/DS) is formed of glucuronic acid 1–4 linked to N-acetylgalactosamine. In HS and DS the glucuronic acids may undergo C5 epimerization to iduronic acid, and then be sulfated at the 2-O position. HS can undergo further O-sulfation at the 3-O and 6-O positions of the glucosamine, while CS/DS can be O-sulfated at the 4-O and 6-O positions of the galactosamine. In HS, the N-acetyl groups can be removed and replaced with a sulfate, or left as free amines. (b) Branched glycans can be divided into N-linked and O-linked glycans. N-linked glycans can be further subdivided into high mannose, hybrid, and complex (such as the example shown) sugars. O-linked glycans can be further subdivided by core type. These subdivisions of branched sugars can differ in the number of branches, and the types of monsaccharide units present and their linkages.

The heterogeneity of glycan structures allows them to interact with a wide variety of proteins. The aforementioned biological effects of glycans are due to their interactions with a variety of enzymes, growth factors, cytokines, cell adhesion molecules, and pathogens. Detailed studies have shown that in most cases, several biological glycans can interact with graded affinities with the same protein [6,9]. The few crystal structures of glycan–protein interactions have shown that while some functional groups on the glycan may be critical for specificity, many others are tertiary to the glycan–protein interaction, allowing for considerable diversity among glycan-binding partners [6]. This structural and functional overlap may be partly necessitated by the stochastic biosynthesis of the glycan molecules, which generates an ensemble of glycotypes for each glycoprotein.

Unlike amino acid chains whose sequences are encoded on a nucleic acid template, glycans are manufactured stochastically based on the coordinated expression of synthetic and modifying enzymes and the substrates present [5,10–12]. These enzymes are expressed in a temporal and tissue-specific manner, allowing a cell to dynamically change its glycoprofile based on its environment [7]. The function of these enzymes has been investigated via recombinant expression and knockout organisms, but the number of synthetic enzymes involved complicates these studies [13–15]. By virtue of the non-template based biosynthesis, and the complexity of the glycan structures, glycans isolated from cells and tissues are always heterogeneous mixtures of chemical structures.

In summary, several challenges present themselves when determining the structure–function relationships for glycans. First, the chemical complexity of each glycan molecule means that it has high information content while simultaneously making it difficult to analyze. Second, the non-template based biosynthesis means that researchers are always dealing with mixtures. The relative abundance of each species is needed, while the easily obtained molecular weight distribution and monosaccharide compositional analysis are not sufficient to describe these mixtures. Third, as the samples are purified from biological sources, with no possible amplification process, the samples will be scarce. Fourth, these glycans act in a complex and often multivalent manner in the extracellular space, so *in vitro* data obtained using purified glycans is often misleading. All of these confounding issues can be dealt with by integrating multiple data sets across various organizational levels [16].

Presently, new tools are being developed to deal with the fundamental challenges of glycan research. New experimental procedures aim to measure structure and sequence aspects of scarce glycans with minimal or no prior separation from a mixture. Simultaneously, new computational tools are becoming available to analyze and integrate these data sets to obtain the structure–function relationships of bioactive glycans. Finally, new frameworks are being created to present these structures in terms of the properties that lead to their functions, to better enable expansion and utilization of this knowledge.

2. ANALYTICAL TECHNIQUES FOR GLYCOBIOLOGY

2.1 Glycan analysis

Any detailed investigation into the structure–function relationship of glycans involves the application of multiple techniques and the integration of multiple data sets. Glycans are chemically so complex that single techniques are usually not capable of sequencing these molecules. These techniques include, but are not limited to, enzymatic or chemical digests, high-performance liquid chromatography (HPLC), capillary electrophoresis (CE), mass spectroscopy (MS), nuclear magnetic resonance (NMR), and "hyphenated" techniques combining the above (Table 1). Combining the information provided by all of the different techniques provides its own set of challenges. Once the structures are known, the problem of relating these structures to their functions can be tackled.

Table 1 Analysis techniques. A brief summary of common analytical techniques used to analyze glycan structure, along with the typical sample amounts required to utilize the techniques

Analytical technique	Use	Sensitivity
MALDI-MS	Mass-composition relationships	~ pg-fg
ESI-MS^n	Mass-composition relationships and structural information from tandem fragmentation patterns	~ ng-pg
CE	Electrophoretic mobility, may identify glycans with properly characterized standards	~ ng-pg
HPLC	Separation and preparation, may identify glycans with properly characterized standards	~ mg-ng
NMR	Monosaccharide composition and linkage information	~ mg-ug
Enzymatic Digests	Structural constraints at cleavage points based on enzyme specificities, may enable exosequencing	
FACE Gels	Enzymatic exosequencing of N-linked glycans	~pmols

One way to deal with mixtures of glycans, and to analyze fine structure, is to separate them during analysis. HPLC and CE have evolved as methods of choice to separate and analyze glycans due to their increased sensitivity over traditional gel electrophoresis and gravity column chromatography [17–19]. Fine structure analysis generally begins with the selection of a single chromogenic or mass peak (if an MS detector is being used), followed by depolymerization or fragmentation and analysis. A limitation of these non-MS analytical methods is that they require a label (radioactive, fluorescent, or chromogenic) on the glycans.

Depolymerization is frequently performed chemically or enzymatically to gain structure information, or to add labels. Some chemical digests will only cleave between certain residues, or will provide chemical handles for labeling, such as nitrous acid degradation of heparan sulfate [20]. Libraries of glycan-degrading enzymes (such as exo-glycosidases or lyases) with various specificities also allow for fine structure analysis, while lyases simultaneously generate double bonds that can be used as chromogenic labels [18,20]. These chemical or enzymatic digests can be complete, to gain a compositional analysis of mixtures or individual glycans, or partial, to gain information about linkages or moieties of a single glycan. Chromatographic shifts indicate various structures depending on the specificities of the chemicals or enzymes used. HPLC has the added advantage of allowing preparation of fractionated samples for later fine structure analysis.

There are other high-throughput techniques that can give information on glycan structures without prior separation. MS has been used extensively to get the mass profile of a mixture of glycans [21–25]. Matrix-assisted laser desorption/ionization MS (MALDI-MS) (see Chapter 7) allows high-throughput glycoprofiling utilizing a minimum amount of sample. Recently, computational tools have been developed to annotate these spectra for N-glycans while incorporating domain knowledge from other sources [26]. Electrospray ionization mass spectrometry (ESI-MS^n) and other tandem MS methods allow for some structural information to be gleaned during the MS process, through isolation and fragmentation of the glycans in the MS after ionization. Computational tools are also being developed to sequence glycans from these spectra [27–31] (Fig. 2).

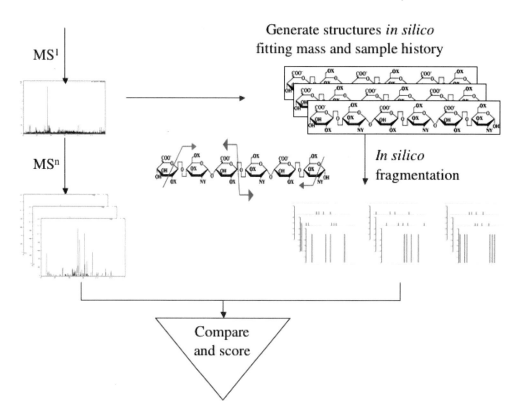

Figure 2. Informatics techniques for MS^n sequencing. A paradigm for determining glycan structures using tandem mass spectrometry (MS^n). All possible glycan structures fitting the mass from the MS^1 spectra and known sample history (composition, modifications, method of generation) are computationally generated. Known fragmentation patterns are then applied to generate theoretical MS^n fragments for each possible structure. These theoretical fragments are compared to those actually observed, and a score is generated for each potential sequence. Computation tools to apply this paradigm include the web accessible GylcoFragment and GlycoSearch MS from the Glycomics Initiative of the German Cancer Research Institute [30], and the EXCEL based Heparin Oligosaccharide Sequencing Tool (HOST) available from researchers at UC Davis [27].

In the case of heparan oligosaccharides, the fragment ions are compared to theoretical fragments generated from known glycan structures, and other information gained from enzymatic digests is also incorporated.

Cutting-edge techniques such as Fourier transform ion cyclotron resonance mass spectrometry (FTICR-MS) and chip-based electrospray interface have also been applied to glycan analysis. The highly sensitive FTICR-MS can characterize glycans, glycopeptides, and glycolipids with exceptional accuracy [32–34]. Nanoscale liquid delivery from chip interfaces reduces the sample consumption by orders of magnitude, and has been coupled to tandem MS and FTICR-MS to gain structural information from minute amounts of glycans [35,36]. Unfortunately, this high-throughput fragmentation often does not allow unambiguous assignment of many complex structures following MS, but much effort is being made to advance this technique.

NMR can be used with larger amounts of glycans to obtain sequence information from mixtures. The monosaccharide contents of glycans can be quantified by combining the information from the one-dimensional proton and carbon spectra along with the coupling constants from the homonuclear (gradient-selected correlation spectroscopy (gCOSY) and total correlated spectroscopy (TOCSY)) and heteronuclear spectra (heteronuclear multiple quantum correlation (HMQC) and heteronuclear multiple bond correlation (HMBC)) [37–39]. These techniques are useful to obtain ratios of sugars such as glucose to galactose to mannose in branched glycans, or iduronic to glucuronic acid in GAGs. Besides characterizing the ratios of specific monosaccharides, reducing-end linkage information can be determined based on the anomeric chemical shifts. This information is highly useful to determine linkage information for terminal sialic acids, which can be $\alpha2$–3- or $\alpha2$–6-linked at the reducing terminus of the chain. To facilitate NMR analysis, the relevant chemical shifts have been catalogued in databases (Table 2).

Bioinformatic techniques have the potential to improve glycan analysis by integrating multiple complementary data sets in order to determine sample fine structure. It generally takes data from several orthogonal methods to uniquely determine a sample's fine structure. To increase throughput, analysis techniques can be combined into so-called "hyphenated" techniques or informatics-based sequencing techniques can be used to efficiently process and combine data. At the collection level, HPLC/MS and CE/MS have both been used successfully to obtain mass profiles of mixtures of glycans [25,40,41]. At the analysis level, several informatics-based methodologies have been developed to use the constraints provided from data gained using multiple methodologies to converge on the unique glycan sequence [8,27,42].

In order to get a structure–function relationship, the fine structure information is combined with data about how the structure is produced and with which proteins it interacts. Many of these analytical techniques are still limited because they remove the glycans from their biological contexts. Frequently glycans are analyzed as mixtures from multiple proteins (or even from whole cell or whole tissue extracts), making interpretation of the biologically relevant functions of these glycans difficult. Glycans may also be fragmented prior to

Table 2 Web databases of glycan information. List of web accessible databases and tools relating to glycan structures and glycan binding proteins. Resources include those available through the large Consortium for Functional Glycomics, Glycomics Initiative of the German Cancer Research Institute, and the Kyoto Encyclopedia of Genes and Genomes

Consortium for Functional Glycomics: CFG http://www.functionalglycomics.org/

Central database http://www.functionalglycomics.org/glycomics/publicdata/home.jsp
- Glycan profiling (catalogues the presence of N- and O-linked glycans in human and mouse tissues)
- Gene microarray (mRNA microarray data from glycogene chip)
- Mouse phenotyping (histology, hematology, immunology, and metabolism data for knockout mice)
- Glycan array (results from high-throughput screening for identifying lectin-ligand interactions)

GBP database http://www.functionalglycomics.org/glycomics/molecule/jsp/gbpMolecule-home.jsp
(an integrated presentation of public database and Consortium information)
Glycan database http://www.functionalglycomics.org/glycomics/molecule/jsp/carbohydrate/carbMoleculeHome.jsp
(database containing structural and chemical information as well as related references)
Glycosyltransferase database http://www.functionalglycomics.org/static/gt/gtdb.shtml
(a graphical interface for navigating the glycoenzyme database)

Glycomics Initiative of the German Cancer Research Institute: http://www.glycosciences.de/

Databases for glycobiology and glycomics http://www.glycosciences.de/sweetdb/index.php
- Structure database (searchable by structures, molecular formula, composition, classification, or motif)
- NMR database (glycan NMR proton and carbon chemical shift database, searchable by atom or by peak)
- MS database (searchable by mass, will compute theoretical fragments)
- Glycan PDB Database (search for and extract carbohydrate containing PDB entries)

Kyoto Encyclopedia of Genes and Genomes Glycan: KEGG Glycan http://www.genome.jp/kegg/glycan/

KEGG glycan database search program: KcaM http://www.genome.jp/ligand/kcam/
(a collection of experimentally determined glycan structures from CarbBank, recent publications, and structures present in KEGG pathways)
KEGG pathway database http://www.genome.jp/kegg/pathway/map/map01170.html
(glycan biosynthesis and metabolism pathways with links to relevant enzymes)
Composite Structure Map: CSM http://www.genome.jp/kegg-bin/draw_csm
(visual representation of all possible glycan structures generated from the KEGG GLYCAN database)

Other glycan databases

Carbohydrate active enzymes: CAZy http://www.cazy.org/CAZY/
  CAZy (database of families of enzymes that degrade, modify, or create glycosidic bonds)
Complex carbohydrate structure database: carbBank http://www.boc.chem.uu.nl/sugabase/carbbank.html
  CarbBank (web interface to carbohydrate structure database)
Sugabase http://www.boc.chem.uu.nl/sugabase/sugabase.html
  Sugabase (glycan NMR proton and carbon chemical shift database)
Animal lectins database http://www.imperial.ac.uk/research/animallectins/
  Information about animal lectins involved in various sugar-recognition processes
GlycoSuite Database: https://tmat.proteomesystems.com/glycosuite/
  Commercial glycan database and analysis tools

analysis, separating epitopes on the same glycan that may influence each other. These two factors make it difficult to determine the biologically relevant structure–function relationships that may require intact proteoglycans, such as multivalent interactions, allosteric interactions, and mixed protein–glycan interactions [43]. Tools that increase sensitivity and throughput (by integrating orthogonal data sets, and utilizing the strengths of each) may allow analysis of glycans from single protein preparations. Computational techniques, especially those that incorporate knowledge of biosynthetic pathways, may also be able to reassemble whole glycan structures from sequenced fragments or from partial sequence data. In these ways, bioinformatic techniques may overcome many of the major limitations of glycan analysis.

2.2 Glycoprotein and glycoprotein-encoding gene analysis

A systems biology approach to glycan structure–function analysis must include the proteins that synthesize or interact with the glycans, and the genes that encode them. Analysis of glycogene expression is being undertaken to try and relate protein levels to glycobiology. At the same time, investigations of glycan–protein interactions are also being completed to understand the structure–function relationships of glycans. Understanding these components is critical to determine the full picture of how glycans are utilized in biological systems, and how cells regulate their influence.

One of the more powerful tools in systems biology is the gene chip, which allows for the simultaneous analysis of the mRNA expression levels of thousands of genes. Although there are advanced commercial gene expression chips available, most notably those from Affymetrix, glycoprotein-encoding gene specific DNA microarrays have been developed. These arrays focus on overcoming the challenges of using whole genome chips, such as limited representations of glycan synthesis enzymes and limited sensitivity of these measurements compared to other proteins [44]. Reverse transcriptase polymerase chain reaction (RT-PCR) techniques can analyze any glycan-related transcripts with the highest specificity and sensitivity. However, this lower throughput technique can only analyze a handful of transcripts simultaneously. Customized glycoprotein-encoding gene microarrays, based on the Affymetrix technology, were designed for the National Institute for General Medical Sciences (NIGMS)-funded Consortium for Functional Glycomics (CFG) and have been used to analyze hundreds of samples [44]. These microarrays can simultaneously provide information on the expression of biosynthetic enzymes and glycan-binding proteins. These data can then be combined with data gained from analyzing the actual glycan structures purified from the same samples.

Since glycans affect biology through their interactions with proteins, it is critical to understand the specificity of glycan–protein interactions. The term glycan-binding protein (GBP) can be used to describe proteins involved in cell adhesion, as well as trafficking and signaling events, that interact with either *N*- or *O*-linked glycans [4,5,45]. GBPs also include foreign proteins that act as receptors for pathogens to attach to host-cell glycans. Three major classes of GBPs are C-type lectins, galectins, and siglecs. Each binding site on these GBPs usually binds to tetrasaccharide or smaller epitopes with relatively low (micro- to millimolar) affinity. Some sulfated GAG-binding proteins can bind to longer epitopes with slightly higher affinities. These low affinities can be overcome *in vivo* by the multivalent binding of glycans by a multimeric association of GBPs. These multivalent units can be soluble or membrane-bound, dispersed or localized [9].

GBPs have been used to analyze the glycans present in a mixture based on their specific affinities. On the other hand, various strategies have been used to construct glycan ligands to investigate the binding affinity of these GBPs. The specificities of recombinant or purified GBPs, including lectins, have been used in column or array format to analyze glycosylation patterns [46]. Solid phase and chemo-enzymatic synthesis techniques have allowed the construction of diverse glycan structures [47,48]. These structures have sometimes been used in assays to determine binding affinities of GBPs to various ligands [9], and to devise inhibitors of the physiological glycan–protein interactions.

The CFG has developed two types of glycan arrays; one array has soluble glycan ligands in a well-based assay and a second array has a solid-phase glycan printed onto an NHS (N-hydroxysuccin)-activated glass slide [49]. The printed array mimics the surface distribution of glycans exhibited on cell surfaces, and may be more physiologically relevant. The GBPs are then presented at high enough concentrations so that they can be present in their multimeric states. Detection schemes for protein–glycan binding include using primary antibodies to the GBPs followed by labeled secondary antibodies, similar to standard ELISA setups, or by directly labeling the proteins (Fig. 3). These glycan arrays have been successfully utilized to investigate the ligand specificities of physiologically important GBPs such as dendritic cell-specific ICAM grabbing non-integrin (DC-SIGN) and DC-SIGN related (DC-SIGNR) protein [50].

When analyzing GBPs and the genes that code for them, one must be careful in determining the relationships between the data and the biology. For example, the connection between relative gene expression changes in synthetic enzymes and actual glycan structures is not proven or deterministic. Also, *in vitro* experiments investigating interactions between GBP and glycans must be carefully designed and accurately scrutinized with regard to how well they reflect biologically relevant interactions. Interaction assay formats that may limit the freedom of movement of glycans or GBPs could misrepresent the biological systems where these components are free to move and interact in a multivalent manner. Our increasing body of knowledge regarding these systems should inform our assay design and data interpretation in the future.

3. BIOINFORMATICS PLATFORMS FOR GLYCOMICS

Multiple data sets, spanning from fine structure analysis to gene expression data to GBP binding data need to be integrated to encompass the biology of glycans. In order to integrate, process, and present these data in the most meaningful manner a bioinformatic framework needs to be developed. One of the most important goals of these tools is to be able to link and integrate orthogonal data sets generated from identical or similar samples, generated at different times or

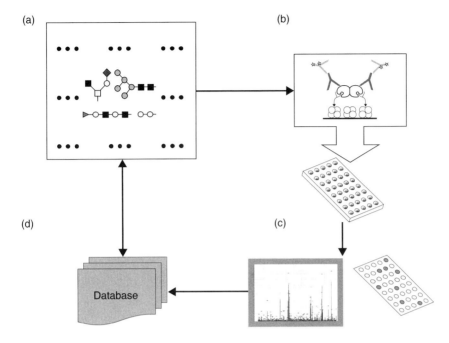

Figure 3. Glycochips developed by the Consortium for Functional Glycomics can be used to determine glycan-binding specificities for GBPs. (a) Libraries of biological and synthesized glycans are generated, with attached linkers. (b) These glycans can be assayed for binding to a GBP presented in monomeric or multimeric forms (pictured as dimers) either in a soluble well-based format or attached to a surface. Glycans with a biotin linker (triangle) can be attached to a well surface through streptavidin molecules (tetrameric circles). In another format, glycans with an amine linker can directly be printed onto an NHS-activated glass slide (not shown). Relative binding can be detected via a primary/labeled secondary antibody system (black "Y" and gray "Y" with stars) or by directly fluorescently labeling the GBP. (c) During data analysis, the data for each well is linked back to the original glycan structure in the database (d), and to other data sets.

by different investigators. While many techniques for analyzing other biopolymers, such as DNA or proteins, have been transferred to analyzing glycans, the chemical complexity and branching nature of some glycans presents unique challenges for integrating and presenting relevant data. The field has been steadily progressing, from early attempts at glycan structure databases and analysis such as the failed Complex Carbohydrate Structures Database (CCSD) to ongoing current efforts such as the Glycosuite database [51], KEGG Glycan database [52], and other tools for representing and analyzing glycan structures [53–55] (see Table 2).

It would be a great aid to investigators to present data in a manner where relationships between different entries in the database are clear. In order to achieve this, an object-based relational database was developed by CFG. The three primary objects in this glycomics database are the biosynthetic enzymes, the glycan structures, and the GBPs. The data sets generated by the aforementioned techniques are then organized into other levels of objects that have defined relationships to each other and to the primary objects. The ontology diagram that encompasses the definitions and relationships is exceedingly complex – but software has been developed to facilitate data input and analysis without the investigators being weighed down by this complexity. The user interface is separated from the back-end database by middleware that communicates between the two. Therefore the investigator's data are automatically organized and integrated after input, allowing investigators to see how different proteins or glycans relate to each other.

An important aspect of the data-integration process is to link orthogonal data sets from identical or similar samples. For example the compositional analysis and NMR data from a GAG sample from a tumor type would need to be integrated and then associated with any gene expression data or immunological profiles from the same tumor type. This integration is critical for understanding the complete picture of how the glycans are influenced or are influencing the pathology or biology of the tumor, and therefore tools need to be developed to integrate these data sets.

Another aid to investigators would be to present the data in useful, familiar, and searchable manner. The CFG has introduced a molecule page, becoming standard in genomics and proteomics [56], that presents information on a single molecule from the molecular to organism level. The CFG molecule pages can automatically acquire information from other public databases, interface with CFG data, and show contributions from a group of experts. This easy-to-use and familiar interface enables faster assimilation of data by the investigator. In addition, the bioinformatics platforms need to allow data mining from the stored data sets. For example, the availability of diverse data sets via relational databases may allow computational identification of the structural determinants that allow multivalent high-affinity glycan interactions with specific GBPs, analogous to computational analysis of genomes to identify promoters or computational analysis of protein sequences to find conserved functional domains.

4. APPLICATIONS

Glycans and glycoconjugates have inherent advantages over protein- or nucleotide-based drugs that give them great potential as therapeutics. In general, glycans are smaller, more stable, and more easily formulated for drug delivery than protein-based drugs [16]. In addition, glycans are potentially less immunogenic than protein- and nucleotide-based therapies [57]. Unfortunately, the manufacture of complex glycans is far more difficult than other biopolymers because glycans cannot be easily synthesized chemically or by using molecular biology tools, and the identification of target glycan-based drugs is hindered by the lack of known structure–function relationships. In order to harness the potential of glycans as therapeutics, these hurdles need to be overcome.

Advancing the use of glycans in biotechnological applications requires the knowledge of their structure–function relationships. Historically, the GAG heparin (see Chapter 9) has been used as an anticoagulant with no understanding of its mechanism of action. But as our understanding of heparin's mode of action has advanced [58], new and better heparinoids have been developed that are more potent and have fewer side-effects than the original unfractionated heparin [59]. More recently, glycan-related genetic disorders (such as Gaucher's disease and mucopolysaccharidosis type 1) have been treated by replacement therapy with enzymes that modify glycan structures. The development of these therapies required a detailed understanding of protein effects on glycan structure and the role of glycan structure in disease. In contemporary drug development, glycoprotein drugs such as the anti-anemia glycoprotein drug erythropoeitin have been modified to change their glycosylation patterns (which resulted in Aranesp) – and these modifications have changed the pharmacokinetic properties of these proteins [60]. These examples illuminate different areas where glycans are involved in drug development – either as standalone saccharides, as targets themselves, or as components of glycoproteins (Table 3).

Table 3 Representative glycan-based therapeutics

Drug name	Disease	Drug	Clinical status	Manufacturer
Lovenox	Thrombosis	LMWH generated by chemical depolymerization	Market	Aventis
Fragmin	Thrombosis	LMWH generated by chemical depolymerization	Market	Pfizer
Arixtra	Thrombosis	Synthetic oligosaccharide	Market	Sanofi
Healon	Cataracts	High molecular weight sodium hyaluronate	Market	Pfizer
Seprafim	Anti-adhesive	Chemically modified sodium hyaluronate and carboxymethyl cellulose	Market	Genzyme
Aranesp	Anaemia	Glycoprotein genetically engineered for increased glycosylation	Market	Amgen
Aldurazyme	MPS I	Lysosomal glycan degrading enzyme	Market	Genzyme
PI-88	Cancer	Heparan sulfate mimetic	Phase II	Progen

Several bioactive saccharides, either synthesized or purified from biological sources, are already in use in the clinic. The first strategy for manufacturing glycan-based drugs involves isolating and modifying natural glycans. This approach is technologically less challenging and scales easily, so it has been widely used to create the largest class of glycan drugs, the low molecular weight heparins (LMWHs) [59]. The currently used LMWHs are derived from unfractionated heparin purified from animal sources, with each LMWH then being degraded into smaller pieces using different degradative methods. While these drugs are more controlled than the starting material, they remain heterogeneous mixtures. In order to understand their composition, to design better LMWHs, and to recreate generics of these drugs, various orthogonal data sets needed to be integrated to come up with a complete description of the mixture. Multiple rationally designed LMWHs are in development to take advantage of our continually improving understanding of the mechanisms of action of these molecules [61]. The first pure glycan pharmaceutical, the pentasaccharide fondaparinux (Arixtra), was created by *de novo* chemo-enzymatic synthesis for use as an anticoagulant [62]. Although synthesis of complex glycans is difficult, new solid-phase synthetic techniques for linear and branched sugars analogous to those in use for peptide and nucleic acids are in continuous development, and have been used already for carbohydrate-based vaccines.

Since most circulating proteins in the body are glycoproteins, most therapeutic proteins derived from antibodies, growth factors, and cytokines are glycoproteins. The glycan component of these proteins has been shown to influence protein function, immunogenicity, and pharmacokinetic properties such as half-life and stability. Since, in general, proteins are expressed with various glycoforms, there is the opportunity to influence the quality of a protein therapeutic by influencing its natural glycosylation [63,64]. The glycan components can also be rationally modified beyond the natural glycoforms. To return to the erythropoietin example, not only does adding glycoslyation sites increase the erythropoietic activity and serum half-life, but removing the capping sialic acids from the glycans decouples broad neuroprotective activities from the erythropoietic activity [60,65].

Researchers have attempted to modulate the glycosylation of proteins by either adding chemicals to influence biosynthesis, by genetically manipulating the cells in which the proteins are produced, by modifying the proteins *in vitro* after their translation, or by modifying the protein backbones themselves. While proteins are being synthesized, the effects of changing cell culture conditions or adding specific natural monosaccharides have been investigated [63,64,66]. More recently non-natural amino acids or monosaccharides with chemical handles, which can be incorporated with the native machinery, have been added for later use in analyzing or in tailoring the glycosylation [67–71]. Mammalian and insect cell lines have been successfully genetically modified to simplify the variety of glycoforms produced, and various efforts have been made to accomplish the gargantuan task of replicating the mammalian glycosylation machinery in yeast cells [72]. Another strategy is to attach specific saccharides after the proteins are made, either chemically or using exogenous glycosyltransferases. Development of chemoselective and chemoenzymatic techniques to modify proteins *in vitro* has demonstrated success [73]. Finally, the gene for the protein itself can be modified to change glycosylation sites, as mentioned before in the erythropoietin example.

New analytical and bioinformatic techniques that advance our understanding of the structure–function relationships in glycans are quintessential in allowing the use of glycan-based therapies in two important ways. First, understanding the structure–function relationships in normal and pathological situations allows for the targeted design of new therapeutics. Second, understanding the structure–function relationships of existing drugs allows the complete description of their activities, enabling the design of generic drugs, as well as the modification of existing drugs to create new tailored molecules with differing activities.

5. CONCLUSIONS

The growing success of glycan and glycoprotein pharmaceuticals, based on increased understanding of the structure–function relationships of specific glycans, is motivating increasing amounts of research into glycobiology at the basic science, transitional, and clinical research levels. Simultaneously, new analytical methods are generating larger and more diverse data sets about glycans and their functions. These developments present challenges and opportunities, including the need to integrate several types of measurements that yield complementary data sets, the need to represent data in a meaningful way, and the need to allow data mining to answer relevant questions. These questions include explaining glycan diversity regulation as a function of its biosynthesis, explaining the basis for specificity and the affect of valency in glycan–protein interactions, and explaining how extracellular glycans regulate cell–cell communication and signal transduction. Answers to these questions should guide researchers aiming to create new and better glycan, glycan-modifying, and glycoprotein research products and pharmaceuticals. Analytical and bioinformatic tools are continually being developed to address these challenges and questions, and a focus needs to be made to perfect, standardize, and encourage adoption of these new tools.

REFERENCES

1. Hacker U, Nybakken K & Perrimon N (2005) *Nat. Rev. Mol. Cell Biol.* 6, 530–541.
2. Sasisekharan R *et al.* (2002) *Nat. Rev. Cancer* 2, 521–528.
3. Rabenstein DL (2002) *Nat. Prod. Rep.* 19, 312–331.
4. Taylor ME & Drickamer K (2006) *Introduction to Glycobiology*, 2nd edn. Oxford University Press, Oxford and New York.
5. Varki A (1999) *Essentials of Glycobiology*. Cold Spring Harbor Laboratory Press, Cold Spring Harbor, NY.
6. Raman R, Sasisekharan V & Sasisekharan R (2005) *Chem. Biol.* 12, 267–277.
7. Lowe JB & Marth JD (2003) *Annu. Rev. Biochem.* 72, 643–691.
8. Venkataraman G *et al.* (1999) *Science* 286, 537–542.
9. Collins BE & Paulson JC (2004) *Curr. Opin. Chem. Biol.* 8, 617–625.
10. Sasisekharan R & Venkataraman G (2000) *Curr. Opin. Chem. Biol.* 4, 626–631.
11. Sugahara K & Kitagawa H (2002) *IUBMB Life* 54, 163–175.
12. Sugahara K *et al.* (2003) *Curr. Opin. Struct. Biol.* 13, 612–620.
13. Homeister JW, Daugherty A & Lowe JB (2004) *Arterioscler. Thromb. Vasc. Biol.* 24, 1897–1903.
14. Martin LT *et al.* (2002) *J. Biol. Chem.* 277, 32930–32938.
15. Kjellen L (2003) *Biochem. Soc. Trans.* 31, 340–342.
16. Shriver Z, Raguram S & Sasisekharan R (2004) *Nat. Rev. Drug Discov.* 3, 863–873.
17. Rudd PM *et al.* (2001) *Proteomics* 1, 285–294.
18. Royle L *et al.* (2002) *Anal. Biochem.* 304, 70–90.
19. Mao W, Thanawiroon C & Linhardt RJ (2002) *Biomed. Chromatogr.* 16, 77–94.
20. Conrad HE (1998) *Heparin-Binding Proteins*. Academic Press, San Diego.
21. An HJ *et al.* (2003) *Anal. Chem.* 75, 5628–5637.
22. Cipollo JF, Costello CE & Hirschberg CB (2002) *J. Biol. Chem.* 277, 49143–49157.
23. Dell A & Morris HR (2001) *Science* 291, 2351–2356.
24. Morelle W & Michalski JC (2005) *Curr. Pharm. Des.* 11, 2615–2645.
25. Henriksen J, Roepstorff P & Ringborg LH (2006) *Carbohydr. Res.* 341, 382–387.
26. Goldberg D *et al.* (2005) *Proteomics* 5, 865–875.
27. Saad OM & Leary JA (2005) *Anal. Chem.* 77, 5902–5911.
28. Ethier M *et al.* (2002) *Rapid Commun. Mass Spectrom.* 16, 1743–1754.
29. Joshi HJ *et al.* (2004) *Proteomics* 4, 1650–1664.
30. Lohmann KK & von der Lieth CW (2004) *Nucleic Acids Res.* 32 (Web Server issue), W261–266.

31. Tang H, Mechref Y & Novotny MV (2005) *Bioinformatics* **21(Suppl 1)**, i431–i439.
32. Park Y & Lebrilla CB (2005) *Mass Spectrom. Rev.* **24**, 232–264.
33. Hakansson K *et al.* (2001) *Anal. Chem.* **73**, 4530–4536.
34. McFarland MA *et al.* (2005) *J. Am. Soc. Mass Spectrom.* **16**, 752–762.
35. Froesch M *et al.* (2004) *Rapid Commun. Mass Spectrom.* **18**, 3084–3092.
36. Zamfir A *et al.* (2004) *Anal. Chem.* **76**, 2046–2054.
37. Manzi AE *et al.* (2000) *Glycobiology* **10**, 669–689.
38. Lopez M *et al.* (1997) *Glycobiology* **7**, 635–651.
39. Guerrini M, Bisio A & Torri G (2001) *Semin. Thromb. Hemost.* **27**, 473–482.
40. Zamfir A *et al.* (2004) *Electrophoresis* **25**, 2010–2016.
41. Thanawiroon C *et al.* (2004) *J. Biol. Chem.* **279**, 2608–2615.
42. Guerrini M *et al.* (2002) *Glycobiology* **12**, 713–719.
43. Paulson JC, Blixt O & Collins BE (2006) *Nat. Chem. Biol.* **2**, 238–248.
44. Comelli EM *et al.* (2002) *Biochem. Soc. Symp.* **69**, 135–142.
45. Sharon N & Lis H (2003) *Lectins*, 2nd edn. Kluwer Academic Publishers, Dordrecht and Boston.
46. Hirabayashi J (2004) *Glycoconj. J.* **21**, 35–40.
47. Hanson S *et al.* (2004) *Trends Biochem. Sci.* **29**, 656–663.
48. Seeberger PH & Werz DB (2005) *Nat. Rev. Drug Discov.* **4**, 751–763.
49. Blixt O *et al.* (2004) *Proc. Natl Acad. Sci. USA* **101**, 17033–17038.
50. Guo Y *et al.* (2004) *Nat. Struct. Mol. Biol.* **11**, 591–598.
51. Cooper CA *et al.* (2003) *Nucleic Acids Res.* **31**, 511–513.
52. Kanehisa M *et al.* (2004) *Nucleic Acids Res.* **32** (Database issue), D277–280.
53. Aoki KF *et al.* (2004) *Bioinformatics* **20(Suppl 1)**, I6–I14.
54. Bohne-Lang A *et al.* (2001) *Carbohydr. Res.* **336**, 1–11.
55. Kikuchi N *et al.* (2005) *Bioinformatics* **21**, 1717–1718.
56. Li J *et al.* (2002) *Nature* **420**, 716–717.
57. Sioud M (2004) *Trends Pharmacol. Sci.* **25**, 22–28.
58. Petitou M, Casu B & Lindahl U (2003) *Biochimie* **85**, 83–89.
59. Linhardt RJ & Gunay NS (1999) *Semin. Thromb. Hemost.* **25(Suppl 3)**, 5–16.
60. Egrie JC & Browne JK (2001) *Br. J. Cancer* **84(Suppl 1)**, 3–10.
61. Sundaram M *et al.* (2003) *Proc. Natl Acad. Sci. USA* **100**, 651–656.
62. Bauer KA (2003) *Chest* **124(Suppl)**, 364S–370S.
63. Baker KN *et al.* (2001) *Biotechnol. Bioeng.* **73**, 188–202.
64. Yang M & Butler M (2002) *Biotechnol. Prog.* **18**, 129–138.
65. Erbayraktar S *et al.* (2003) *Proc. Natl Acad. Sci. USA* **100**, 6741–6746.
66. Senger RS & Karim MN (2003) *Biotechnol. Prog.* **19**, 1199–1209.
67. Wieser JR *et al.* (1996) *FEBS Lett.* **395**, 170–173.
68. Luchansky SJ, Goon S & Bertozzi CR (2004) *Chembiochem* **5**, 371–374.
69. Kiick KL *et al.* (2002) Proc. Natl Acad. Sci. USA **99**, 19–24.
70. Kayser H *et al.* (1992) *J. Biol. Chem.* **267**, 16934–16938.
71. Zhang Z *et al.* (2004) *Science* **303**, 371–373.
72. Hamilton SR *et al.* (2003) *Science* **301**, 1244–1246.
73. Ritter TK *et al.* (2003) *Angew. Chem. Int. Ed. Engl.* **42**, 4657–4660.

SECTION 4
Cell biology of carbohydrates and glycoproteins

CHAPTER 9

Some observations on the biology of cell surface heparan sulfate proteoglycans

John T. Gallagher

1. INTRODUCTION

Heparan sulfate (HS) was first discovered in 1950 in the discarded by-products of the industrial manufacture of heparin [1]. It was initially named heparin monosulfuric acid and, as the name suggests, it had approximately one sulfate group per disaccharide unit. This polysaccharide entity was something of a curiosity because although it had some heparin-like characteristics it was considerably less sulfated than pharmaceutical heparin and its apparent lack of anticoagulant activity meant that it had little therapeutic value. Despite its low standing in the commercial world HS attracted the interest of several academic groups and over the next 20 years papers from the laboratories of Cifonelli, Linker, and Dietrich revealed that HS was widely distributed in animal tissues, including primitive organisms, and that its sulfate content and chain length varied depending on the source [2].

A major finding in the early 1970s was the report by Kramer that HS was present in trypsin extracts of mammalian cell cultures, implying its close association in proteoglycan form with the plasma membrane [3]. This important development suggested that HS was in some way directly influencing cell growth and cellular interactions with the microenvironment, although at the time there was no clear understanding of how HS could regulate the behavior of cells. It is an interesting coincidence that the extracellular matrix (ECM) protein fibronectin was discovered at around the same time as Kramer published his findings on cell surface HS. It is now well-established that one of the key interactions of HS that has a profound influence on cell adhesion and migration, is with pericellular fibronectin [4] (see section 6).

Direct evidence for the significance of HS in cell growth was published by Yayon and by Rapraeger in 1991, who both showed that basic fibroblast growth factor (bFGF or FGF2) required cell surface HS for the efficient binding to FGF2 receptors and for eliciting a mitogenic response [1,6]. Rapraeger and colleagues also found that other FGF family members required an HS "coreceptor" or cofactor for stimulating cell growth [6].

Although it is not unexpected that the FGFs share a common cell activation mechanism, what is surprising and intriguing is that many other growth factors, often structurally unrelated, such as hepatocyte growth factor/scatter factor (HGF/SF), vascular endothelial growth factor (VEGF), and glial cell-derived neurotrophic factor (GDNF), have now been shown to utilize an HS coreceptor as a means of delivering signals to cells [7]. Cell growth and cell adhesion are closely integrated processes and HS, in its proteoglycan form (see below), may play a vital role in coordinating the transmembrane signals elicited by the ECM and by soluble growth factors, morphogens, etc.

2. THE MOLECULAR STRUCTURE OF HEPARAN SULFATE

In common with other sulfated glycosaminoglycans (GAGs) HS is a linear, polyanionic molecule composed of from 50 to 150 repeating disaccharide units of the general structure $(A-U)_n$ where A is an amino sugar and U a uronic acid. The amino sugar is *N*-acetylated or *N*-sulfated glucosamine (GlcNAc or GlcNS) and the uronic acid is present in the form of glucuronic acid (GlcA) or its C-5 epimer iduronic acid (IdoA) (Fig. 1). Occasionally the glucosamine is *N*-unsubstituted. In addition to the *N*-sulfates, ester-linked sulfates (*O*-sulfates) are common at C-2 of IdoA and C-6 of the amino sugars. Less frequent modifications include C-3 sulfation of GlcNS and C-2 sulfation of GlcA [8]. In principle, this wide variety of modified sugar residues could give rise to immense structural diversity but constraints are imposed on variations in monosaccharide sequence and sulfation patterns by the tight substrate specificities of the HS modifying enzymes.

The defining feature of HS that distinguishes it from other GAGs is its domain structure in which the vast majority of the sulfated sugars are localized in a series of "composite sulfated regions" or CSRs (Fig. 2) that are positioned at regular intervals along the polymer chain [9]. Between them are sections of the chain with little or no sulfation that consist mainly of GlcNAcα1-4GlcAβ1-4 repeat units (NAc domains) interspersed with the occasional *N*-sulfated disaccharide. Domains of variable sulfation and functional importance are also found in other sulfated GAGs (see Chapter 5) but HS appears to be the only GAG in which tandemly repeating sulfated regions can be readily identified.

The CSRs are, on average, about 14–16 disaccharides in length. They can be prepared from HS by treatment with the enzyme K5 lyase that attacks the chain exclusively in the non-sulfated sections [9]. The CSRs consist of two subregions, the S-domains and the transition zones or T-zones. The S-domains are the most anionic sections of HS with a dominant repeat sequence of the disulfated disaccharide GlcNS α1-4 IdoA,2S α1-4 [10,11]. S-domains, which can be excised by the enzyme heparinase III, vary in length from 2 to 7/8 disaccharide units, though most fall in the range of 3–5. T-zones are regions of intermediate sulfation

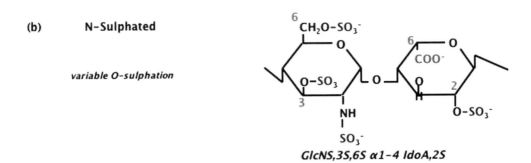

Figure 1. Disaccharide units in heparan sulfate. The disaccharides in HS are composed of either N-acetylated or N-sulfated glucosamine linked to uronic acid. Glucuronate (GlcA) is always linked to C-1 of GlcNAc whereas in the N-sulfated units the uronate is most often present as iduronate (IdoA), a C-5 epimer of GlcA. Common positions of O-sulfation in the N-sulfated disaccharides are C-2 of IdoA and C-6 of GlcNS. C-3 sulfation is rare but functionally significant (see text). C-6 sulfation of the GlcNAc residue only occurs in the T-zones (see Fig. 2) where the N-acetylated and N-sulfated disaccharides are in alternate sequences.

positioned around the S-domains. They are composed of alternating N-sulfated and N-acetylated disaccharides and make up an appreciable fraction (~25%) of a typical HS chain. C-6 sulfation occurs on glucosamine residues in both the S-domains and the T-zones. Variations in the frequency and position of 6-sulfated glucosamines in the CSRs contribute significantly to the overall fine structural heterogeneity in polymer sulfation [8].

The rare 3-O-sulfates are present in the S-domains of some species of HS (e.g. from endothelial cells), where they amplify structural diversity and confer very important recognition properties on specific S-domain type sequences. The best examples of this are the well-described antithrombin (AT)-binding pentasaccharide sequence and the binding site for the fusogenic gD protein of herpes simplex virus (HSV) that is essential for virus entry into the host cell. In the latter case the 3-O-sulfated group is linked to an N-unsubstituted glucosamine [12]. These 3-O-sulfated sequences are involved in protein activation mechanisms. The AT-binding pentasaccharide is a unique sequence motif of structure:

GlcNAc,6S–GlcA–GlcNS,3S,6S–IdoA,2S–GlcNS,6S

It was elucidated over 20 years ago by Lindahl's group [13] and it still remains the best example of specificity in GAG–protein recognition. Polysaccharides that contain this sequence (therapeutic heparin is enriched in it) activate AT by an allosteric mechanism that accelerates its interaction with the procoagulant proteases Factor Xa and thrombin. This sequence is responsible for the anticoagulant activity of heparin.

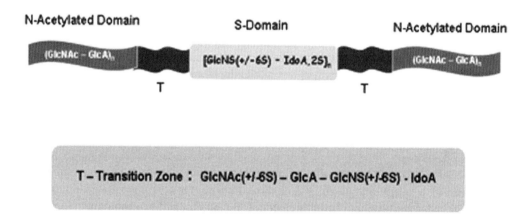

Figure 2. Composite sulfated region (CSR) of heparan sulfate. The CSRs are distributed at fairly regular intervals along the HS chain. They consist of a central S-domain flanked by transition zones (T-zones) that separate the S-domains from the non-sulfated N-acetylated regions. S-domains can be excised from the HS chain using the enzyme heparinase III and intact CSRs can be prepared by exploiting the specificity of K5 lyase for the N-acetylated regions (see text for details).

In contrast to the AT–heparin/HS interaction there is little molecular detail available on recognition of the HSV gD protein by HS. However it is assumed that HS with the GlcNH$_3^+$,3S residue in the appropriate sequence will induce a conformational change in the gD protein that drives the fusion reaction between the viral and cell membranes.

3. SACCHARIDE CONFORMATION AND ORIENTATION OF SULFATE GROUPS

Although no direct studies have been done on HS we can make certain assumptions about the conformation of the S-domains based on the solution structures of heparin. Heparin is mainly produced by connective tissue mast cells where it is packaged in a pseudo-crystalline array in secretory granules with biogenic amines and cationic proteases [14]. Although heparin has the same basic disaccharide structure as HS it lacks a well-defined domain structure. It is very highly sulfated and GlcNS,6S α1–4 IdoA,2S is the main disaccharide unit (>75%). NMR spectroscopy has revealed a helical structure for the heparin chain with clusters of sulfate groups in the disaccharide repeats positioned on opposite faces along the helical axis [15]. Heparin is a chemical analog of the S-domains of HS and it is reasonable to assume that S-domains will adopt a heparin-like helical structure.

The molecular geometry of the T-zones and N-acetylated regions is unclear but they are likely to be significantly more flexible than the S-domains. The evidence for this is indirect and based largely on HS-binding studies with dimeric or oligomeric proteins such as interleukin 8 (IL-8), macrophage inflammatory protein 1 alpha (MIP-1α), interferon gamma (IFNγ) and platelet factor 4 (PF4) (see, for example, [16,17]). In their quarternary structures these proteins are bivalent and bind simultaneously to two S-domains spaced along the HS chain. Molecular modeling suggests that the T-zone and the NAc domains that maintain continuity between the S-domains must be flexible to allow the S-domains to align and dock in the favored orientation with two separate recognition sites on the bound proteins [16,17].

4. THE BIOSYNTHESIS OF HEPARAN SULFATE

The biosynthesis of HS takes place in the Golgi cisternae and begins with formation of a non-sulfated precursor called heparan or N-acetylheparosan. Heparan is assembled directly on protein cores at specific serine residues primed with the tetrasaccharide linkage sequence (Ser)-Xyl-Gal-Gal-GlcA that is also found in chondroitin sulfate/dermatan sulfate (CS/DS) (see Chapter 5). The synthesis of HS rather than CS is favored where there are two or three contiguous or closely spaced Ser-Gly repeats and where the local peptide environment contains a hydrophobic amino acid and a cluster of acidic residues [18]. The tertiary structure of protein acceptors also influences GAG chain preferences [19].

The polymerization of the GlcNAc-GlcA repeat units of the heparan precursor is carried out by two HS polymerases that are now known to be equivalent to the *ext1* and *ext2* genes (exostosis genes) [20]. Mutations in either of these two genes gives rise to the human genetic disease called hereditary multiple exostosis (HME), characterized by benign orthogonal outgrowths of cartilage capped bone that become malignant in some cases [21]. These dominantly acting mutations cause lesions mainly in the growth plate of long bones. At present little information is available on the deficit in HS synthesis in these patients nor is it known why *ext* gene mutations appear to only affect bone growth. This disease does indicate a weak tumor suppressor activity for the *ext* genes.

Both EXT1 and EXT2 polymerases have GlcNAc and GlcA transferase activity essential for the synthesis of heparan [22]. However, the efficiency of polymerization is significantly enhanced when they combine as a functional heterodimer or larger aggregates [23]. It is also interesting that the formation of the EXT1/EXT2 complex is an early event after their synthesis in the endoplasmic reticulum and is essential for their intracellular transfer to the Golgi cisternae [24].

The transformation of heparan to HS begins with the conversion of clusters of GlcNAc residues to GlcNS brought about by the dual-function *N*-deacetylase/*N*-sulfotransferase (NDST) enzymes that combine both *N*-deacetylase and *N*-sulfotransferase activities in a single protein. There are four genetically distinct isoforms of the NDSTs but the widely distributed NDST-1 is the major enzyme involved in the biosynthesis of HS [25,26]. It is assumed that this enzyme establishes the basic domain structure of HS although it is not known how its activity is directed onto localized areas of the chain. The substrate specificities of all the other polymer-modifying enzymes restrict their activities to regions of *N*-sulfation.

Following *N*-sulfation the next step in the synthesis of HS is the conversion of GlcA to IdoA by an HS C-5 epimerase and this is closely coupled to the C-2 sulfation of IdoA by an HS 2–*O*-sulfotransferase. There is only one form of each of these two enzymes and, like the EXT polymerases, they appear to be cotransported from the endoplasmic reticulum (ER) to the Golgi [61]. Synthesis of HS is completed by the action of C-6 and C-3-*O*-sulfotransferases that transfer sulfate groups to glucosamine residues at C-6 and C-3 respectively. The 6-*O*-sulfotransferases and the 3-*O*-sulfotransferases comprise multigene families that generate much of the observed variation in structure of HS derived from different sources [27,28].

Ledin and colleagues observed that HS isolated from various tissues and cell cultures showed consistent and specific differences in chain length and polymer sulfation [29]. This is an interesting finding because although there is no known template that directs the synthesis of HS these results show that the activities of the HS-biosynthetic enzymes are tightly regulated.

5. "EDITING" THE STRUCTURE OF HEPARAN SULFATE

Newly synthesized HS is transported as an HS-proteoglycan from the Golgi to the cell surface or secreted into the ECM. Here its sulfation can be modified or "edited" by the action of two closely related endosulfates (sulfs) that desulfate the polymer at C-6 of GlcNS residues. Sulfs were first detected in quail (Qsulf) and were subsequently found in all vertebrates examined to date and in model organisms such as *Caenorhabditis elegans* and *Drosophila* [30,31]. The Qsulf enzyme is quite selective in its action, targeting the S-domains rather than the T-zones [30,32].

Although little is known about how the catalytic activity of these plasma membrane-associated enzymes is regulated they are clearly important in embryonic development and indeed they were first identified in a genetic screen for target genes of the Hedgehog protein [31].

Endosulfatases have significant effects on the biological activities of a number of morphogens and cytokines. For example the sulfs decrease the affinity of the Wnt ligand for HS, enabling it to bind efficiently to its signaling receptor frizzled [30,31]. The association of the bone morphogenetic protein (BMP) inhibitor Noggin with cell surface HS is also strongly reduced by endosulfatase action [32]. The sulf enzymes negatively regulate FGF2 signaling [34]. This is an interesting effect because the binding of FGF2 to HS is not dependent on 6-O-sulfates [8]. In this case sulfatase action is probably removing one or more key functional groups essential for the association of HS–FGF2 complexes with FGF2 signaling receptors [8].

In addition to the editing action of the sulfs, cell surface HS can be partially degraded to large fragments by an endo-β-glucuronidase or heparanase (hpa1) that acts on glucuronic acid residues at the T-zone/S-domain junction. It does not degrade HS in the N-acetylated regions where the majority of the GlcA residues are found [35]. Heparanase is highly enriched in platelets, where it is stored in α-granules and released on platelet activation. The enzyme is synthesized as an inactive monomeric protein that is activated by proteolytic removal of an ~8 kDa internal peptide sequence to form a functional heterodimer composed of a large (50 kDa) and small (8 kDa) subunit [36].

Normal nucleated cells release little heparanase – it remains inside the cell and probably initiates the first steps in the degradation of HS. In contrast, many tumor cells secrete heparanase and the level of expression and/or secretion of the enzyme correlates strongly with tumor angiogenesis and metastasis [37]. Because it has only a very limited degradative effect on HS, the enzyme is unlikely to diminish interactions of HS with growth factors and other protein effectors. Its apparent ability to stimulate tumor growth and invasion could be because it releases saccharide–growth factor complexes from heparan sulfate proteoglycans (HSPGs) and such complexes may be less constrained in their ability to engage cognate signaling receptors.

6. CELL SURFACE HEPARAN SULFATE PROTEOGLYCANS

The presence of HS in considerable abundance on cell surfaces is due mainly to its covalent association with two families of proteins, the transmembrane syndecans and the glycosylphosphatidylinositol (GPI)-anchored glypicans (Fig. 3); a variant form of CD44 harboring the V3 exon is also glycanated with HS [5,8].

6.1 Syndecans

In mammals there are four members of the syndecan family (20–45 kDa) that share sequence homology in the transmembrane and short cytoplasmic domains [38]. Three sites for attachment of HS are located in a short homologous region

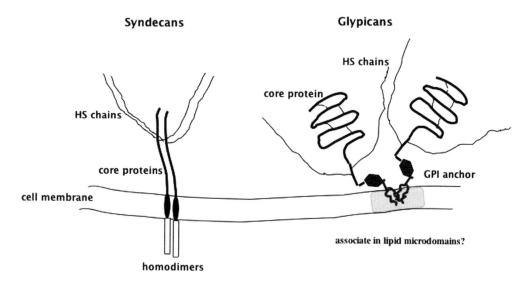

Figure 3. Cell surface heparan sulfate proteoglycans (HSPGs). The major cell surface HSPGs are the transmembrane syndecans and the GPI-anchored glypicans comprising four and six family members respectively. The syndecans are type I transmembrane proteins that are tightly associated as homodimers with regions of high homology in the membrane-spanning and cytoplasmic domains. The HS chains are clustered towards the N-terminus. The tertiary structure of the glypicans is stabilized by 14 highly conserved disulfide bonds and the HS chains are attached to the stem region close to the GPI anchor. Glypicans are monomeric proteins but may associate in lipid microdomains in the plasma membrane.

towards the N-terminus of the ectodomain. The syndecans are very stable homodimers and this is believed to be due to strong self-affinity interactions in the hydrophobic membrane-spanning regions [39] and the cytoplasmic tail [40]. Common intracellular binding partners for the syndecans are the bivalent PDZ-domain proteins syntenin and CASK. These proteins recognize the EFYA tetrapeptide sequence at the C-terminus of the conserved C2 region [41,42]. PDZ proteins are often scaffolding proteins with several interaction domains for association with intracellular signaling and membrane transporting systems [43].

Syntenin and CASK influence membrane targeting of the syndecans and in the case of syntenin evidence has emerged to show that it regulates the endocytosis and intracellular routing of cell surface syndecans [44]. It was recently reported that one of the PDZ-domains of syntenin binds to phosphatidylinositol 4,5-bisphosphate (PtdIns(4,5)P_2) [45] and this novel interaction may have the effect of localizing the bound syndecans to specialized phosphoinositide-rich membrane microdomains.

The syndecans provide a clear illustration of the concerted actions of core proteins and HS chains of HSPGs in controlling cell behavior. A thorough review of this theme has been published recently [46] but some key points in connection with syndecan 2 and syndecan 4 merit consideration here.

Syndecan 4 and focal adhesions

Syndecan 4 is essential for the formation of focal adhesions in fibroblasts attaching to ECM substrates such as fibronectin. These highly complex structures are assembled as a result of cooperative binding of a membrane integrin (e.g. $\alpha_5\beta_1$) to the RGD cell-binding region of fibronectin and HS chains to fibronectin Hep II region [46,47]. The mechanism of syndecan clustering in focal adhesions is unclear but some form of transmembrane communication between HS in the ectodomain and the intracellular cytoplasmic tail seems probable. The cytoplasmic domain plays a critical role in syndecan 4-mediated cell adhesion, signaling, and morphogenesis. One of its key molecular interactions is with α-actinin, one of the linker proteins that connects the actin cytoskeleton with the inner surface of the plasma membrane [48].

In all the syndecans, the short cytoplasmic regions of approximately 30–35 amino acids can be subdivided into conserved C1 and C2 regions that lie respectively proximal and distal to the plasma membrane and a central variable or V-region which probably specifies distinct functions of each syndecan (Fig. 4). In its dimeric form the V-region of syndecan 4 (KKPIYKK) binds to PtdIns(4,5)P$_2$ with lysine residues (K) making important contacts with phosphate groups of the inositol moiety [46,49]. This complex can then bind and activate protein kinase C alpha (PKCα) [50], which is known to be an important regulator of cell shape and

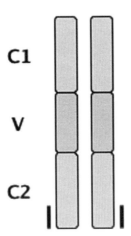

Figure 4. Cytoplasmic domains of the syndecans. The syndecan cytoplasmic domains (approx 35 amino acids) consist of a C1 conserved region proximal to the cell membrane, a central variable (V) region and a second conserved sequence (C2) at the C-terminus. The V-regions confer specific interactions on the individual syndecans. A tetrapeptide sequence EFYA at the end of the C2 regions (black lines) is a docking site for the PDZ proteins syntenin and CASK. These scaffold proteins influence the membrane localization and intracellular routing of the syndecans following endocytosis.

cell adhesion, possibly as a result of activating members of the Rho family of guanosine triphosphatases (GTPases). Syndecan-mediated binding and activation of PKCα is regulated by the action of PKCδ that impairs the interaction with PtdIns(4,5)P$_2$ by phosphorylating the one serine residue in the C1 region of the cytoplasmic tail [51]. Thus an interplay between two PKC isoforms controls the signaling potential of syndecan 4 and this will have an important bearing on cell shape, adhesion, and morphology.

Syndecan 2 in cell development

Intriguing developmental and morphogenetic roles have been identified for syndecan 2. In cultured neuronal cells it is important for dendritic spine formation where it is associated with and phosphorylated by EphB2 receptor tyrosine kinase [52,53]. The interaction with ephrin receptors probably triggers syndecan 2 clustering that parallels the extension of dendritic spines.

In *Xenopus* embryos it has been shown that syndecan 2 is essential for left–right asymmetry in the developing embryonic heart [54]. This is a well-orchestrated and complex process in which syndecan 2, phosphorylated on a pair of unique serines in its V-region, binds and presents vitellogenin 1 (Vg1), a member of the transforming growth factor beta (TGFβ) family, to migratory mesodermal cells. The syndecan 2 is found in the embryonic ectoderm but only cells in "right side" ectoderm contain the phosphorylated serines as a result of the action of PKCγ [55]. The Vg1 is complexed to the HS chains of the phosphorylated syndecan 2 and although it is attractive to speculate that the phosphorylated serines somehow influence HS sulfation, and hence its capacity to bind Vg1, no evidence for this has been published to date.

6.2 The glypicans

In contrast to the syndecan HSPGs, the glypicans are quite large monomeric proteins that range in size from 60 to 70 kDa. Although overall the amino acid sequence similarities are rather weak, all the mammalian glypicans contain 14 conserved cysteine residues that probably dictate a common tertiary structure. A short stem region that contains 2–3 HS glycanation sites, links the glypicans to a GPI membrane anchor [56]. The glypicans are widely expressed in animal organs and tissues, particularly in the developing embryo, where they play important roles in cell growth control and in the formation and stabilization of morphogen gradients [28]. Loss-of-function mutations in glypican 3 (GPC3) give rise to the dysmorphic and overgrowth syndrome Simpson Golabi Behmel in humans [57].

At present the mode of action of GPC3 in regulating morphology and organ size is unclear, although evidence from GPC3 knockout mice suggests that it suppresses cell proliferation and regulates cell survival [58]. Genetic evidence from studies in *Drosophila* and zebrafish implicate glypicans in the regulation of cell migration and cell fate through interactions with signaling proteins such as Wingless/Wnt and BMP. Three of the glypicans (GPC1, GPC3, and GPC4) have an internal proteolytic cleavage site in the cysteine-rich region of the core protein. Cleavage at this site produces ~30 and ~ 40 kDa subunits that remain associated in a heterodimer cross-linked by disulfide bonds. David and colleagues have demonstrated that in transfected MCF7 and CHO cells endoproteolytic cleavage of the GPC3 ectodomain by proprotein convertases is necessary for interaction with Wnt5a and for subsequent GPC3-facilitated Wnt5a cell signaling [59]. This study also demonstrated that convertase "activation" of GPC3 was essential for convergent/extension movements in zebrafish embryos. However, the necessity for GPC3 to be processed by convertases is not an absolute requirement for Wnt signaling through the canonical β-catenin pathway in hepatocellular carcinoma cells [60].

At the present time it is unclear why proteolytic processing of GPC3 is necessary for activity only in some cellular contexts. More studies are needed on the structure of the Wnt interaction domain in GPC3 and on the transfer of Wnt from GPC3 to its signaling receptor system.

7. SUMMARY

A clear picture is now emerging of the functional diversity of HSPGs. Genetic manipulation combined with detailed structural analysis of HS chains and their core proteins has yielded new insights on the molecular mechanisms underpinning the signal transduction properties of HSPGs. We still need more information on the cooperative modes of action of the HS and protein components of HSPGs in regulating cell growth and development. HSPGs are normally conceived as being involved in "outside-in" signaling, with information flowing from HS chains (in complexes with one of their many client proteins) to the cell interior via membrane-associated PG core proteins. However, the mode of action of syndecan 2 in the development of the *Xenopus* heart, where HS chain structure is influenced by PKCγ, is an example of "inside-out" signaling, and this outward directional flow of information may prove to be more common than is currently realized.

HSPGs still harbor many secrets and unravelling them will be a challenging but rewarding task, not only because we will gain a deeper understanding of fundamental cellular mechanisms but also because new opportunities will emerge for targeting PG-dependent mechanisms in human disease.

REFERENCES

1. Jorpes JE, Gardell S (1948) *J. Biol. Chem.* **176**, 267–276.
2. Gallagher JT, Walker A (1985) *Biochem. J.* **230**, 665–674.
3. Kramer PM (1971) *J. Biol. Chem. Biochem.* **10**, 1437–1444.
4. Couchman JR, Chen L, Woods A (2001) *Int. Rev Cytol.* **207**, 113–150..
5. Yayon A, Klagsbrun M, Esko JD, Leder P, Ornitz D (1991) *Cell* **64**, 841–848.
6. Rapreager AC, Krufka A, Olwin BB (1991) *Science* **252**, 1705–1708.
7. Gallagher JT, Lyon M (2000) In: *Proteoglycans: Structure, Biology and Molecular Interactions*, pp. 27–60. Edited by R Iozzo. Marcel Dekker New York.
8. Gallagher JT (2001) *J. Clin. Invest.* **108**, 349–355.
9. Murphy K, Merry CLR, Lyon M, Thompson JE, Roberts IS, Gallagher JT (2004) *J. Biol. Chem.* **279**, 27239–27245.

10. Merry CLR, Lyon M, Deakin J, Hopwood JJ, Gallagher JT (1999) *J. Biol. Chem.* **274**, 18455–18462.
11. Kreuger J, Salmirvirta M, Sturiale L, Gimenez-Gallego G, Lindahl U (2001) *J. Biol. Chem.* **276**, 30744–30752.
12. Shukla D, Liu J, Blaiklock P *et al.* (1999) *Cell* **99**, 13–22.
13. Lindahl U, Thunberg L, Backstrom G, Riesenfeld J, Nordling K, Bjork I (1984) *J. Biol. Chem.* **259**, 12368–12376.
14. Kolset S, Gallagher JT (1990) *Biochim. Biophys. Acta* **1032**, 191–211.
15. Mulloy B, Forster MJ (2000) *Glycobiology* **10**, 1147–1156.
16. Lortat-Jacob H, Turnbull JE, Grimaud JA (1995) *Biochem. J.* **310**, 497–505.
17. Stringer S, Gallagher JT (1997) *J. Biol. Chem.* **272**, 20508–20514.
18. Zhang L, Esko JD (1994) *J. Biol. Chem.* **269**, 19295–19299.
19. Chen RL, Lander AD (2001) *J. Biol. Chem.* **276**, 7507–7517.
20. Senay C, Lind T, Muguruma K *et al.* (2000) *EMBO Rep.* **1**, 282–286.
21. Kim BT, Kiagawa H, Tamura J, Saito T, Kusche-Gulberg M, Lindahl U (2001) *Proc. Natl Acad. Sci. USA* **98**, 7176–7181.
22. McCormick C, Leduc Y, Martindale D *et al.* (1998) *Nat. Genet.* **19**, 158–161.
23. Izumikawa T, Egusa N, Taniguchi F, Sugahara K, Kitgawa H (2006) *J. Biol. Chem.* **281**, 1929–1934.
24. McMormick C, Duncan G, Goutsos KT, Tufaro F (2000) *Proc. Natl Acad. Sci. USA* **97**, 668–673.
25. Kjellen L (2001) *Biochem. Soc. Trans.* **31**, 340–342.
26. Ringvall M, Ledin J, Holmberg K *et al.* (2000) *J. Biol. Chem.* **275**, 25926–25930.
27. Esko JD, Lindahl U (2001) *J. Clin. Invest.* **108**, 169–173.
28. Esko JD, Selleck SB (2002) *Annu. Rev. Biochem.* **71**, 435–471.
29. Ledin J, Staatz W, Jin-Ping L *et al.* (2004) *J. Biol. Chem.* **279**, 42733–42741.
30. Ai X, Do A, Lozynska G, Kusche-Gullberg M, Lindahl U, Emerson C (2003) *J.Cell Biol.* **162**, 341–351.
31. Dhoot GK, Gustafsson MK, Ai X, Sun W, Standiford DM, Emerson CP (2001) *Science* **293**, 1663–1666.
32. Viviano BL, Paine-Saunders S, Gasiunas IN, Gallagher JT, Saunders S (2004) *J. Biol. Chem.* **279**, 5604–5611.
33. Ai X, Kusche-Gullberg M, Lindahl U, Lu K, Emerson CP (2006) *J. Biol. Chem.* **281**, 4969–4976.
34. Uchimura K, Morimoto-Tomita M, Bistrup A *et al.* (2006) *BMC Biochemistry* **7**, 1–13.
35. Vlodavsky I, Friedmann Y (2001) *J. Clin. Invest.* **147**, 99–108.
36. McZenzie E, Young K, Hiscock M *et al.* (2003) *Biochem. J.* **373**, 423–435.
37. Zetser A, Bashenk Y, Miao HQ, Vlodavsky I, Ilan N (2003) *Cancer Res.* **63**, 7733–7741.
38. Bernfield M, Gotte M, Park PW *et al.* (1999) *Annu. Rev. Biochem.* **68**, 729–777.
39. Carey D (1998) *Biochem. J.* **327**, 1–16.
40. Choi S, Lee E, Kwon S *et al.* (2005) *J. Biol. Chem.* **280**, 42573–42579.
41. Zimmermann P, Tomatis D, Rosas M *et al.* (2001) *Mol. Biol.Cell.* **13**, 339–350.
42. Hung AY, Sheng M (2002) *J. Biol. Chem.* **277**, 5699–5702.
43. Fanning AS, Anderson JM (1999) *J. Clin. Invest.* **103**, 767–772.
44. Zimmermann P, Meerschaert K, Reeksmans G *et al.* (2002) *J. Mol. Cell.* **9**, 1215–1225.
45. Zimmermann P, Zhang S, Degeest G *et al.* (2005) *Dev. Cell* **9**, 377–388.
46. Couchman JC (2003) *Nat. Rev.* **4**, 926–937.
47. Woods A Couchman JR (2001) *Curr. Opin. Cell. Biol.* **13**, 578–583.
48. Greene DK, Tumova S, Couchman JR, Woods A (2003) *J. Biol. Chem.* **278**, 7617–7623.
49. Oh E-S, Woods A, Lim S-T, Thelbert AW, Couchman JR (1998) *J. Biol. Chem.* **273**, 10624–10629.
50. Lim S-T, Longley RL, Couchman JR, Woods A (2003) *J. Biol. Chem.* **278**, 13795–13802.
51. Murakami M, Horowitz A, Tang S, Ware JA, Simons M (2003) *J. Biol. Chem.* **278**, 7617–7623.
52. Ethell IM, Yamaguchi Y (1999) *J. Cell Biol.* **144**, 575–586.
53. Ethell IM, Irie F, Kalo MS, Couchman JR, Pasquale EB, Yamaguchi Y (2001) *Neuron* **31**, 1001–1013.

54. Kramer KL, Yost HJ (2002) *Dev. Cell* **2**, 115–124.
55. Kramer KL, Barnette JE, Yost HJ (2002) *Cell* **111**, 981–990.
56. Filmus J, Selleck SB (2001) *J. Clin. Invest.* **108**, 497–501.
57. Pilia G, Hughes-Benzie RM, MacKenzie A *et al.* (1996) *Nat. Genet.* **12**, 241–247.
58. Paine-Saunders S, Viviano Bl, Zupicich J, Skarnes WC, Saunders S (2000) *Dev. Biol.* **225**, 179–187.
59. De Cat B, Muyldermans S-Y, Coomans C *et al.* (2003) *J. Cell. Biol.* **163**, 625–635.
60. Capurro MI, Shi W, Sandal S, Filmus J (2005) *J. Cell Biol.* **280**, 41201–41206.
61. Pinhal MAS, Smith M, Olsen S, Aikawa J, Kimata K & Esko JD (2001) *Proc. Natl. Acad. Sci. USA* **98**, 12984–12989.

CHAPTER 10
Galectins: effective modulators of cytoskeletal organization and cellular growth – focus on galectin-8 as a model system

Yehiel Zick

1. INTRODUCTION

Extracellular matrix (ECM) proteins have important functions in providing structural integrity to tissues, and in presenting proper environmental cues for cell adhesion, migration, growth, and differentiation [1,2]. These functions rely on spatiotemporal expression of adhesive as well as antiadhesive components of the ECM proteins [3]. ECM proteins like fibronectin [4], collagen [5], and laminin [6] are best characterized, though other types of proteins, including mammalian lectins, also function as ECM proteins. Selectins mediate cell–cell interactions [7] through calcium-dependent recognition of sialylated glycans [8], whereas galectins, animal lectins that specifically bind β-galactoside residues [9], were implicated as modulators of cell–matrix interactions. While lacking a signal peptide and found mainly in the cytosol, galectins are externalized by an atypical secretory mechanism [10] to modulate cell growth, cell transformation, embryogenesis, and apoptosis (reviewed in [11]). In accordance with their proposed functions, galectins enhance or inhibit cell–matrix interactions (see [12] for review).

In this chapter the role of galectins as ECM proteins will be discussed, placing special emphasis on galectin-8. In many aspects, the effects of galectin-8 on cytoskeletal rearrangement and cell growth reflect the mode of action of other galectins; nonetheless, several features of galectin-8 single it out of the entire galectin family. These common and unique features are discussed here.

2. CHARACTERISTIC FEATURES OF THE GALECTIN FAMILY

The galectins are a family of carbohydrate-binding proteins that are widely distributed in the animal kingdom from lower invertebrates to mammals [13].

Members of the galectin family are defined by two properties: affinity for β-galactoside-containing glycoconjugates and conserved amino acid sequence in the carbohydrate-binding domain (CRD) [9,14].

In mammals, 15 galectins have been identified to date [15]. They can be structurally classified into three types [16]: (1) The prototype galectins (-1, -2, -5, -7, -10, -11, -13, -14, and -15] [15] that contain a single CRD. Members of this group can act either as monomers or as dimers, with subunit molecular mass of ~14 kDa [17]. (2) The chimera type has one CRD and another type of domain. In vertebrates, the only known member of this type is galectin-3, composed of a long N-terminal domain, connected to the CRD [17]. (3) The tandem repeat type galectins (-4, -6, -8, -9, and -12) are composed of two CRDs, joined by a linker peptide of variable length. The specificity and binding affinity of the two CRDs are not necessarily the same and they can probably bind different glycoconjugates [13,17,18].

Several lines of evidence suggest that galectins play a role in development [17], cell adhesion [12], immunity [15,19], cancer [15,20], and apoptosis [15,21]. However, little is known about the exact roles of galectins in these processes.

3. GALECTIN-8 – A TANDEM REPEAT GALECTIN

Galectin-8, a member of the tandem repeat type galectins, is a 34 kDa protein made of two homologous CRDs of about 140 amino acid each, with 38% identity among them. The two CRDs are joined by a linker peptide of variable length [16]. Potential differences in sugar binding between the domains are predicted from a conserved difference in their sequence (WGXEXI vs. WGXEXR at the N- and C-terminal CRDs, respectively) [18,22,23]. Galectin-8 was initially isolated from rat liver, but it is also expressed in other tissues including lung, kidney, brain, hind limb, and cardiac muscle [22]. Prostate carcinoma tumor antigen 1 (PCTA-1), which is highly expressed in prostate carcinomas, was the first human galectin-8 isoform to be discovered [24,25]. In humans, at least six isoforms of galectin-8 are known. Two isoforms are, in fact, prototype galectins because they contain only the N-terminal domain, fused to "hinge" regions of different length. The four isoforms that are of the tandem repeat type also differ in the size of their "hinge region", which varies in length from 24 to 74 amino acids. The variable length of the "hinge region" is likely to affect the repertoire of the glycoproteins which interact with galectin-8, because the two CRDs spaced at different distances are likely to bind different spatially oriented carbohydrates. This may affect the function of galectin-8 [18] and may account for the presence of its different isoforms.

Although galectin-8 is a cytoplasmatic protein, it is not uniformly spread. Similar to other galectins [26–28] it shows a micro-clustering pattern reminiscent of that seen with proteins associated with mitochondria, the Golgi or trans-Golgi membranes. Furthermore, galectin-8 is a secreted protein [29]. However, like other galectins, it lacks signal peptide suitable for endoplasmic reticulum (ER)/Golgi-mediated secretion [16]. Therefore, galectins (including galectin-8) are assumed to be secreted by an unconventional, yet unknown, mechanism. At least four potential mechanisms for unconventional protein export are known to mediate secretion of cytosolic proteins into the extracellular space. Two of these involve secretion of intracellular vesicles of the endocytic membrane system such as secretory lysosomes and exosomes [30]. Two alternative unconventional secretory mechanisms are characterized by a direct translocation of cytosolic proteins across the plasma membrane using either resident transporters or a process called "membrane blebbing" [30]. Atypical secretion is not a unique property of galectins, because other cytoplasmic proteins like thioredoxin, interleukin-1β and basic fibroblast growth factor (FGF) lack a signal sequence, yet are externalized and function extracellularly [30].

4. GALECTIN-8 AS A MEDIATOR OF CELL ADHESION

Studies of galectin-8 revealed that it positively or negatively regulates cell adhesion, depending on the extracellular context [18,29,31–35]. When immobilized onto matrix, galectin-8 can be classified as an ECM protein equipotent to fibronectin in promoting cell adhesion, spreading, and migration [31]. In contrast, excess soluble galectin-8 interacts both with cell surface integrins and with other soluble ECM proteins, such as fibronectin, and inhibits cell–matrix interactions [18,29].

4.1 Integrins as mediators of cell adhesion to galectin-8

Cell–matrix interactions depend to a large extent upon the engagement of specific ligands, like galectin-8, with a diverse class of cell surface $\alpha\beta$ heterodimeric receptors known as integrins [36]. Integrins mediate cell adhesion, migration, and invasion, and were implicated in the regulation of many cellular functions including embryonic development, tumor cell growth, programmed cell death, leukocyte homing, bone resorption, clot retraction, and the response of cells to mechanical stress [37]. Integrins also have a multitude of intracellular effects both on cytoskeletal organization, and as triggers of an intricate network of signaling pathways [1,2,36].

Cell adhesion to galectin-8 involves its interaction with a selected subgroup of cell surface integrins that include α_M [33], α_1, α_5 [35], α_3, α_6, and β_1 [29], while it interacts to a very limited extent with α_4, β_3 [29], and α_2 [35] integrins. Moreover, galectin-8 interactions with integrins involves its binding to sugar moieties, rather than the ligand-binding RGD site on the extracellular domain of the integrin molecules. Indeed, cell adhesion to galectin-8 is completely insensitive to the presence of RGD peptides [31]. Still, the ability of EDTA and Mn^{2+} to inhibit or potentiate, respectively, cell adhesion to galectin-8 suggests that the metal-bound conformation of integrins is the preferred conformation that promotes adhesion to galectin-8.

Interaction of galectin-8 with integrins is of physiological relevance. First, integrin-blocking antibodies decrease adhesion mediated by galectin-8 [29,31,33,35]. Furthermore, galectin-8 and $\alpha_3\beta_1$ complexes, formed in naïve intact cells, can be isolated by precipitation with specific α_3-antibodies [29]. These observations suggest that a secreted form of galectin-8 presumably acts in an autocrine or a paracrine fashion and binds to the extracellular regions of cell surface integrins. These findings implicate integrins as major cellular components through which a secreted galectin-8 manifests its biological activities. Other galectins also interact with integrins. Galectin-3 [38] and galectin-1 [39] interact with $\alpha_1\beta_1$ and $\alpha_7\beta_1$, integrins, respectively, whereas $\alpha_M\beta_2$ integrin is a major receptor for galectin-1 [40] and -3 [41]. Hence, different galectins might interact with selected subsets of integrins.

While immobilized galectin-8 promotes cell adhesion, this process is inhibited by exogenously added soluble galectin-8 [18,29]. In that respect galectin-8 resembles other soluble ECM proteins like laminin [6] and fibronectin [4] as well as other galectins [12] that inhibit cell adhesion upon binding to integrins and inducing a steric hindrance. Alternatively, similar to galectin-3 [42], galectin-8 could induce the internalization of cell surface integrins and in such a way impair cell adhesion. Galectin-8 also binds soluble fibronectin present in the serum [29]. Hence, the antiadhesive effects of galectin-8 could be mediated either upon direct binding of excess soluble galectin-8 to cell surface integrins, or alternatively, upon binding and recruitment to the cell surface of other soluble ECM proteins such as fibronectin that could exert an antiadhesive effect of their own [4]. This function requires the occupancy of both CRDs of the native galectin-8, and might account for the inability of the truncated monovalent soluble N-galectin-8 to inhibit cell adhesion [18]. The inhibitory effects of galectin-8 were reversed by manganese ions, a powerful activator of integrins, suggesting that ligation by soluble galectin-8 may stabilize integrins in a low affinity state [29]. Altogether, the function of soluble galectin-8 seems to be dictated by the combinatorial arrangement of available cell surface and extracellular ligand.

Due to its pro- and antiadhesive functions galectin-8 might be considered as a member of adhesion-modulating proteins such as SPARC, thrombospondin, tenascin, hevin, and disintegrins, collectively known as matricellular proteins [43]. Still, it should be emphasized that the action of matricellular proteins depends upon protein–protein interactions, while galectin-8's function depends upon sugar–protein interactions. Such usage of the glyco-code adds a novel biological role for lectin–carbohydrate interactions.

4.2 Signal transduction cascades triggered upon cell adhesion to galectin-8

Cell adhesion onto galectin-8 induces signaling cascades that are being utilized by integrins upon ligation by other ECM proteins. Most prominent among them is the focal adhesion kinase (FAK), which undergoes integrin-stimulated autophosphorylation. Tyr-phosphorylated FAK recruits Grb2-Sos complexes which activate the Ras-MAPK signaling pathway. FAK also phosphorylates p130^{Cas} that binds Crk and generates further signals through C-jun N-terminal kinase (JNK). P-Tyr^{397} of FAK serves as a binding site for the SH2 domain of p85α, the regulatory subunit of phosphatidylinositol 3-kinase (PI3K) that propagates signals to protein kinase B (PKB) and p70S6K (reviewed in [1,37]). Stimulation of integrins also activates the Rho-family GTPases Rho, Rac, and Cdc42, which mediate the formation of stress fibers, lamellipodia, and filopodia, respectively [44].

Phosphorylation of FAK and p130^{Cas} which takes places during early stages of cell–matrix interactions is a common signal emitted upon cell adhesion to fibronectin or galectin-8 [32]. Thereafter, bifurcation of signals mediated by fibronectin and galectin-8 presumably takes place. While cell adhesion to fibronectin leads to transient activation of mitogen-activated protein kinase (MAPK), engagement of integrins by galectin-8 leads to sustained activation of ERK-1 and ERK-2, and a more robust and sustained activation of PKB and p70S6K [18,32,34,35]. PKB has already been implicated as a mediator of cell adhesion [45], but there is little evidence that p70S6K is required for the processes of cell adhesion. Activation of p70S6K was shown to be independent of pathways that regulate formation of focal adhesions [46], therefore p70S6K might modulate integrin activation by selective ECM proteins such as galectin-8, although the direct targets of this kinase within cell adhesion complexes remain to be determined.

How is a similar extent of activation of FAK and p130^{Cas} by galectin-8 or fibronectin translated into differences in the state of activation of their downstream effectors (Ras, ERK-1,2, PKB, and p70S6K)? One possibility is that a different set of integrins is ligated by galectin-8 or fibronectin. While galectin-8 ligates a selected subset of integrins, the repertoire of integrins ligated by fibronectin is much broader. As a result, the composition of signaling complexes formed between the different cytoplasmic tails of integrins and their downstream effectors might differ, depending on whether integrins were clustered by galectin-8 or fibronectin.

The robustness and duration of the activation of a given signaling pathway has far-reaching biological consequences. For example, it is well established that transient activation of the MAPK cascade (e.g. by epidermal growth factor (EGF)) leads to enhanced growth of PC-12 cells, while stimulation of these cells with nerve growth factor (NGF), induces sustained activation of the MAPK cascade, which leads to cellular differentiation [47]. Accordingly, the sustained and robust activation of the MAPK and PI3K signaling pathway upon cell adhesion to galectin-8 might account for the unique cytoskeletal organization and biological functions of cells adherent to this lectin.

4.3 Cytoskeletal organization triggered upon cell adhesion onto galectin-8

The unique signaling pattern triggered upon cell adhesion to galectin-8 translates to a faster cell spreading and a distinctive organization of cytoskeletal elements. Prominent stress fibers that traverse the cell body are readily observed in cells adherent to fibronectin, but they are less abundant in cells adherent to galectin-8. Instead, adhesion to galectin-8 triggers sustained formation of F-actin microspikes (Fig. 1). This is rather a general phenomenon observed in a number of, though not all, cell lines [31,32]. Similarly, formation of focal contacts is limited when cells adhere onto galectin-8 [31]. These adhesion sites contain minimal amounts of vinculin and paxillin, which correlates with a reduced Tyr phosphorylation of paxillin in cells adherent to galectin-8 [31]. These findings suggest that other adhesion complexes presumably act in conjunction with integrins to mediate cell adhesion to galectin-8. This idea conforms to the view that adhesion complexes show extraordinary structural and molecular diversity, and sites of cell adhesion to the ECM can be mediated by a variety of matrix molecules and integrin proteins [48].

(a) **Fibronectin** **(b)** **Galectin-8**

Actin Staining   Actin Staining

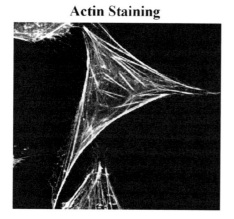 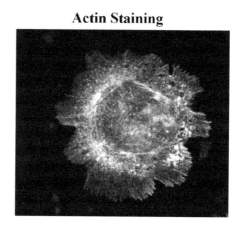

Figure 1. Cytoskeletal organization of cells adherent to (a) fibronectin or (b) galectin-8. The experiment was carried out as described in [31]. CHO cells were allowed to adhere for 2 h on fibronectin (10 μg mL^{-1}) or galectin-8 (25 μg mL^{-1}) precoated coverslips at 37°C. Cells were then fixed, incubated with TRICT-Phalloidin for actin staining, and were examined under confocal microscope.

Hence, formation of protein–protein complexes upon binding of integrins to fibronectin, vs. the formation of protein–sugar complexes between galectin-8 and integrins offers a molecular aspect for the differences in cytoskeletal organization and signaling induced by these two matrices. Indeed, the less-developed pattern of actin filaments and focal contacts observed in cells seeded on galectin-8 resembles the appearance of cells whose integrins were aggregated in the absence of a ligand (e.g. RGD peptide) [49], suggesting that galectin-8 presumably fails to occupy the protein–ligand binding site of integrins, while it effectively induces aggregation of these receptors. The possibility that immobilized galectin-8 induces integrin clustering is consistent with the fact that a truncated form of galectin-8, which contains only its N-terminal half with a single CRD, is much less efficient (about 5-fold) in functioning as an ECM protein that promotes cell adhesion [18,31].

These findings suggest that ligation of cell surface integrins is necessary but insufficient to trigger the biological functions of immobilized galectin-8, and that receptor clustering, in addition to receptor occupancy, is required to promote the adhesive effects of galectin-8. In that respect, galectin-8 resembles "classical" ECM proteins that induce integrin aggregation to trigger cell adhesion and to initiate the signaling cascaded generated thereof.

Inhibitors of the PI3K pathway impair cell adhesion and spreading on galectin-8 and the formation of microspikes, while having no effects on cells adherent to fibronectin, indicating that downstream effectors of PI3K selectively regulate cytoskeletal rearrangements which occur when cells adhere and spread on galectin-8. Indeed, overexpression of PKB, a downstream effector of PI3K potentiates the formation of microspikes whereas overexpression of PKB or p70S6K accentuates the sensitivity of cells adherent and spread on galectin-8 to inhibitors of PI3K [32]. Hence, the differences in cytoskeletal organization observed when cells adhere to galectin-8 or fibronectin can be attributed to differences in the robustness and duration of the PI3K-mediated signals emitted upon adhesion to the two matrices.

Cellular attachment and spreading on fibronectin involves initial requirement for Cdc42 in the formation of filopodial protrusions and subsequent involvement of both Cdc42 and Rac during cell spreading and organization of the actin cytoskeleton [50]. In galectin-8-adherent cells, focal contacts poorly assemble [31] and microspikes containing F-actin are formed instead. These structures have been functionally implicated in cell migration [51], which is readily induced by galectin-8 [31]. Microspikes are readily formed when cells adhere to a variety of other ECM proteins such as thrombospondin-I [51], laminin-5 [52], and tenascin-C splice variants [53]. On fibronectin, Cdc42- and Rac-dependent formation of microspikes is involved in early steps of cell adhesion, but these events are transient and microspikes are rapidly replaced by focal contacts [51]. In contrast, microspikes are stabilized when cells adhere to galectin-8, and the cells do not proceed to form highly developed focal contacts [31]. The microspikes formed when cells adhere to galectin-8 are short and radial, and in that respect resemble microspikes formed when C2C12 cells, overexpressing a constitutively active Rac, adhere onto thrombospondin-I [54]. Indeed, Rac-1 has been implicated in mediating cell adhesion to galectin-8 [35]. We can therefore suggest that formation of microspikes upon cell adhesion to galectin-8 presumably involves Rac activation. Still, additional signaling elements, induced by galectin-8, like the PI3K/PKB pathways, are likely to be involved.

PKB, the downstream effector of PI3K, activates a number of kinases including the Ser/Thr kinase p21-activated protein kinase (PAK) [55], which has been implicated as playing a role in actin organization. PAK inhibits the activity of coffilin (reviewed in [56]) and in such a way may inhibit actin depolymerization and promote formation of F-actin microspikes induced by galectin-8.

4.4 Modulation of cell adhesion and signaling by the interplay between the two CRDs of galectin-8

Being a tandem-repeat type galectin, a key question is whether each of its two CRDs is functionally independent. This is a highly relevant issue in view of the fact that prototype galectins, having a single CRD, are active either as monomers or as dimers. For that purpose we deleted regions and mutated amino acids that were implicated in sugar binding of galectins. Then we explored the ability of the mutated/truncated forms of galectin-8 to bind sugars; to induce intracellular signaling cascades and to modulate cell adhesion. The results of the above analysis led us to conclude that the two CRDs of galectin-8 are not functionally independent [18]. A proper orientation of the two CRDs, determined by the length of the linker peptide, is required for the proper function of this lectin. The results further suggest that while sugar binding activity is a key feature required for the proper function of galectin-8, other structural elements, not involved in sugar binding, affect the signaling capacity and adhesive properties of this lectin.

Several lines of evidence support these conclusions. First, we could show that mutations that have only a minor impact on the sugar-binding capacity of galectin-8 have profound effects on its ability to promote transmembrane signaling and cell adhesion (Table 1). One example is the E88Q mutant, whose sugar-binding activity is essentially identical to that of the wild-type galectin-8, while its ability to stimulate the PI3K pathway and promote cell adhesion is severely impaired. The dichotomy between sugar-binding and signaling activity suggests that structural elements not directly involved in sugar binding might also mediate the signaling capacity of galectin-8. These elements, most likely, are involved in protein–protein interactions between galectin-8 and its cell surface receptors. Still, the importance of sugar binding for the proper function of galectin-8 should not be dismissed. This was evident by the fact that mutations

Table 1 Activities of wild-type (WT) and mutated/truncated forms of galectin-8. The activity of wild-type galectin-8 was taken as 100%. The ability of the various galectin-8 mutants to bind to lactosyl-Sepharose beads, agglutinate red blood cells, promote or inhibit cell adhesion and trigger signaling pathways is presented in a qualitative manner, to enable overall comparison of the different mutants

Construct	Lactose binding	Agglutination	Signaling	Cell adhesion	Antiadhesive activity
Galectin-8 (wild-type)	(100%)	+	+	+	+
GST-Galectin-8	+	+	+	+	+
I90R	+	+	+	+	+
R253I	+	+	+	nd	nd
GST-Δ-hinge	–	+	–	–	–
NT	–	None	–	–	None
E88Q	+	+	–	–	–
E251Q	–	+	+	+	None
GST-W248Y	–	+	–	–	None
GST-CT	–	+	None	None	None
GST-W85Y	–	+	None	None	None
GST-W*2Y	– –	None	None	None	None

+, Normal activity (equivalent to WT).
–, Activity reduced by less than 50% of the WT activity.
– –, Activity reduced by more than 50% of the WT activity.
nd, Not determined.
None, no activity.
Data taken from [18].

that severely impaired sugar binding of galectin-8 such as the double-mutant W85/248Y resulted in complete loss of its signaling and adhesive properties [18].

A major thrust of this study was aimed at elucidating the functional autonomy of the two CRDs of galectin-8. The results clearly indicated that the isolated domains of galectin-8 are functionally impaired. This translates into impaired sugar binding; adhesive and antiadhesive effects; reduced signaling capacity and altered cytoskeletal organization. The impairment is better manifested by the isolated C-CRD, which is practically devoid of adhesive and antiadhesive capabilities, while the isolated N-CRD maintains impaired adhesive functions but no antiadhesive capacities [18,57]. These findings clearly indicate that the presence of both CRDs is necessary for the proper function of galectin-8.

Proper function of galectin-8 also depends upon the proper orientation of its two CRDs, determined by the length of the linker or "hinge" region [18]. This was an unexpected result in view of the fact that the model structure of galectin-8, which was based upon the crystal structure of galectin-1 dimers, clearly indicated that the interactions of the two CRDs occurs along interphases that do not involve the linker peptide [16]. This apparently was not the case as a Δ-hinge mutant was impaired in its capacity to promote the rate of cell spreading and signaling capabilities [18], although the extent of cell adhesion to this mutant did not differ from that of the wild-type lectin [18,58]. These results differ from findings related to galectin-9, where it was demonstrated that shortening of the "hinge" domain had little impact on the function of this lectin [59]. Hence, galectin-8 and -9 presumably evolved differently with regard to their capacity to coordinate the functions of their isolated domains.

In conclusion, the above results are consistent with the hypothesis that galectin-8 must be viewed as a single functional entity whose two CRDs must be properly oriented and act in concert to elucidate the adhesive and signaling functions of this lectin. This contrasts with the more prevailing view of an "antibody model" in which each CRD of a tandem repeat type galectin can function independently. Because galectin-8 can exist in several isoforms, which vary in the length of their "hinge" domain, further studies are required to determine how such variations affect the overall function of this lectin.

5. GALECTIN-8 AS A MEDIATOR OF CELL GROWTH

Proliferation of animal cells is a highly conserved process tightly controlled by the interplay between growth-promoting and growth-limiting signals whose operation results in a timed progression through the cell cycle [60]. Signals which limit cell cycle advance are critically important for the control of cell number and the maintenance of tissue homeostasis both through restraints on cell proliferation and through the induction of programmed cell death [61]. The molecular machinery that controls cell cycle progression is based on the sequential activity of a family of protein kinases known as cyclin-dependent kinases or CDKs [62]. Considering the importance of CDKs in cellular proliferation, it is not surprising that their activity is exquisitely regulated. One of the key players in this process is the cyclin-dependent kinase inhibitor p21 [63]. This protein and its related counterparts p27 and p57, block progression through the cell cycle at the G_1/S and G_2/M checkpoints by forming ternary complexes with CDKs, thus inhibiting their enzymatic activity [64]. Increasing evidence now suggests that p21 confers apoptosis protection [65] and might be important for cell survival [66]. Accordingly, a complex mechanism regulates the cellular content of p21. In addition to transcriptional induction by p53-dependent and -independent mechanisms, both ubiquitin-mediated as well as ubiquitin-independent degradation processes regulate the levels of p21 (reviewed in [67]).

Cyclin-dependent kinase inhibitors (CDKIs) are induced in response to DNA damage through the action of the tumor supressor protein p53 [68]. Growth factors can upregulate p21 in some cells but this occurs in a p53-independent manner [69]. Other receptor–ligands including Fas ligands [70], interferons [71], TGFβ [72], and galectins [73–75] can also act as negative regulators of cell cycle progression. However, the molecular basis for the growth inhibitory effects of galectins in general, and of galectin-8 in particular, remains largely obscure.

Galectin-8 seems to inhibit cellular growth by promoting the accumulation of the CDK inhibitor p21. The accumulation of p21 is mediated, at least in part, through activation of JNK and PKB, whose phosphorylation is markedly increased in galectin-8-treated cells [34]. When cells treated with galectin-8 fail to accumulate p21, then they are subjected to an accelerated apoptotic process [34]. The cytostatic effects of galectin-8 are antagonized by growth factors such as insulin, whose receptors, when overexpressed, enable cells to accommodate to high concentrations of galectin-8 without undergoing apoptosis. These findings implicate galectin-8 as a modulator of cell growth, whose action is controlled by the availability of selected growth factors [34].

Several lines of evidence support such a model. First, failure to stably overexpress significant amounts of galectin-8 in several cell lines that otherwise readily overexpress a variety of other proteins, indicates that the overexpressed lectin exerts growth-inhibitory effects [29,34]. This could be accounted for by an autocrine effect of the secreted lectin that interacts with cell surface integrins. Alternatively, galectin-8 could act intracellularly and inhibit cell growth by an as yet undefined mechanism. Galectin-8 promotes the accumulation of the CDK inhibitor p21. Galectin-8 promotes the transcription of p21 mRNA by activating JNK [34] that induces the activity of NFκB, leading to p21 gene transcription [76]. Accordingly, activation of JNK in response to galectin-8 precedes the increase in cellular content of p21. JNK can be activated by other galectins as well. JNK mRNA is increased when T-cells are incubated with galectin-1 [77], or when galectin-7 is overexpressed in HeLa and DLD-1 cells [78].

A causal link between activation of JNK and the induction of p21 is provided by the fact that SP600125, a selective inhibitor of JNK, effectively inhibits p21 accumulation induced by galectin-8. Furthermore, introduction into H1299 cells of dominant-inhibitory mutant of SEK1 (MKK4), a dual-specificity kinase which activates JNK [79], inhibits the ability of galectin-8 to activate JNK and to induce the accumulation of p21. The second signaling pathway utilized by galectin-8 to induce the accumulation of p21 is PKB [34]. Soluble galectin-8 effectively stimulates PKB that promotes the accumulation of p21 by phosphorylating glycogen synthase kinase 3 beta (GSK3β). Such phosphorylation inhibits the activity of GSK3β that otherwise can phosphorylate p21 and tag it for proteosomal-mediated degradation [80]. This effect of PKB may compensate for, and even prevail over the destabilizing effect of PKB on p21–proliferating cell nuclear antigen (PCNA) complexes [81] associated with the direct phosphorylation of p21 by PKB *in vivo* [82]. Support for the role of PKB in the induction p21 is provided by the fact that wortmannin, a selective inhibitor of PI3K, the upstream regulator of PKB, partially blocks the accumulation of p21 induced by galectin-8 [34].

Activation of JNK and PKB by galectin-8 could be mediated by integrins, the cellular receptors of galectin-8 [29]. Indeed, both JNK and PKB are downstream effectors of integrins [36], some like $\alpha_6\beta_4$ and $\alpha_v\beta_8$ induce the expression of p21 [83,84]. These observations set a link between integrin ligation by galectin-8, activation of PKB and JNK, and the cellular accumulation of p21. Still, JNK, rather than PKB, seems to play the major role in the mechanism underlying the accumulation of p21 because inhibition of JNK activity completely inhibits p21 accumulation, even when PKB is still active. This suggests that the attenuated rate of degradation of p21, induced by PKB, is of a lesser impact.

Other galectins also induce growth arrest. Galectin-1 functions as a cytostatic factor for murine embryo fibroblasts [85], galectin-3 induces cell cycle arrest of human breast epithelial cells [74], while galectin-12 inhibits growth of HeLa cells [75]. Galectin-1 interacts with $\alpha_5\beta_1$ integrin to restrict carcinoma cell growth via induction of p21 and p27 [86], whereas galectin-3-mediated G_1 arrest involves downregulation of cyclin E and cyclin A and upregulation of p21 and p27 [87]. Hence, galectin-8, -1, and -3 can induce growth arrest by a mechanism that involves the accumulation of CDK inhibitors.

The accumulation of p21 induced by soluble galectin-8 protects the cells from the potential pro-apoptotic effects of this lectin. This is evident by the fact that galectin-8 effectively potentiates apoptosis when the accumulation of p21 is prevented [34]. Furthermore, the ability of galectin-8 to induce apoptosis almost doubles in HTC$^{p21-/-}$ cells, which lack p21. Hence, the ability of galectin-8 to drive cells to apoptosis is masked by its ability to induce p21 that diverts the cells from the apoptotic pathway into growth arrest. The induction of growth arrest as a means to escape apoptotic process is well established, but the apoptotic machinery has a complicated relationship to cell cycle control. Upregulation of p21 and its binding to CDKs may trigger growth arrest on one hand, whereas binding of p21 to caspase-3 may inhibit the function of the latter and inhibit pro-apoptotic processes [88]. Similarly, accumulation of cytosolic p21, which occurs during cellular differentiation [89], triggers formation of complexes between p21 and the apoptosis signal regulating kinase 1 (ASK1) [90]. This results in the inhibition of ASK1 activity that is otherwise required for the induction of programmed cell death [89]. The mechanism by which galectin-8 promotes apoptosis involves activation of caspase-9 and caspase-3 [34], although other galectins implicated as inducers of apoptosis [91–94] seem to utilize somewhat different mechanisms. Galectin-9 induces apoptosis through the calcium–calpain–caspase-1 pathway [94]. Likewise, apoptosis induced by galectin-1 involves a significantly reduction in the cellular content of Bcl-2 [93], while galectin-8 fails to affect the cellular content of this protein. Finally, galectin-7 is associated with p53-dependent onset of apoptosis [95], while galectin-8 can induce apoptosis in a p53-independent manner [34]. These observations indicate that different galectins can induce apoptotic processes by somewhat distinct mechanisms.

Why does immobilized galectin-8 serve as an ECM protein that promotes cell adhesion, spreading, and growth [31,32], whereas the soluble or the overexpressed lectin acts as a cytostatic factor? The opposing effects of galectin-8 could be attributed, for example, to the different concentrations of the lectin experienced by the cells. When galectin-8 is present at low concentrations, as an immobilized ligand, it interacts only with high-affinity receptors of the integrin family [29] that promote cell migration and growth. In contrast, when galectin-8 is present at high enough concentrations as a soluble ligand, or when it is overexpressed, it can interact with low-affinity receptors that trigger its cytostatic effects. These receptors could either be other members of the integrin family or different cell surface receptors altogether. Support for this model is provided by the fact that binding of galectin-8 to low- and high-affinity receptors results in a different repertoire of signals. When cells adhere to immobilized galectin-8 and only the high-affinity receptors are engaged, it triggers robust and sustained activation of the PI3K and MAPK pathways [31,32]. In contrast, when applied at high enough doses as a soluble ligand, galectin-8 triggers, in addition to PKB, a delayed response that involves activation of stress-activated kinases like JNK, expression of p21, and induction of cytostatic effects [34].

In summary, it appears that galectin-8 can act in three different modes, depending on the cellular context and the extracellular environment (Fig. 2). When it is immobilized in the presence of growth factors, it interacts with high-affinity receptors to promote cell adhesion, spreading, and cell migration. When it is present at high concentrations as a soluble ligand, or when it is overexpressed and secreted, it interacts with low-affinity receptors that induce the accumulation of CDK inhibitors, represented by p21 that attenuate the rate of DNA synthesis and induce a cytostatic effect. The third, pro-apoptotic mode of action of galectin-8 is exhibited under conditions which prevent the accumulation of p21, or following a sustained deprivation of growth factors. These different modes of action of galectin-8 are mediated by different signaling pathways, with the PI3K and MAPK being the predominant mediators of cell motility induced by immobilized galectin-8; with JNK and p21 being key players in mediating the cytostatic effects, and with JNK and caspases contributing to the pro-apoptotic effects of this lectin. In view of the cytostatic effects of galectin-8, no wonder that a number of tumor cells [96,97], such as malignant colon tissues [98], attenuate the expression of this protein. Still, an interesting question is how galectin-8 is beneficial to invasive prostate carcinomas [24] and other tumors [99] that highly express this lectin. Further studies are therefore required to elucidate the cellular cues that dictate which mode of action of galectin-8 is operative under physiological or pathological conditions.

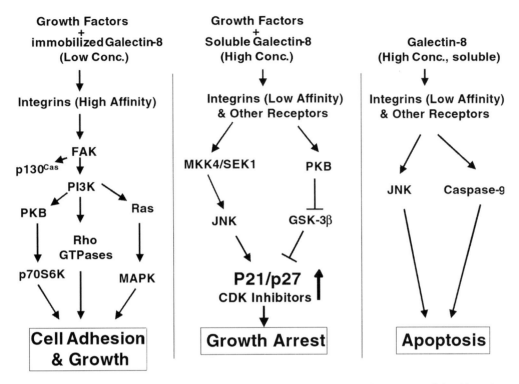

Figure 2. Different modes of action of galectin-8 when presented as a soluble or an immobilized ligand, in the absence or presence of growth factors. Modified from [34].

6. FUTURE PERSPECTIVES

Since the discovery of galectin-8, progress has been made in unraveling its structure, localization, and some of its physiological functions. Still, we are far beyond achieving a clear understanding of its biological role. Better elucidation of the signaling cascades emitted upon cell adhesion to galectin-8, combined with studies aimed at better understanding its role as regulator of cell growth might be revealing. Generation of transgenic mice that overexpress galectin-8, and production of mice that lack the galectin-8 genes are obvious avenues that should be avidly pursued. Ligation of integrins by ECM proteins triggers transcription of a unique set of genes. Accordingly, identification of genes whose expression is modulated upon cell adhesion to galectin-8 may help in elucidating its biological activity. Collectively, these possible experimental approaches are likely to converge into a major undertaking aimed to enlighten novel aspects related to the mode of action of galectin-8 under physiological and pathological conditions.

Acknowledgment

Y.Z. is an incumbent of the Marte R. Gomez Professorial Chair. This work was supported by grants from the CaPCure Israel Foundation; The Moross Center for Cancer Research and the Israel Cancer Association.

REFERENCES

1. Geiger B, Bershadsky A, Pankov R & Yamada KM (2001) *Nat. Rev. Mol. Cell. Biol.* 2, 793–805.
2. Miranti CK & Brugge JS (2002) *Nat. Cell. Biol.* 4, E83–90.
3. Chiquet-Ehrismann R (1995) *Curr. Opin. Cell Biol.* 7, 715–719.
4. Woods ML, Cabanas C & Shimizu Y (2000) *Eur. J. Immunol.* 30, 38–49.
5. Heino J (2000) *Matrix Biol.* 19, 319–323.
6. Calof AL, Campanero MR, O'Rear JJ, Yurchenco PD & Lander AD (1994) *Neuron* 13, 117–130.
7. Juliano RL (2002) *Annu. Rev. Pharmacol. Toxicol.* 42, 283–323.
8. Feizi T & Galustian C (1999) *Trends Biochem. Sci.* 24, 369–372.
9. Barondes SH, Castronovo V, Cooper DN et al. (1994) *Cell* 76, 597–598.
10. Hughes RC (1999) *Biochim. Biophys. Acta* 1473, 172–185.
11. Rabinovich GA, Rubinstein N & Fainboim L (2002) *J. Leukoc. Biol.* 71, 741–752.
12. Hughes RC (2001) *Biochimie* 83, 667–676.
13. Cooper DN (2002) *Biochim. Biophys. Acta* 1572, 209–231.
14. Liu FT, Patterson RJ & Wang JL (2002) *Biochim. Biophys. Acta* 1572, 263–273.
15. Liu FT & Rabinovich GA (2005) *Nat. Rev. Cancer* 5, 29–41.
16. Zick Y, Eisenstein M, Goren RA, Hadari YR, Levy Y & Ronen D (2004) *Glycoconj. J.* 19, 517–526.
17. Leffler H, Carlsson S, Hedlund M, Qian Y & Poirier F (2004) *Glycoconj. J.* 19, 433–440.
18. Levy Y, Auslender S, Eisenstein M et al. (2006) *Glycobiology* in press.
19. Rabinovich GA, Baum LG, Tinari N et al. (2002) *Trends Immunol.* 23, 313–320.
20. Lahm H, Andre S, Hoeflich A et al. (2004) *Glycoconj. J.* 20, 227–238.
21. Hernandez JD & Baum LG (2002) *Glycobiology* 12, 127R–136R.
22. Hadari YR, Paz K, Dekel R, Mestrovic T, Accili D & Zick Y (1995) *J. Biol. Chem.* 270, 3447–3453.
23. Ideo H, Seko A, Ishizuka I & Yamashita K (2003) *Glycobiology* 13, 713–723.
24. Su Z.-Z, Lin J, Shen R, Fisher PE, Goldstein NI & Fisher PB (1996) *Proc. Natl Acad. Sci. USA* 93, 7252–7257.
25. Gopalkrishnan RV, Roberts T, Tuli S, Kang D, Christiansen KA & Fisher PB (2000) *Oncogene* 19, 4405–4416.
26. Hadj SY, Seve AP, Doyennette MM et al. (1996) *J. Cell Biochem.* 62, 529–542.
27. Sarafian V, Jadot M, Foidart JM et al. (1998) *Int. J. Cancer* 75, 105–111.
28. Maldonado CA, Castagna LF, Rabinovich GA & Landa CA (1999) *Invest. Ophthalmol. Vis. Sci.* 40, 2971–2917.
29. Hadari YR, Goren R, Levy Y et al. (2000) *J. Cell Sci.* 113, 2385–2397.
30. Nickel W (2005) *Traffic* 6, 607–614.
31. Levy Y, Arbel-Goren R, Hadari YR et al. (2001) *J. Biol. Chem.* 276, 31285–31295.
32. Levy Y, Ronen D, Bershadsky AD & Zick Y (2003) *J. Biol. Chem.* 278, 14533–14542.
33. Nishi N, Shoji H, Seki M et al. (2003) *Glycobiology* 13, 755–763.
34. Arbel-Goren R, Levy Y, Ronen D & Zick Y (2005) *J. Biol. Chem.* 280, 19105–19114.
35. Carcamo C, Pardo E, Oyanadel C et al. (2006) *Exp. Cell Res.* 312, 374–386.
36. Schwartz MA & Ginsberg MH (2002) *Nat. Cell Biol.* 4, E65–E68.
37. Hood JD & Cheresh DA (2002) *Nat. Rev. Cancer* 2, 91–100.
38. Ochieng J, Leite BM & Warfield P (1998) *Biochem. Biophys. Res. Commun.* 246, 788–791.
39. Gu M, Wang W, Song WK, Cooper DN & Kaufman SJ (1994) *J. Cell Sci.* 107, 175–181.

40. Avni O, Pur Z, Yefenof E & Baniyash M (1998) *J. Immunol.* **160**, 6151–6158.
41. Dong S & Hughes RC (1997) *Glycoconj. J.* **14**, 267–274.
42. Furtak V, Hatcher F & Ochieng J (2001) *Biochem. Biophys. Res. Commun.* **289**, 845–850.
43. Bornstein P & Sage EH (2002) *Curr. Opin. Cell Biol.* **14**, 608–616.
44. Ridley AJ (2001) *Trends Cell Biol.* **11**, 471–477.
45. Chou MM & Blenis J (1996) *Cell* **85**, 573–583.
46. Malik RK & Parsons JT (1996) *J. Biol. Chem.* **271**, 29785–29791.
47. Qui MS & Green SH (1992) *Neuron* **9**, 705–717.
48. Zamir E, Katz M, Posen Y *et al.* (2000) *Nat. Cell Biol.* **2**, 191–196.
49. Miyamoto S, Teramoto H, Coso OA *et al.* (1995) *J. Cell. Biol.* **131**, 791–805.
50. Price LS, Leng J, Schwartz MA & Bokoch GM (1998) *Mol. Cell Biol.* **9**, 1863–1871.
51. Adams JC & Schwartz MA (2000) *J. Cell Biol.* **150**, 807–822.
52. Kawano K, Kantak SS, Murai M, Yao CC & Kramer RH (2001) *Exp. Cell Res.* **262**, 180–196.
53. Fischer D, Tucker RP, Chiquet ER & Adams JC (1997) *Mol Biol Cell* **8**, 2055–2075.
54. Adams JC (1995) *J. Cell Sci.* **108**, 1977–1990.
55. Manser E, Leung T, Salihuddin H, Zhao ZS & Lim L (1994) *Nature* **367**, 40–46.
56. Ridley AJ (2001) *J. Cell Sci.* **114**, 2713–2722.
57. Patnaik SK, Potvin B, Carlsson S, Sturm D, Leffler H & Stanley P. (2006) *Glycobiology* **16**, 305–317.
58. Nishi N, Itoh A, Fujiyama A *et al.* (2005) *FEBS Lett.* **579**, 2058–2064.
59. Sato M, Nishi N, Shoji H *et al.* (2002) *Glycobiology* **12**, 191–197.
60. Coffman JA (2004) *Dev. Cell* **6**, 321–327.
61. Evan GI & Vousden KH (2001) *Nature* **411**, 342–348.
62. Sherr CJ & Roberts JM (1999) *Genes Dev.* **13**, 1501–1512.
63. Ball KL (1997) *Prog. Cell Cycle Res.* **3**, 125–134.
64. Jacks T & Weinberg RA (1998) *Science* **280**, 1035–1036.
65. Dotto GP (2000) *Biochim. Biophys. Acta* **1471**, M43-56.
66. Lawlor MA & Rotwein P (2000) *Mol. Cell. Biol.* **20**, 8983–8995.
67. Gartel AL & Tyner AL. (1999) *Exp Cell Res* **246**, 280–289.
68. Agarwal ML, Taylor WR, Chernov MV, Chernova OB & Stark GR (1998) *J. Biol. Chem.* **273**, 1–4.
69. Macleod KF, Sherry N, Hannon G *et al.* (1995) *Genes Dev.* **9**, 935–944.
70. Muschen M, Warskulat U & Beckmann MW (2000) *J. Mol. Med.* **78**, 312–325.
71. Sangfelt O, Erickson S & Grander D (2000) *Front. Biosci.* **5**, D479–D487.
72. Amati B (2000) *Nat. Cell Biol.* **3**, E112–E113.
73. Perillo NL, Marcus ME & Baum LG (1998) *J. Mol. Med.* **76**, 402–412.
74. Kim HR, Lin HM, Biliran H & Raz A (1999) *Cancer Res.* **59**, 4148–4154.
75. Yang RY, Hsu DK, Yu L, Ni J & Liu FT (2001) *J. Biol. Chem.* **276**, 20252–20260.
76. Kobayashi K & Tsukamoto I (2001) *Biochim. Biophys. Acta* **1537**, 79–88.
77. Rabinovich GA, Alonso CR, Sotomayor CE, Durand S, Bocco JL & Riera CM (2000) *Cell Death Differ.* **7**, 747–753.
78. Kuwabara I, Kuwabara Y, Yang RY *et al.* (2002) *J. Biol. Chem.* **277**, 3487–3497.
79. Sanchez I, Hughes RT, Mayer BJ *et al.* (1994) *Nature* **372**, 794–798.
80. Rossig L, Badorff C, Holzmann Y, Zeiher AM & Dimmeler S (2002) *J. Biol. Chem.* **277**, 9684–9689.
81. Cayrol C & Ducommun B (1998) *Oncogene* **17**, 2437–2444.
82. Rossig L, Jadidi AS, Urbich C, Badorff C, Zeiher AM & Dimmeler S (2001) *Mol. Cell. Biol.* **21**, 5644–5657.
83. Lundberg AS & Weinberg RA (1999) *Eur. J. Cancer* **35**, 531–539.
84. Cambier S, Mu DZ, O'Connell D *et al.* (2000) *Cancer Res.* **60**, 7084–7093.
85. Adams L, Scott GK & Weinberg CS (1996) *Biochim. Biophys. Acta* **1312**, 137–144.
86. Fischer C, Sanchez-Ruderisch H, Welzel M *et al.* (2005) *J. Biol. Chem.* **280**, 37266–37277.
87. Yoshii T, Fukumori T, Honjo Y, Inohara H, Kim HR & Raz A (2002) *J. Biol. Chem.* **277**, 6852–6857.
88. Suzuki A, Tsutomi Y, Akahane K, Araki T & Miura M (1998) *Oncogene* **17**, 931–939.
89. Asada M, Yamada T, Ichijo H *et al.* (1999) *EMBO J.* **18**, 1223–1234.

90. Ichijo H, Nishida E, Irie K *et al.* (1997) *Science* **275**, 90–94.
91. Perillo NL, Pace KE, Seilhamer JJ & Baum LG (1995) *Nature* **378**, 736–739.
92. Hotta K, Funahashi T, Matsukawa Y *et al.* (2001) *J. Biol. Chem.* **276**, 34089–34097.
93. Novelli F, Allione A, Wells V, Forni G & Mallucci L (1999) *J. Cell Physiol.* **178**, 102–108.
94. Kashio Y, Nakamura K, Abedin MJ *et al.* (2003) *J. Immunol.* **170**, 3631–3636.
95. Bernerd F, Sarasin A & Magnaldo T (1999) *Proc. Natl Acad. Sci. USA* **96**, 11329–11334.
96. Danguy A, Rorive S, Decaestecker C *et al.* (2001) *Histol. Histopathol.* **16**, 861–868.
97. Nagy N, Bronckart Y, Camby I *et al.* (2002) *Gut* **50**, 392–401.
98. Nagy N, Legendre H, Engels O *et al.* (2003) *Cancer* **97**, 1849–1858.
99. Lahm H, Andre S, Hoeflich A *et al.* (2001) *J. Cancer Res. Clin. Oncol.* **127**, 375–386.

CHAPTER 11

Intracellular lectin involvement in glycoprotein maturation and quality control in the secretory pathway

Gerardo Z. Lederkremer

1. INTRODUCTION

During the biogenesis and maturation of membrane and secretory proteins they undergo modifications in the form of glycosylation, processing of these sugar chains, specific cleavages of the precursor proteins, folding, and assembly. Most proteins traversing the secretory pathway in eukaryotic cells are glycoproteins and are subjected to an elaborate quality control of these processing reactions, mostly before their exit from the endoplasmic reticulum (ER). Proteins that fail to reach proper folding, cleavage, or assembly in a given time-frame are sent to ER-associated degradation (ERAD). Recent findings suggest a stepwise processing of N-linked sugar chains followed by recognition of intermediate species by a series of lectins (reviewed in [1–5]). These processes seem to determine the fate of the glycoprotein towards productive maturation and exit to the Golgi or delivery to the cytosolic proteasomes for degradation. Here I review the current knowledge in this developing field, focusing on the role of intracellular lectins, and outline future research directions.

2. GLYCOPROTEIN BIOSYNTHESIS AND MATURATION

Glycoprotein biosynthesis starts with mRNA translation in cytosolic ribosomes and targeting by a signal peptide and signal recognition particle (SRP) to the ER membrane, followed by translocation into the ER lumen. This process can be posttranslational in *Saccharomyces cerevisiae* but is mostly cotranslational in higher eukaryotes, where N-linked glycosylation also takes place cotranslationally when the protein encounters the luminal active site of oligosaccharyltransferase (OST) [6]. N-Glycosylation involves transfer by OST of the precursor oligosaccharide $Glc_3Man_9GlcNAc_2$ from the intermediate lipid dolichol to an asparagine residue on the target protein (in an Asn-X-Ser (or Thr) motif, where X may be any amino acid except for Pro). The precursor oligosaccharide is universally conserved with few exceptions [7]. After transfer of this precursor to the protein, a glucose residue is removed by glucosidase I and the two remaining ones by glucosidase II [2,8] (see Fig. 1). Mannose residues can then be removed by α1,2-mannosidases, of which there is only one in S. cerevisiae but several in higher eukaryotes, in the ER and the Golgi complex [9]. Usually, after proper folding, a glycoprotein will exit from the ER to the Golgi with its sugar chains presenting the structures $Man_9GlcNAc_2$, $Man_8GlcNAc_2$ and possibly $Man_7GlcNAc_2$ (see Fig. 2). Occasionally though, it exits with glucose residues, structures that are converted

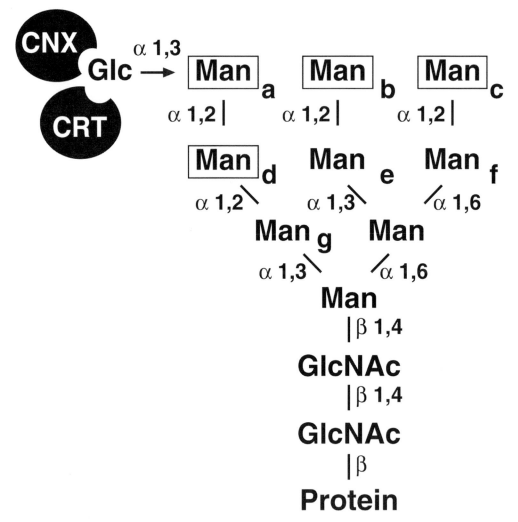

Figure 1. Structure of the $Glc_1Man_9GlcNAc_2$ N-linked oligosaccharide. Structure of the oligosaccharide which is the ligand for glycoprotein binding to the chaperone-lectins calnexin (CNX) or calreticulin (CRT), highlighting the α1,2-linked mannose (Man) residues (a–d) (boxed) that can undergo trimming in the ER and the position (arrow) of the glucose (Glc) residue on the original precursor (after its two outer glucoses were excised) or after its re-addition by the folding-sensor glucosyltransferase UGGT. Note that trimming of mannose-c and/or mannose-b does not affect reglucosylation but loss of mannose-a does.

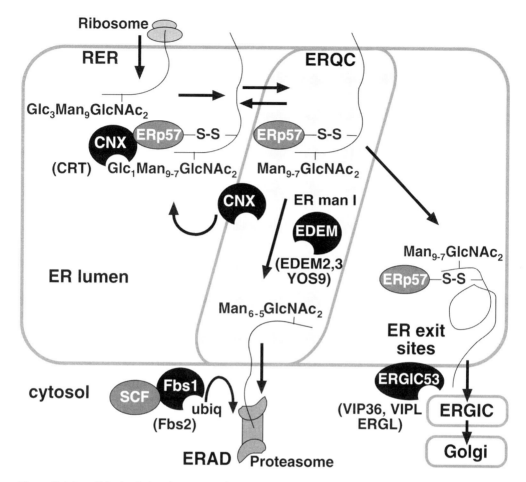

Figure 2. Intracellular lectin involvement in glycoprotein maturation and in quality control and delivery to ER-associated degradation (ERAD). After cotranslational transfer of the precursor oligosaccharide $Glc_3Man_9GlcNAc_2$ to a nascent protein in the mammalian rough ER (RER), trimming of two glucose residues takes place leading to binding of calnexin (or calreticulin). Calnexin recruits the oxidoreductase ERp57 and the complex enters the ER-derived quality control compartment (ERQC). Calnexin folding cycles then start, where the glycoprotein is deglucosylated by glucosidase II, releasing calnexin and subject to trimming of mannose residues by ER mannosidase I (ER man I) and possibly other $\alpha 1,2$-mannosidases. This is followed by reglucosylation by UGGT, reassociation with calnexin, deglucosylation and again mannose trimming. During these cycles the glycoprotein can acquire proper folding, it is no longer recognized as a substrate by the folding sensor UGGT and travels to ER exit sites and from there to the ERGIC and Golgi and from there to its final destination, sometimes with the help of ERGIC-53 (or other lectins, VIP36, VIPL, ERGL). If the glycoprotein cannot fold successfully, the critical mannose-a residue (see Fig. 1) is then removed, leading to release from the calnexin cycles because of absence of the acceptor for glucose transfer by UGGT. In this case, binding to the putative lectin EDEM (or other lectins, EDEM 2, EDEM 3, YOS9) commits the glycoprotein to ERAD. It is then retrotranslocated to the cytosol where it encounters E2 ubiquitin conjugating enzymes and E3 ubiquitin ligases (some of them with lectin components, like Fbs1 or 2) that ubiquitinate it, resulting in binding to cytosolic chaperones like p97/VCP and targeting to proteasomal degradation.

in the Golgi to $Man_8GlcNAc_2$ by an endomannosidase [5]. In the Golgi it will then lose the remaining $\alpha 1,2$-mannoses to produce $Man_5GlcNAc_2$. A GlcNAc is then added, followed by removal of two additional mannose residues to produce $GlcNAcMan_3GlcNAc_2$. A series of sugars are then added, fucose, N-acetylglucosamine, galactose, and sialic acid, to create the sugar chain structures that will be present on the native glycoprotein. This elaborate scheme of reactions involves many specific enzymes and energy-consuming steps [10]. Until recently it has been a mystery why such a large lipid-linked oligosaccharide precursor had to be built if it was subsequently disassembled for addition of different sugar residues. As we will see below, information gathered in recent years starts to provide sense to this process as it links glycoprotein maturation and quality control stages to the processing of its sugar chains.

Before exit to the Golgi, the secondary, tertiary, and quaternary structures of the protein are usually completed and failure of folding and assembly results in retention in the ER. A series of ER-resident chaperones protect the newly synthesized protein from aggregation and help in the folding process. The best characterized are the HSP70 family members BiP (GRP78) and GRP170, in conjunction with a series of DnaJ-like cofactors (ERdj1-5) [11]. Another ER luminal protein involved in the process is the HSP90 member GRP94 [12]. Two chaperones that are unique in their properties, as they are lectins that associate to glycoprotein sugar chains, are calnexin and calreticulin. I will describe them in detail later. In addition there are a series of accessory proteins that participate in disulfide bonding and isomerization of these bonds on the nascent protein, a process that is intimately linked to folding to the native structure. These are PDI [13], ERp57 [14] and other less well-characterized ER-resident proteins [15,16]. ERp57 seems to be specific for N-linked glycoproteins as it acts in complexes with calnexin or calreticulin [17]. When the glycoprotein is ready for export to the Golgi it is recruited to COPII vesicles in discrete ER exit sites for vesicular traffic [18]. At this stage there are proteins required for the transport to the Golgi, the best characterized being ERGIC-53 [19], which resides mainly in the ER-to-Golgi intermediate compartment (ERGIC).

3. ER QUALITY CONTROL OF GLYCOPROTEINS

The ER quality control machinery identifies improperly processed proteins and initiates a process that involves rescue attempts such as refolding. If these attempts fail it targets the aberrant protein to degradation. The main feature recognized as a sign of improper or incomplete folding or assembly is the exposure of hydrophobic patches that when folding is correct become buried inside the structure of the native protein. N-linked oligosaccharides provide large hydrophilic groups that affect folding and help stabilize the native conformation of glycoproteins. In many cases inhibition of N-glycosylation or lack of individual sugar chains irreversibly compromises their folding [2]. The same chaperones that participate in protein maturation in the ER are also responsible for the quality control. Chaperones like BiP associate to the exposed hydrophobic regions of misfolded or incompletely folded proteins and retain them in the ER. Interestingly, in the case of calnexin and calreticulin they do not seem to recognize directly the hydrophobic patches, but these are recognized by a sugar chain-modifying enzyme that creates the structure recognized by the chaperone lectins as I describe later. This is a main checkpoint in the ER quality control of glycoproteins.

The nature of the ER quality control retention process is still not well understood. It seems to involve true retention (rather than recycling from the Golgi) [20]. Subsequently, misfolded proteins are delivered to a specialized pericentriolar subcompartment termed the ER-derived quality control compartment (ERQC) [21,22]. Of several ER resident proteins studied, only a few, prominently calnexin and calreticulin, concentrate in this compartment [22]. In contrast, BiP and PDI do not accumulate in the ERQC and remain distributed throughout the ER [22]. UDP-glucose glycoprotein glucosyltransferase (UGGT, discussed later) and the oxidoreductase ERp57 do not accumulate in the ERQC either, trafficking instead to ER exit sites and the ERGIC [21–23]. ER quality control could involve recycling through the ER and the ERQC as will become more clear later on.

The ER quality control process provides a time interval for the protein to reach its native folding, processing, and assembly. Failure to do so after this time leads to delivery of the aberrant protein for disposal by cytosolic proteasomes in a process called ER-associated degradation.

4. ER-ASSOCIATED DEGRADATION OF GLYCOPROTEINS

ERAD targets not only normal newly made proteins that fail to achieve proper folding in the time-frame given by ER quality control but also mutated misfolded proteins that arise in many genetic diseases [24,25]. This mechanism is also exploited by viruses, as in the case of human immunodeficiency virus (HIV), which targets CD4 for proteasomal degradation through interaction with viral VpU, or cytomegalovirus (CMV), which possesses several proteins that associate with major histocompatibility complex (MHC) class I molecules and sends them to degradation [26–28].

Although ERAD was initially thought to take place in the lumen of the ER, it was later determined that degradation of aberrant membrane and secretory proteins that fail to pass the ER quality control checkpoints takes place in the cytosol. This requires their delivery from the ER lumen, in a process known as retrotranslocation or reverse translocation (reviewed in [23,29]). The retrotranslocated proteins are ubiquitinated and then degraded by cytosolic proteasomes. Several proteins are known to participate in the retrotranslocation, including a cytosolic protein, member of the family of AAA ATPases, p97 (also called Cdc48 or VCP) [30,31] and membrane-bound Derlin 1-3, VIMP, and Sec61β [32–34]; in complexes that have so far been better delineated in *S. cerevisiae* [35–37]. However, much remains to be understood. For instance, does Sec61β, a subunit of the ER translocation channel, also integrate a channel along the route for retrotranslocation [38]? p97 comprises a ring of ATPases that is anchored to the cytosolic side of the membrane via a complex with Derlin-1 and

VIMP and is able to attach to membrane-associated poly-ubiquitinated substrates and shuttle them to the proteasomes aided by additional factors [23,34,39,40]. This complex binds ERAD substrates after their exposure to the cytosol and is necessary for their delivery to the proteasomes. Movement to the cytosolic side is mandatory for ubiquitination of luminal proteins, before their degradation by the proteasomes. For membrane proteins, cytosolic exposure of the luminal domains seems to precede ubiquitination as well, even though theoretically they could be ubiquitinated on their cytosolic tails before retrotranslocation [41,42]. A portion of the ubiquitin-proteasome machinery is localized at, or near to, the ER membrane. A growing number of E2 ubiquitin conjugating enzymes, E3 ubiquitin ligases, ubiquitin-binding proteins and delivery proteins have been identified on the cytoplasmic side of the membrane [5,23,43,44].

The most widely studied E3 involved in ERAD is Hrd1 (Der3), found in mammals and yeast, where it participates in a putative retrotranslocation complex [35–37]. HRD1 is complemented by additional membrane-associated ligases such as Doa10 in yeast and gp78 in mammals. Other participating E3 ubiquitin ligases are cytoplasmic, yet recruited to the ER membrane for ubiquitination of particular proteins, for example Rsp5 (yeast) and Parkin (mammals) [44,45].

After ubiquitination the protein is extracted from the membrane, delivered to the proteasome and degraded. Glycoprotein sugar chains are removed by the cytoplasmic peptide:*N*-glycanase (PNGase [46]). Fewer oligosaccharides are released from cytosolic glycoproteins when proteasomes are inhibited [47], suggesting that the retrotranslocation and deglycosylation must be coupled to proteasomal degradation. Moreover, for most ERAD substrates proteasomal inhibition leads to inhibition of their retrotranslocation [22,32,41]. PNGase was found to interact with p97, which in turn associates with the proteasome [39,48]. PNGase also associates with the proteasome-binding HR23 (Rad23) [49]. These interactions suggest a series of coupled events that lead to proteasomal degradation.

5. LECTINS IN GLYCOPROTEIN MATURATION

A link between glycoprotein maturation in the ER and processing of their sugar chains is provided by a series of lectins. The first ER resident proteins to be characterized as lectins were calnexin and calreticulin in mammalian cells [50,51]. Calnexin is a type I ER membrane protein and calreticulin is a soluble ER protein with homology to calnexin. Both are monomeric, calcium-binding proteins, and have ER localization signals. They have two separate domains, one is a globular beta-sandwich domain with homology to legume lectins and the other is a long extended hairpin fold corresponding to a proline-rich domain (the P-domain). The globular domain contains a concave and a convex beta-sheet with the carbohydrate recognition domain (CRD) located on the concave surface partially shielded by the P-domain [2,52,53].

As mentioned before, calnexin and calreticulin bind transiently to newly synthesized membrane and secretory proteins. Removal by glucosidase I of the glucose residue from the non-reducing end of the oligosaccharide precursor on nascent glycoproteins and of the second glucose residue by glucosidase II creates the monoglucosylated ligand recognized specifically by these lectins, $Glc_1Man_9GlcNAc_2$ [50,51,54]. In addition to the single terminal β1,3-linked glucose, which is essential, three mannoses in the same branch (a, d, and g) contribute to the binding (see Fig. 1). On the other hand mannose residues on the other branches are less important for the binding. Calnexin and calreticulin associate with the oxidoreductase ERp57, which can then bind to the target glycoprotein and participate in its folding by aiding disulfide bond formation (Fig. 2). Cycles of removal of the remaining glucose residue by glucosidase II, release from the chaperones, reglucosylation by the glucosyltransferase UGGT, and reassociation with calnexin and calreticulin take place until the glycoprotein achieves proper folding (reviewed in [8,55]). This process is known as the calnexin folding cycle. It is unique because the folding status of the glycoprotein is not recognized, as is usually the case, by the chaperones themselves but by UGGT that tags the glycoprotein for a new chaperone binding and folding cycle if proper folding has not been achieved. *S. cerevisiae* lacks this cycle as it has no UGGT nor calreticulin and it possesses a quite different calnexin homolog with unclear function.

Correct folding in mammalian cells results in UGGT no longer recognizing the glycoprotein as only incompletely folded glycoproteins are substrates for this glycosyltransferase (reviewed in [2,55]). The released glycoprotein, dissociated from calnexin, still remains associated to ERp57 that escorts it to ER exit sites [21]. At the ER exit sites the glycoprotein encounters ERGIC-53. ERGIC-53 (also called LMan1 or p58) is a type I membrane protein that recycles between the ER, the ERGIC, and the Golgi [56]. It has a CRD with homology to legume lectins [57] and it forms hexamers. Like in calnexin, the CRD is located in a concave beta-sheet that is packed together with a convex beta-sheet in a beta-sandwich structure [58]. ERGIC-53 can function as a "cargo receptor," binding glycoproteins bearing high mannose sugar chains in the ER, recruiting them into coated vesicles and releasing them in the Golgi. This function, accomplished with the aid of a protein called MCFD2 [59,60] was shown so far only for a few glycoproteins acting as cargo [61,62]. Interestingly, mutations in ERGIC-53 or in MCFD2 strongly affect the secretion of coagulation factors V and VIII, causing an autosomal recessive bleeding disorder [63,64]. ERGIC-53 binds to unglucosylated high mannose chains [62] although its exact carbohydrate specificity is not known. ERGIC-53 binds its glycoprotein cargo in a calcium-dependent manner in the ER and releases it in the Golgi, apparently due to the slightly lower pH in this organelle, which reduces the affinity for the ligand [65]. ERGIC-53 binds to a ligand oligosaccharide only when the latter is associated with a surface-exposed peptide β-hairpin loop on the cargo glycoprotein, implying that folding must be complete for the recognition to take place [66]. *S. cerevisiae* possesses an ERGIC- 53 homolog called Emp47 [67].

A series of other lectins with homology to ERGIC-53 have been identified that are resident in the early secretory pathway in mammals, VIP36, VIPL, ERGL, all of them type I membrane proteins [19,68,69]. Not much is known about their function; they might represent cargo receptors for different classes of glycoproteins. VIP36 has been shown to reside in the ERGIC and *cis*-Golgi, possibly recycling between them and to bind unglucosylated high mannose oligosaccharides, although it is not calcium-dependent [19,70]. VIPL was reported to reside in the ER [71] or recycle between the ER and the *cis*-Golgi and its knockdown using siRNA retards the trafficking of a subset of glycoproteins [68].

6. LECTINS IN ER QUALITY CONTROL AND ERAD

Similar to the participation of the calnexin folding cycle in normal glycoprotein maturation, it is also central to the quality control retention of misfolded glycoproteins. As we have seen above, calnexin and/or calreticulin (depending on the specific glycoprotein) bind to monoglucosylated molecules, after recognition and glucose transfer by UGGT. Glucosidase II removes the glucose and the molecule is tested again by UGGT that transfers a new glucose, allowing reassociation to the chaperone lectins. If the misfolded glycoprotein cannot be repaired the cycles would repeat endlessly. At some point the cell must decide that the glycoprotein is hopelessly misfolded and should be removed from these futile reglucosylation cycles and sent to ERAD. This decision is dependent on another sugar chain-processing reaction, trimming of mannose residues. Many papers in the last 10 years showed trimming of mannose residues as necessary for delivery of misfolded glycoproteins to ERAD. Inhibitors of class I α1,2-mannosidases block almost completely their degradation. A model was proposed by which a slow acting mannosidase would function as a timer for ERAD [72]. Removal of mannose residues by this mannosidase would convert the oligosaccharide $Man_9GlcNAc_2$, the ideal substrate for UGGT when linked to a misfolded glycoprotein, into shorter oligosaccharides that are less efficient glucose acceptors [8]. Thus, if after a certain time period the glycoprotein is not properly folded, mannose trimming would delay its reglucosylation and re-entry into the calnexin folding cycles, exposing it to the ERAD machinery.

It was later proposed that removal of one specific mannose residue (mannose-b in Fig. 1) to generate $Man_8GlcNAc_2$ is the key to slower cycles of deglucosylation and reglucosylation [73]. After trimming of this mannose residue, reglucosylation by UGGT is slower, as explained above. Deglucosylation by glucosidase II is also slower after trimming of mannose-b. This could predict a slower calnexin binding cycle for the ERAD substrate. However, it would not prevent its perpetual reglucosylation and re-entry into this refolding cycle instead of being targeted for ERAD. Recent reports show that mannose trimming on ERAD substrates can proceed to produce $Man_6GlcNAc_2$ and even $Man_5GlcNAc_2$ [74–77]. This extensive trimming is what commits a substrate to ERAD, whereas the removal of mannose residues progresses only very slowly on stable ER-retained glycoproteins [76]. Trimming of one or two residues yields $Man_8GlcNAc_2$ and $Man_7GlcNAc_2$, species that can still be reglucosylated (see Fig. 1). However, trimming of additional α1,2-linked mannose residues, to $Man_6GlcNAc_2$, effectively removes the glycoprotein from the calnexin folding cycles, as it lacks the specific mannose acceptor for glucose transfer.

Mannose-a is then the crucial residue, as its removal sends the glycoprotein to ERAD. This is further supported by the fact that in mutant cells that transfer a specific $Man_5GlcNAc_2$ isomer (containing mannose-a and lacking mannose-b) to protein instead of Glc$_3$Man$_9$GlcNAc$_2$, removal of mannose-a residue is still required for proteasomal degradation to proceed [78]. A model was then proposed where progressive trimming and reglucosylation provide a time interval until UGGT cannot recognize the glycoprotein, even if it is still not properly folded [3]. The time interval would be imparted by slow mannose trimming due to the fact that after removal of each mannose the glycoprotein can be reglucosylated and re-enter the calnexin folding cycle. This creates a window of opportunity for rescue of the glycoprotein from ERAD until its sugar chains are trimmed to Man$_6$GlcNAc$_2$ when it can no longer be rescued (Fig. 1).

As mentioned above, in *S. cerevisiae* there is only one ER mannosidase responsible for removal of one mannose residue. This pointed to ER mannosidase I, homolog of the yeast mannosidase as responsible for the mannose trimming in mammals. Consistent with this, ERAD is blocked by inhibitors of ER mannosidase I. However, in mammals there are several other α1,2-mannosidases sensitive to the same inhibitors, which might recycle from the Golgi to the ER, namely Golgi mannosidases I A (also called Man9-mannosidase), I B, and I C [9]. These enzymes could act alone or in combination with ER mannosidase I. A hint that ER mannosidase I is indeed involved comes from studies where its overexpression caused accelerated ERAD [75,79]. Initially, ER mannosidase I was thought to remove only mannose-b and generate a precursor to ERAD [73]. However, it was found that *in vitro*, at high concentrations, the enzyme is capable of removing three more α1,2-linked mannose residues in addition to mannose-b [80]. Can these high concentrations be achieved in the cell? Possibly, by subcompartmentalization. In fact, we have found that ER mannosidase I is localized to the ERQC (our unpublished results).

Compartmentalization of ER mannosidase I to the ERQC would help explain the slow action of the enzyme, as its substrates would only be available for a limited time in each cycle. For this one can envisage transport of the glycoprotein substrate together with calnexin and calreticulin to the ERQC and recycling to the ER upon release (Fig. 2). Seclusion of the mannosidase in the ERQC could explain not only how nascent, still unfolded glycoproteins are not degraded but also how ER-resident glycoproteins are spared, as their proper folding would release them from the calnexin cycle. What happens after removal of mannose-a? Another putative lectin, EDEM (Htm1 in yeast) takes over. EDEM is homologous to ER mannosidase I and other mannosidases but does not show enzymatic activity when tested *in vitro* [81] though it seems to accelerate mannose trimming *in vivo* [82]. It is thought to be a lectin although this has not been formally proven yet. The exact role of EDEM is not clear but it is required for ERAD of glycoproteins; its overexpression accelerates ERAD and its deletion in yeast or knock-down in mammalian cells inhibits the degradation [2,5]. Membrane-bound EDEM could possibly participate in the retrotranslocation of the ERAD substrate glycoprotein, but it undergoes cleavage to a soluble form [83], which like two recently identified soluble luminal homologs, EDEM 2 and 3 [83,84], would require for that purpose a membrane-bound adaptor.

Interestingly, in the cytosolic ubiquitination step, some newly found E3 ubiquitin ligase components, Fbs1 and Fbs2, are also lectins and able to recognize the glycans still present on retrotranslocated substrates [85–88]. The newest addition to the family of lectins in the secretory pathway is YOS9, recently identified in *S. cerevisiae* [89-92]. Two candidate mammalian homologs are OS9 or XTP3-B [90,93]. YOS9 is a soluble, luminal ER-retained glycoprotein with a mannose 6-phosphate receptor homology domain. It binds misfolded glycoproteins through their high mannose sugar chains and participates in ERAD in a still unknown manner, possibly as a member of a retrotranslocation complex [35-37]. Its deletion in yeast stabilizes ERAD substrate glycoproteins.

7. CONCLUDING REMARKS

Studies in recent years have delineated pathways for glycoprotein maturation and quality control in the secretory pathway that are defined by a series of lectins. These lectins are central to the mechanism of folding and to the timing and decision-making process in quality control. The recent discovery of novel participating lectins with still unclear function suggests their role in building a manufacturing chain where glycoproteins are escorted and sorted from their initial translocation into the ER, through their delivery to the Golgi, or their quality control retention and refolding, transport from the ER to a specialized quality control compartment and their delivery, if everything else fails, to proteasomal degradation in the cytosol. Future studies should determine the mechanism of participation of several of these lectins: Are EDEM and YOS9 final acceptors of aberrant glycoproteins after sugar trimming events or do they escort them during the quality control process? Do the EDEM and ERGIC-53 homologs simply reflect the redundancy of the system or are they specialized in subsets of substrates? What are the exact sugar chain structures that are recognized by each lectin, what mannosidases are involved in their creation and in which exact subcellular location?

We have just started to understand why the creation of a large precursor *N*-linked oligosaccharide followed by its extensive trimming is not a waste of energy but an investment by the cell, to be able to achieve proper glycoprotein folding and maturation involving stepwise recognition by intracellular lectins.

Acknowledgments

Research related to this work was supported by grant 638/02 from the Israel Science Foundation.

REFERENCES

1. Hebert DN, Garman SC & Molinari M (2005) *Trends Cell. Biol.* **15**, 364-370.
2. Helenius A & Aebi M (2004) *Annu. Rev. Biochem.* **73**, 1019-1049.
3. Lederkremer GZ & Glickman MH (2005) *Trends Biochem. Sci.* **30**, 297-303.
4. Moremen KW & Molinari M (2006) *Curr. Opin. Struct. Biol.* **16**, 592-599.
5. Spiro RG (2004) *Cell. Mol. Life Sci.* **61**, 1025-1041.
6. Yan A & Lennarz WJ (2005) *J. Biol. Chem.* **280**, 3121-3124.
7. Parodi AJ (2000) *Biochem. J.* **348**, 1-13.
8. Parodi AJ (2000) *Annu. Rev. Biochem.* **69**, 69-93.
9. Herscovics A (2001) *Biochimie* **83**, 757-762.

10. Kornfeld R & Kornfeld S (1985) *Annu. Rev. Biochem.* **54**, 631–664.
11. Ma Y & Hendershot LM (2004) *J. Chem. Neuroanat.* **28**, 51–65.
12. Argon Y & Simen BB (1999) *Semin. Cell Dev. Biol.* **10**, 495–505.
13. Freedman RB, Hirst TR & Tuite MF (1994) *Trends Biochem. Sci.* **19**, 331–336.
14. Oliver JD, van der Wal FJ, Bulleid NJ & High S (1997) *Science* **275**, 86–88.
15. Anken E & Braakman I (2005) *Crit. Rev. Biochem. Mol. Biol.* **40**, 269–283.
16. Hosoda A, Kimata Y, Tsuru A & Kohno K (2003) *J. Biol. Chem.* **278**, 2669–2676.
17. Oliver JD, Roderick HL, Llewellyn DH & High S (1999) *Mol. Biol. Cell* **10**, 2573–2582.
18. Tang BL, Wang Y, Ong YS & Hong W (2005) *Biochim. Biophys. Acta* **1744**, 293–303.
19. Hauri H, Appenzeller C, Kuhn F & Nufer O (2000) *FEBS Lett.* **476**, 32–37.
20. Shenkman M, Ayalon M & Lederkremer GZ (1997) *Proc. Natl Acad. Sci. USA* **94**, 11363–11368.
21. Frenkel Z, Shenkman M, Kondratyev M & Lederkremer GZ (2004) *Mol. Biol. Cell* **15**, 2133–2142.
22. Kamhi-Nesher S, Shenkman M, Tolchinsky S, Fromm SV, Ehrlich R & Lederkremer GZ (2001) *Mol. Biol. Cell* **12**, 1711–1723.
23. Ahner A & Brodsky JL (2004) *Trends Cell Biol.* **14**, 474–478.
24. Helenius A (2001) *Philos. Trans. R Soc. Lond. B Biol. Sci.* **356**, 147–150.
25. Ellgaard L & Helenius A (2003) *Nat. Rev. Mol. Cell Biol.* **4**, 181–191.
26. Lilley BN & Ploegh HL (2005) *Immunol. Rev.* **207**, 126–144.
27. Meusser B & Sommer T (2004) *Mol. Cell* **14**, 247–258.
28. Schubert U, Anton LC, Bacik I, Cox JH, Bour S, Bennink JR, Orlowski M, Strebel K & Yewdell JW (1998) *J. Virol.* **72**, 2280–2288.
29. Romisch K (2005) *Annu. Rev. Cell Dev. Biol.* **21**, 435–456.
30. Rabinovich E, Kerem A, Frohlich KU, Diamant N & Bar-Nun S (2002) *Mol. Cell. Biol.* **22**, 626–634.
31. Ye Y, Meyer HH & Rapoport TA (2001) *Nature* **414**, 652–656.
32. Hirsch C, Jarosch E, Sommer T & Wolf DH (2004) *Biochim. Biophys. Acta* **1695**, 215–223.
33. Lilley BN & Ploegh HL (2004) *Nature* **429**, 834–840.
34. Ye Y, Shibata Y, Yun C, Ron D & Rapoport TA (2004) *Nature* **429**, 841–847.
35. Carvalho P, Goder V & Rapoport TA (2006) *Cell* **126**, 361–373.
36. Denic V, Quan EM & Weissman JS (2006) *Cell* **126**, 349–359.
37. Gauss R, Jarosch E, Sommer T & Hirsch C (2006) *Nat. Cell. Biol.* **8**, 849–854.
38. Kalies KU, Allan S, Sergeyenko T, Kroger H & Romisch K (2005) *EMBO J.* **24**, 2284–2293.
39. McNeill H, Knebel A, Arthur JS, Cuenda A & Cohen P (2004) *Biochem. J.* **384**, 391–400.
40. Richly H, Rape M, Braun S, Rumpf S, Hoege C & Jentsch S (2005) *Cell* **120**, 73–84.
41. Bubeck A, Reusch U, Wagner M *et al.* (2002) *J. Biol. Chem.* **277**, 2216–2224.
42. Furman MH, Loureiro J, Ploegh HL & Tortorella D (2003) *J. Biol. Chem.* **278**, 34804–34811.
43. Hampton RY (2002) *Curr. Opin. Cell Biol.* **14**, 476–482.
44. Kostova Z & Wolf DH (2003) *EMBO J.* **22**, 2309–2317.
45. Tanaka K, Suzuki T, Hattori N & Mizuno Y (2004) *Biochim. Biophys. Acta* **1695**, 235–247.
46. Suzuki T, Park H & Lennarz WJ (2002) *FASEB J.* **16**, 635–641.
47. Karaivanova VK & Spiro RG (2000) *Glycobiology* **10**, 727–735.
48. Katiyar S, Li G & Lennarz WJ (2004) *Proc. Natl Acad. Sci. USA* **101**, 13774–13779.
49. Kim I *et al.* (2006) *J. Cell Biol* (2006) **172**, 211–219.
50. Peterson JR, Ora A, Van PN & Helenius A (1995) *Mol Biol. Cell* **6**, 1173–1184.
51. Ware FE, Vassilakos A, Peterson PA, Jackson MR, Lehrman MA & Williams DB (1995) *J. Biol. Chem.* **270**, 4697–4704.
52. Schrag JD, Bergeron JJ, Li Y, Borisova S, Hahn M, Thomas DY & Cygler M (2001) *Mol. Cell* **8**, 633–644.
53. Schrag JD, Procopio DO, Cygler M, Thomas DY & Bergeron JJ (2003) *Trends Biochem. Sci.* **28**, 49–57.
54. Zapun A, Petrescu SM, Rudd PM, Dwek RA, Thomas DY & Bergeron JJM (1997) *Cell* **88**, 29–38.
55. Trombetta ES & Parodi AJ (2003) *Annu. Rev. Cell. Dev. Biol.* **19**, 649–676.

56. Kappeler F, Klopfenstein DRC, Foguet M, Paccaud JP & Hauri HP (1997) *J. Biol. Chem.* **272**, 31801–31808.
57. Fiedler K & Simons K (1994) *Cell* **77**, 625–626.
58. Velloso LM, Svensson K, Schneider G, Pettersson RF & Lindqvist Y (2002) *J. Biol. Chem.* **277**, 15979–15984.
59. Nyfeler B, Michnick SW & Hauri HP (2005) *Proc. Natl Acad. Sci. USA* **102**, 6350–6355.
60. Zhang B, Kaufman RJ & Ginsburg D (2005) *J. Biol. Chem.* **280**, 25881–25886.
61. Appenzeller C, Andersson H, Kappeler F & Hauri HP (1999) *Nat. Cell. Biol.* **1**, 330–334.
62. Cunningham MA, Pipe SW, Zhang B, Hauri HP, Ginsburg D & Kaufman RJ (2003) *J. Thromb. Haemost.* **1**, 2360–2367.
63. Nichols WC, Seligsohn U, Zivelin A, Terry VH, Hertel CE, Wheatley MA, Moussalli MJ, Hauri HP, Ciavarelli N, Kaufman RJ & Ginsburg D (1998) *Cell* **93**, 61–70.
64. Zhang B, McGee B, Yamaoka JS, Guglielmone H, Downes KA, Minoldo S, Jarchum G, Peyvandi F, de Bosch NB, Ruiz-Saez A, Chatelain B, Olpinski M, Bockenstedt P, Sperl W, Kaufman RJ, Nichols WC, Tuddenham EG & Ginsburg D (2006) *Blood*, **107**, 1903–1907.
65. Appenzeller-Herzog C, Roche AC, Nufer O & Hauri HP (2004) *J. Biol. Chem.* **279**, 12943–12950.
66. Appenzeller-Herzog C, Nyfeler B, Burkhard P, Santamaria I, Lopez-Otin C & Hauri HP (2005) *Mol. Biol. Cell* **16**, 1258–1267.
67. Sato K & Nakano A (2002) *Mol. Biol. Cell* **13**, 2518–2532.
68. Neve EP, Svensson K, Fuxe J & Pettersson RF (2003) *Exp. Cell Res.* **288**, 70–83.
69. Yerushalmi N, Keppler-Hafkemeyer A, Vasmatzis G *et al.* (2001) *Gene* **265**, 55–60.
70. Kamiya Y, Yamaguchi Y, Takahashi N *et al.* (2005) *J. Biol. Chem.* **280**, 37178–37182.
71. Nufer O, Mitrovic S & Hauri HP (2003) *J. Biol. Chem.* **278**, 15886–15896.
72. Helenius A (1994) *Mol. Biol. Cell* **5**, 253–265.
73. Cabral CM, Liu Y & Sifers RN (2001) *Trends Biochem. Sci.* **26**, 619–624.
74. Foulquier F, Duvet S, Klein A, Mir AM, Chirat F & Cacan R (2004) *Eur. J. BioChem.* **271**, 398–404.
75. Hosokawa N, Tremblay LO, You Z, Herscovics A, Wada I & Nagata K (2003) *J. Biol. Chem.* **278**, 26287–26294.
76. Frenkel Z, Gregory W, Kornfeld S & Lederkremer GZ (2003) *J. Biol. Chem.* **278**, 34119–34124.
77. Kitzmuller C, *et al.* (2003) *Biochem. J.* **376**, 687–696.
78. Ermonval M, Kitzmuller C, Mir AM, Cacan R & Ivessa NE (2001) *Glycobiology* **11**, 565–576.
79. Wu Y, Swulius MT, Moremen KW & Sifers RN (2003) *Proc. Natl Acad. Sci. USA* **100**, 8229–8234.
80. Herscovics A, Romero PA & Tremblay LO (2002) *Glycobiology* **12**, 14G–15G.
81. Hosokawa N, Wada I, Hasegawa K, Yorihuzi T, Tremblay LO, Herscovics A, Nagata K (2001) *EMBO Rep.* **2**, 415–422.
82. Olivari S, Cali T, Salo KE, Paganetti P, Ruddock LW & Molinari M (2006) *Biochem. Biophys. Res. Commun.* **349**, 1278–1284.
83. Olivari S, Galli C, Alanen H, Ruddock L & Molinari M (2005) *J. Biol. Chem.* **280**, 2424–2428.
84. Mast SW, Diekman K, Karaveg K, Davis A, Sifers RN & Moremen KW (2005) *Glycobiology* **15**, 421–436.
85. Mizushima T, Hirao T, Yoshida Y, *et al.* (2004) *Nat. Struct. Mol. Biol.* **11**, 365–370.
86. Yoshida Y, Adachi E, Fukiya K, Iwai K & Tanaka K (2005) *EMBO Rep.* **6**, 239–244.
87. Yoshida Y, Chiba T, Tokunaga F *et al.* (2002) *Nature* **418**, 438–442.
88. Yoshida Y, Tokunaga F, Chiba T, Iwai K, Tanaka K & Tai T (2003) *J. Biol. Chem.* **278**, 43877–43884.
89. Buschhorn BA, Kostova Z, Medicherla B & Wolf DH (2004) *FEBS Lett.* **577**, 422–426.
90. Bhamidipati A, Denic V, Quan EM & Weissman JS (2005) *Mol. Cell* **19**, 741–751.
91. Kim W, Spear ED & Ng DT (2005) *Mol. Cell* **19**, 753–764.
92. Szathmary R, Bielmann R, Nita-Lazar M, Burda P & Jakob CA (2005) *Mol. Cell* **19**, 765–775.
93. Baek JH, Mahon PC, Oh J *et al.* (2005) *Mol. Cell* **17**, 503–512.

SECTION 5
Carbohydrate-binding proteins

CHAPTER 12

The diversity of sialic acids and their interplay with lectins

Roland Schauer

1. INTRODUCTION

Sialic acids are electronegatively charged monosaccharides in higher animals and some microorganisms [1,2]. They contribute to the enormous structural diversity of complex carbohydrates, which are major constituents mostly of proteins and lipids of cell membranes and secreted macromolecules. Sialic acids are prominently positioned, usually at the outer end of these molecules. In this way they are extremely well suited to interact with molecules of the cell environment including other cells and pathogens. The diversity of glycan chains is increased even more by the biosynthesis of various kinds of sialic acids [3,4], thus potentiating the complexity of the "third language of life" (after nucleic acids and proteins as first and second languages, respectively) used by glycoconjugates.

Sialic acids are regularly found in the deuterostome branch of the animal kingdom, and they play important regulatory and protective roles in cell biology. Prominent examples are ion transport, stabilization of protein conformation, protection from proteolytic attack, regulation of innate and acquired immune responses, modulation of receptor affinity and transmembrane signaling, involvement in fertilization, differentiation, aging, and apoptosis, component of receptors or masking of recognition sites thus regulating molecular and cellular interactions, and antioxidative effects [1–4]. It is therefore believed that the appearance of these monosaccharides has facilitated the evolution of higher organisms. They are indispensable for the life of higher organisms. Inhibition of their biosynthesis during early development of mice (see Chapter 19) is lethal, as was shown by inactivation of UDP-*N*-acetylglucosamine-2-epimerase/*N*-acetylmannosamine kinase, the key enzyme of sialic acid biosynthesis [5]. There is almost no biological event in mammals in which these "promiscuous" compounds are not involved. Therefore, errors in their biosynthesis or degradation have dramatic biological consequences and may lead to diseases such as autoimmune reactions and cancer.

In addition, microorganisms have developed various strategies that use the easily accessible cell surface sialic acids as anchors for adhesion and subsequent infection of the cell or that destroy the acidic protective sialic acid shield by secreting hydrolytic enzymes. Thus, apart from their "internal" tasks, sialic acids permanently interact with the environment of cells and whole organisms. This "battle" is fought in conjunction with carbohydrate moieties bearing sialic acid residues. It is believed that this defense barrier has continuously and individually been adapted to new conditions, which may also have driven the evolution of higher organisms [1,4]. In all these biological and pathophysiological events the interactions of sialic acids with lectins play a dominant role.

2. THE DISTRIBUTION OF SIALIC ACIDS IN NATURE AND THEIR CHEMICAL DIVERSITY

Sialic acids are a family of monosaccharides comprising about 50 members, which are derivatives of neuraminic acid (5-amino-3,5-dideoxy-D-*glycero*-D-*galacto*-non-2-ulopyranosonic acid, Neu). They carry various substituents at the amino or hydroxyl groups (Fig. 1) [1,2]. The amino group of neuraminic acid in most cases is

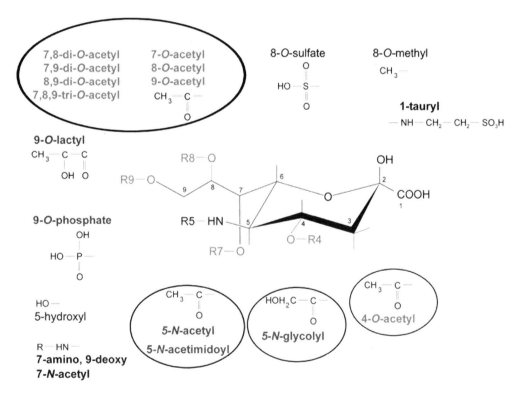

Figure 1. The naturally occurring sialic acids. The residues on the neuraminic acid molecule may be combined depending on the species or tissue. The 5-hydroxyl is characteristic for 2-keto-3-deoxy-nononic acid (Kdn). The 5-*N*-acetimidoyl, 7-*N*-acetyl, and 9-deoxy groups are typical for the microbial legionaminic acid (Leg) derivatives [15], which also belong to the sialic acid family.

acetylated or glycolylated, while one or various acetyl groups may occur at all non-glycosidic hydroxyl residues. Usually, there is only one O-acetyl group, mostly at C-9, but di- and tri-O-acetylated sialic acids are known, especially in mucins from bovine submandibular gland and human colon. Lactyl or phosphoryl residues may occur at C-9 and methyl or sulfate groups at C-8. All these different substituents may be combined (e.g. 8-O-methyl with 9-O-acetyl and N-glycolyl) yielding the many types of sialic acids found throughout the animal kingdom. Apart from these types, unsaturated sialic acids as well as anhydro, lactone, and lactam forms have also been identified in various biological sources. 5-Desamino-5-hydroxy-neuraminic acid or 2-keto-3-deoxy-nononic acid (Kdn) is increasingly being found in microorganisms and animals. Neuraminic acid itself (Neu), the de-N-acetylated product of N-acetylneuraminic acid (Neu5Ac), is a further component of mammalian glycoconjugates.

The natural sialic acids are listed, along with their main sources and abbreviations, in [1,2]. These sialic acids, except unsaturated sialic acids (2-deoxy-2,3-didehydro sialic acids) and N-acetylneuraminic acid-9-phosphate (Neu5Ac9P), usually occur in glycosidic linkages of oligosaccharides, polysaccharides (polysialic acids), glycoproteins, gangliosides, and lipopolysaccharides [1,2]. In most cases they are bound to galactose (Gal) or N-acetylgalactosamine (GalNAc) groups via α2,3 or α2,6 bonds or to sialic acid itself by α2,8 or, rarely, α2,9 linkages. Neu5Ac9P is an intermediate in sialic acid biosynthesis.

N-Acetylneuraminic acid (Neu5Ac), N-acetyl-9-O-acetylneuraminic acid (Neu5,9Ac$_2$) and N-glycolylneuraminic acid (Neu5Gc) are the three most frequently occurring members of the sialic acid family, probably followed by N-acetyl-9-O-lactylneuraminic acid (Neu5Ac9Lt). Only Neu5Ac is ubiquitous, while the others are not found in all species. The best investigated example next to Neu5Ac is Neu5Gc, which occurs frequently in the animal kingdom, but not in healthy human tissues – only in some tumors in minute quantities [6] – and has not been detected in bacteria. It is also missing in many bird and reptile species as well as in Australian monotremes [7]. In contrast, O-acetylated sialic acids are more abundant, occurring in humans, birds, reptiles, and bacteria.

No studies have found all kinds of sialic acids in a single cell or organism. The distribution depends on the animal and cell species as well as on the function of a cell and seems to be strongly regulated at the gene level. The animal with the highest number of different sialic acids known so far is the cow. In its submandibular gland mucin 15 types were detected, most of them O-acetylated on the sialic acid side-chain, but also 9-O-lactylated as well as N-acetylated or N-glycolylated [1,2]. In humans, the number of sialic acids types is much smaller, with Neu5Ac prevailing and followed by derivatives which are O-acetylated and O-lactylated at the sialic acid side-chain. Remarkably, in echinoderms, best studied in *Asterias rubens*, many sialic acid types are known, some of them representing the well-known monosaccharides of higher animals, such as Neu5Ac, Neu5Gc, and O-acetylated types, as well as dehydro-derivatives in traces. In *A. rubens*, a large percentage of these sialic acids are additionally methylated at O-8, and in other echinoderms this position can be sulfated [1,2,8]. This wide abundance of sialic acids types in echinoderms gave rise to the theory that the corresponding enzymes leading to the various sialic acid types were first expressed at least 500 million years ago when the deuterostomes evolved. Sialic acids are only rarely found in animals lower than echinoderms [1,4]. Furthermore, genes homologous to those involved in mammalian sialic acid biosynthesis have been found in plants [9]. A sialyltransferase-like protein expressed in rice plants (*Oryza sativa*) is capable of transferring sialic acid from CMP-Neu5Ac to terminal galactose of a glycan [10].

3. BIOSYNTHESIS OF SIALIC ACIDS

N-Acetylneuraminic acid is the maternal molecule of all other Neu derivatives. It is synthesized in the cytosol by condensation of phosphoenol-pyruvate with *N*-acetylmannosamine-6-phosphate for Neu5Ac9P or with mannose-6-phosphate for Kdn9P, followed by dephosphorylation (for reviews see [1,2,4]). The crucial regulatory enzyme in the long reaction sequence from glucose is the epimerase reaction converting UDP-*N*-acetylglucosamine to *N*-acetylmannosamine-6-phosphate. After dephosphorylation, the free monosaccharide is activated to the CMP-glycoside in the nucleus. En route, CMP-Neu5Ac can be modified to CMP-Neu5Gc in the cytosol by CMP-Neu5Ac hydroxylase (Fig. 2). In the hydroxylase reaction electrons are enzymatically transferred from NAD(P)H to cytochrome b_5, which is a cofactor of the specific hydroxylase oxidizing CMP-Neu5Ac with the aid of molecular oxygen. This enzyme has been isolated from starfish and various mammals and its primary structure has been elucidated, showing that it is highly conserved in the deuterostome branch. Neu5Gc expression is missing in humans due to a defect at the N-terminus of the hydroxylase.

CMP-Neu5Ac and CMP-Neu5Gc are then transported by a specific carrier into the Golgi apparatus, where they are transferred by sialyltransferases onto nascent glycoproteins and glycolipids. There is a large family of sialyltransferases [9]. The various enzyme members differ in their specificity towards both the acceptor glycan and the type of sialic acid linkage formed. They seem, however, to be less specific with regard to the type of sialic acid transferred.

The subcellular site and mechanism of the *O*-acetylation of sialic acids seem to be even more complex and have not yet been fully elucidated (Fig. 2). The enzymes involved, the acetyl-CoA:sialate-4-*O*-acetyltransferase and the corresponding 7(9)-*O*-acetyltransferase, are firmly attached to the Golgi membranes and have only recently been solubilized by detergents and partially purified from bovine submandibular gland (Lrhorfi et al., Biol. Chem., in press) and guinea-pig liver [11]. CMP-Neu5Ac has been shown to be the best substrate for both the 4- and 7(9)-*O*-acetyltransferases [11,12]. The most probable hypothesis, fitting best to the older model [13], is that in a protein complex the CMP-Neu5Ac (CMP-Neu5Gc) transporter, the *O*-acetyltransferases esterifying CMP-Neu5Ac and the sialyltransferases transferring the *O*-acetylated product are cooperating in the Golgi apparatus. Since the primary enzyme reaction product is Neu5,7Ac$_2$ [12] and Neu5,9Ac$_2$ is mainly found in tissues, enzymatic isomerization by a "migrase" seems to be involved in the *O*-acetylation of glycans. Spontaneous isomerization

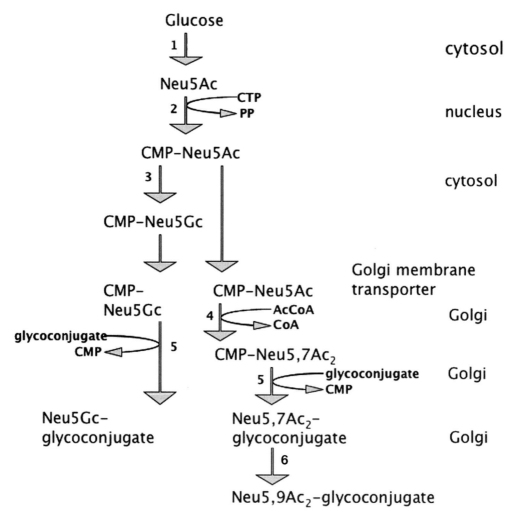

Figure 2. Schematic presentation of the pathway and subcellular site of the biosynthesis of N-acetylneuraminic acid, N-glycolylneuraminic acid, and O-acetylated species. **1**, ten enzymatic steps lead to Neu5Ac; **2**, CMP-Neu5Ac synthase; **3**, CMP-Neu5Ac hydroxylase; **4**, CMP-Neu5Ac 7-O-acetyltransferase; **5**, sialyltransferase; **6**, hypothetical sialic acid O-acetyl isomerase.

at physiological pH would be too slow [14]. Rapid O-acetyl migration within the sialic acid glycerol side-chain is also believed to enable oligo-O-acetylation (e.g. in human colon).

Although the substrate specificities and kinetic properties of these O-acetyltransferases have been studied extensively, knowledge about their structural properties is scant, and expression cloning and mutagenesis experiments have failed to reveal the gene structures of mammalian O-acetyltransferases [15]. Although sialate-O-acetyl genes have been cloned in several microorganisms, such as group B *Streptococcus*, *Escherichia coli*,

Legionella pneumophila, and *Neisseria meningitides* [16–19], protein expression and catalytic properties have not yet been reported. Study of the regulation of gene expression of sialate-*O*-acetyltransferases is most important, since this sialic acid modification is involved in many physiological and pathological cellular events. These will be discussed below, because sialic acid *O*-acetylation has a significant influence on recognition processes.

Evidence is accumulating that sialic acids are de-*N*-acetylated and re-*N*-acetylated when bound to glycoproteins and especially gangliosides [1,2,20]. It is assumed that this variable modification is involved in transmembrane signaling. Further research is necessary to prove this hypothesis and the enzymology responsible for this modification. Not much is known of the enzymology and molecular biology of the enzymes involved in 9-*O*-lactylation and 8-*O*-sulfation of sialic acids. Only the enzyme responsible for sialic acids 8-*O*-methylation has been isolated and characterized from subcellular membranes of the gonads of *Asterias rubens* [21]. The biological significance of this chemical diversity of sialic acids appears to regulate the turnover of sialic acids and correspondingly of glycoconjugates [2,3]. Furthermore, the various sialic acid derivatives influence recognition and other cellular processes, as was best studied with Neu5Gc and *O*-acetylated sialic acids (see below).

4. DEGRADATION OF SIALIC ACIDS

Since sialic acids are involved in so many biological processes, their turnover rate deserves special attention. The decomposition of sialylated glycoconjugates usually starts with the hydrolysis of sialic acid by extracellular or intracellular sialidases. The sialidases (neuraminidases) are a large family, comprising hydrolases from viruses, bacteria, protozoa, and animals [2,22–24]. Some trypanosome species express trans-sialidases which can transfer sialic acids from one glycosidic linkage to another. Common features of sialidases are approximately 12 conserved amino acids involved in catalysis, two motifs ("Asp-boxes" and a FRIP region) as well as a six-bladed propeller structure. Modified sialic acids are more or less resistant to enzymatic hydrolysis, as has been best studied with *O*-acetylated and *N*-glycolylated sialic acids. For example, an *O*-acetyl group at C-4 renders the corresponding sialic acid completely resistant towards all sialidases tested, with the exception of influenza virus sialidases. Therefore, these impeding substituents have to be removed by specific enzymes before sialidase action (e.g. *O*-acetyl groups must be hydrolyzed by corresponding esterases that occur in many tissues, bacteria, and viruses) [2,24,25]. In cellular metabolism this de-*O*-acetylation mainly occurs in lysosomes, where sialo-glycoconjugates are desialylated. Liberated sialic acids are transported into the cytosol mediated by the protein "sialin." There, they can be reutilized by activation with CTP and retransported into the Golgi compartment or they are degraded by cytosolic aldolase, the sialate-pyruvate lyase, yielding pyruvate and acylmannosamines [2,24]. This enzyme also acts best on Neu5Ac, while a *N*-glycolyl group and especially *O*-acetyl groups impede the cleavage process.

5. SIALIC ACIDS REGULATE MOLECULAR AND CELLULAR INTERACTIONS

The external position of sialic acids on glycoproteins and gangliosides, either alone or in oligo- or polymeric form, and correspondingly on the outer cell membranes implies a strong influence in cell biology. These acidic monosaccharides can most easily interact with the components of other cell surfaces, extracellular substances, and effector molecules. Evidence is increasing that they are involved in a multiplicity of cell signaling events. When considering their functions, sialic acids may be divided into more general classes, irrespective of the variability of their structures, and into those exerted by chemical modifications of these acidic monosaccharides [1,2,24]. The biology of sialic acids may also be viewed from their dual role, that is that they either mask recognition sites, or, in contrast, represent a biological target, allowing recognition by a receptor protein, a lectin, thus representing a ligand or counterreceptor [26,27]. The latter role may be modulated or even abolished by sialic acid substituents, most effectively by *O*-acetyl groups. Thus, esterification of sialic acids can mask their ligand function.

These multiple possibilities make thorough understanding of the sialic acids difficult. This is even more pronounced, since the environment of these monosaccharides and the nature of the molecule to which they are bound may influence the biological effects. However, the present knowledge in this area allows the prediction of biological events if the amount of sialic acids on cells increases or decreases, or if the nature of the sialic acids typical for a given cell or tissue changes. In the following, some light will be shed on this situation, showing that sialic acids are ideal mediators of fine-tuning of cell behavior. However, they are also ideal ligands for the binding of and infection of mammalian cells by microorganisms.

The antirecognition effects of sialic acids are explained by the negative charge in combination with a bulky, hydrophilic molecule. This is a very important research field concerning the repulsion of cells and molecules [2,26,27], such as erythrocytes, by the negatively charged sialic acid residues, the masking of penultimate sugars (mostly galactose) which nature has designated to be recognized by receptors, such as galactose, the shielding of antigenic sites in macromolecules either soluble or bound to cell membranes, the antiproteolytic effect in glycoproteins, and the hindrance of the action of some endoglycosidases. Desialylation may lead to binding to galactose-specific lectins or to recognition of macromolecules and cells by the complement and immune system. Demasking of galactose can target molecules and cells to specific sites, often in order to promote degradation, which may be of physiological or pathological importance.

The first reported and ground-breaking example of this phenomenon was the uptake of desialylated serum glycoproteins by hepatocytes [28]. A similar function was observed with the interaction between liver, spleen, and peritoneal macrophages and erythrocytes. After sialidase treatment of mammalian erythrocytes and reinjection, most of the modified cells disappear from the bloodstream within a few hours, although their normal survival time in humans is about 120 days. They are phagocytosed by liver Kupffer cells, spleen, or peritoneal macrophages [29] (Fig. 3). Sialidase-treated thrombocytes and lymphocytes experience the same fate, although the latter are released after one day of incubation, probably due to resialylation of the cell surface [2,30].

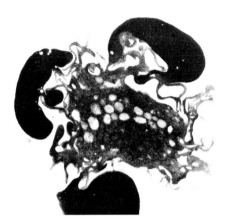 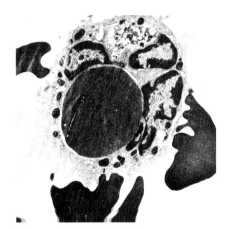

Figure 3. Binding of three sialidase-treated rat erythrocytes to homologous peritoneal macrophages with microvilli (left) and phagocytosis (right). Electron micrograph, magnification × 5000 (left) and × 6000 (right).

On these blood cells sialic acid is bound to galactose or N-acetylgalactosamine, which can be recognized by corresponding lectins after the enzymatic release of the sialic acid moieties. The red blood cells bind via their demasked galactose residues to a galactose-specific receptor of phagocytes and ultimately are taken up and degraded ("lectinophagocytosis" [30,31]). Correspondingly, trapping of sialidase-treated erythrocytes can be inhibited by galactose or more efficiently by galactosides. This mechanism can work without the involvement of immunoglobulin or complement and appears to be involved in the sequestration process of aged red cells [32]. It was demonstrated that old erythrocytes expose more galactose residues than younger cells. The galactose-recognizing receptor belonging to the C-type lectins responsible for the binding of blood cells was isolated from rat peritoneal macrophages by Kelm *et al.* [33] and represents one of the first characterized mammalian galactose receptors after the hepatocyte receptor. Since then large gene families encoding for the galectins and other proteins recognizing galactose have been found in microorganisms, plants, and animals [34].

Malignant cells can also be eliminated in this way by macrophages, and it is understandable that oversialylation, often as sialyl Lewis^x, frequently observed in such cells [35], protects them from humoral and cellular defense systems and thus increases their malignancy and metastatic potential. The strategy to increase sialylation over physiological levels has been used to create a "hyper-erythropoietin" by insertion of more glycosylation sites, which exhibits better pharmacokinetic properties due to an extended life-time in blood serum [36]. In this regard, polysialic acids found on some bacteria and on mammalian cells are strongly antiadhesive, thus regulating cellular contacts and movements [1,37]. Polysialic acids enhance virulence in, for example, *Escherichia coli*. In higher animals they play a crucial role in development, especially of neuronal tissues and are involved in the maintenance of neuronal plasticity and the regulation of circadian rhythms. They also facilitate tumor growth and spreading.

Sialylation of microorganisms follows a similar antirecognition strategy, allowing better survival in the host organisms and thus enhancing virulence. Similar to the binding of blood cells after demasking of their galactose residues, many, and often pathogenic, bacteria attach to mucous endothelia via galactose-recognizing lectins, for example in the oro-gastrointestinal tract, after desialylation of endothelial cells and exposure of penultimate galactose residues [27]. Such bacteria secrete sialidase, which removes the protecting sialic acid moieties and can thus be considered as a virulence factor enabling colonization and spreading. Desialylation of endothelia can also occur by viral infection (e.g. influenza) which leads to bacterial colonization and consequently to aggravation of the disease.

A second group of examples for the masking effect of sialic acid is the production of autoantibodies after cell membrane desialylation by bacterial or viral sialidases. This exposure of cellular antigens may lead to chronic diseases such as neuronal disorders or glomerulonephritis [38]. Carbohydrates, including sialic acids, may be antigenic determinants, especially in gangliosides, but they more often shield antigenic sites and thus weaken the immunoreactivity. Sialic acids render cells as "self," not allowing recognition by the immune system or by macrophage lectins, for example. The loss of these monosaccharides makes these cells "non-self" and therefore vulnerable. Therefore sialic acids can be considered as members of the innate immune system.

In contrast to masking, sialic acid molecules take directly part in numerous recognition processes (vertebrate cells, protozoa, bacteria, mycoplasma, viruses, hormones, toxins, plant lectins, enzymes, and antibodies). This may be the most important role of these monosaccharides. It was first noted in microorganisms, which in most cases use cell surface carbohydrates for binding to their host cells [27,34,39]. Sialic acids appear to be the most frequent ligands for pathogenic and non-pathogenic viruses, bacteria, and protozoa. The best known and longest studied example is the influenza viruses. Hirst observed in 1942 [40] the binding of influenza A virus to human erythrocytes and mucins of the respiratory tract. The attachment, mediated by a viral lectin called hemagglutinin, was reversible and accompanied by the release of an acidic sugar-like substance (later identified as Neu5Ac) by a viral "receptor-destroying enzyme," later named neuraminidase (sialidase). The infection mechanism includes binding of the virus to endothelial cells of the respiratory tract, followed by penetration, multiplication, and exocytosis of the virus.

Viral sialidase seems to facilitate the spreading of viruses in tissues by preventing their further attachment to the cells from where they originated and to mucus layers protecting epithelia of the respiratory tract. Using this knowledge it was possible to synthesize a very strong inhibitor (2,3-didehydro-2,4-dideoxy-4-guanidinyl-N-acetylneuraminic acid; 4-guanidinyl-Neu2en5Ac) of the viral sialidase based on a natural sialidase inhibitor (Neu2en5Ac) and exact knowledge of the three-dimensional structure of the enzyme [41]. This substance and analogs thereof are also active *in vivo* and are being successfully used for the treatment of influenza.

Transmission of influenza virus from animals (e.g. birds) to humans or other mammals involves the specificity of viral hemagglutinins towards the type of sialic acids linkage (α2,3 or α2,6). Glycan arrays are now available to exactly test the hemagglutinins of various virus types [42].

Many other viruses also attach to cells via sialylated glycans and the number of known examples is growing (see [27] for a review and [43]). The most prominent ones are adeno-, corona-, herpes, orthomyxo- (including influenza viruses), papova-, paramyxo-, parvo-, picorna-, rhabdo-, rheo-, and rotaviruses. On infection with HIV viruses sialylated glycans both from the virus and cellular receptors are involved. Most of them bind to Neu5Ac, but some virus species have highest affinity to modified sialic acids. Human influenza C virus and salmon isavirus as well as some coronavirus, torovirus, and rotavirus species recognize Neu4,5Ac$_2$, Neu5,9Ac$_2$, or di-*O*-acetylated sialic acids (Table 1). These viruses express various specific esterases as receptor-destroying enzymes [25], which may have a function similar to that of sialidase in influenza A and B viruses. Binding of these influenza viruses is hindered by sialic acids-*O*-acetylation. Viruses can also specifically bind to Neu5Gc, such as the transmissible gastroenteritis virus of pig [44] and some rotavirus species [45].

Table 1 Occurrence of lectins preferring binding to *N*-glycolylated or *O*-acetylated sialic acids

Preferred sialic acid	Lectin occurrence
N-Glycolylneuraminic acid	Murine and ape CD22 (siglec-2)
	Crabs *Scylla serrata* and *Carcinoscopus rotunda*
	Clam *Anadara granosa*
	Snail *Pila globasal*
	Mulberry *Morus alba*
	Mushrooms *Hericium erinaceum* and *Chlorophyllum molybdites*
	Malaria parasite *Plasmodium reichenowi*
	Escherichia coli strain K99
	Transmissible gastroenteritis virus
	Rotavirus strains
N-Acetyl-9-*O*-acetylneuraminic acid	Crab *Cancer antennarius*
	Crustacean *Liocarcinus depurator*
	Snails *Achatina fulica* ("achatinin-H") and *Cepaea hortensis*
	Influenza C virus
	Strains of coronavirus[a], isavirus[a], rotavirus, and torovirus

[a]Some strains also recognize *N*-acetyl-4-*O*-acetylneuraminic acid.

Bacteria produce carbohydrate-specific adhesins which are frequently located on their fimbriae or pili. Prominent examples of colonization via sialic acids are pathogenic bacteria such as strains of *E. coli*, streptococci, and *Helicobacter pylori*, and the list is rapidly growing (for reviews see [27,34,39]). *H. pylori*, which is often found in human stomach and is responsible for gastric inflammation and possibly also cancer, is especially intensively studied. Mahdavi et al. [46] have published the structure of its sialic acid-recognizing adhesin and Johansson et al. have carried out a detailed study on the interaction of these bacteria with sialylated carbohydrates [47]. In most cases Neu5Ac has highest affinity to the bacterial adhesins. An exception is the *E. coli* strain K99 which prefers Neu5Gc [27].

Knowledge about such adhesion mechanisms enables a biology-based therapy of diseases caused by *H. pylori* and other pathogenic microorganisms and may save the use of antibiotics [48]. Receptor–ligand interactions are inhibited with soluble ligands such as sialylated oligosaccharides or glycoproteins. Such oligosaccharides occur in a relatively high concentration in milk, especially the colostrum. The milk from each mammal contains different and more or less sialylated glycans [49]. These carbohydrates are considered to regulate the species-specific colonization of the intestine with microorganisms and to hinder the attachment of pathogenic bacteria such as *E. coli* and *Helicobacter* strains or pathogenic viruses, for example rotaviruses, which is most important in newborns. Correspondingly, the application of sialo-oligosacchairdes in soluble forms or as dendritic polymers is a new strategy to fight microbial infections [50]. Whether the high degree of *O*-acetylation observed in mucins from human colon or bovine submandibular gland similarly acts against Neu5Ac-mediated adhesion of microorganisms (e.g. in the stomach of the ruminant cow) has not yet been investigated. Bacterial toxins, such as cholera, tetanus, diphtheria, and botulinus [51] toxin, firmly adhere to sialic acids, mostly of gangliosides, as the basis of their pathophysiological activity (summarized in 27]).

Some (pathogenic) protozoa can also interact with sialic acids on mammalian cells during the infection process. The most prominent are malaria-causing plasmodia [27]. While the human pathogen *Plasmodium falciparum* prefers Neu5Ac as ligand, *P. reichenowi*, which infects chimpanzees, prefers Neu5Gc as ligand of the erythrocyte-binding antigen-175 of the plasmodia [52]. This difference in susceptibility to malaria mirrors the genetic loss of Neu5Gc in humans. It may hint to the emergence of the human pathogenic *P. falciparum* only after this mutation about 3 million years ago.

A variety of sialic acid-recognizing lectins have been isolated from plants, mushrooms, and animals (reviewed in [4,27,30,34]). Some of these are specialized for *O*-acetylated sialic acids or Neu5Gc (see Table 1). Some snail and crab or crustacean species express lectins recognizing Neu5,9Ac$_2$. An example used for the analysis of *O*-acetylated sialic acids in natural sources (e.g. leukemic cells) is achatinin H [53]. Neu5Gc-binding proteins also occur in the hemolymph of crab species. Table 1 shows that corresponding lectins have been isolated from plant, fungus, and mushroom species. A recently described example is the mushroom *Chlorophyllum molybdites* [54]. These lectins are thought to be involved in mechanisms of innate immunity that defend against sialylated microorganisms such as fungi or bacteria or against herbivores.

Probably the most important sialic acid-recognizing receptors in higher animals are the "siglecs" (see Chapter 4), which belong to the immunoglobulin superfamily (IgSF) with repeating extracellular domains of variable length. Thirteen members of this family have been discovered in the past few years [1,27,30,34,55,56]. The first well-characterized species were macrophage sialoadhesin, involved in binding and nursing of maturing blood cells, CD22 on B-lymphocytes, responsible for the "cross-talk" of these immune-competent cells with T-lymphocytes, and myelin-associated glycoprotein (MAG) on oligodendrocytes and Schwann cells, participating in the growth and myelination of neurites. Most of the other siglecs are involved in the regulation of the functions of various white blood cell types and macrophages. CD22 is the best studied member of this family; it mediates signaling and other functions in B-cell regulation [56–58]. CD33-related siglecs undergo rapid evolution. They are involved in the regulation of the innate immune system, trigger apoptosis, calcium flux, and can provide inhibitory signals [56]. The fish ortholog MAG (siglec-4), involved in the maintenance of a nervous system with myelinated axons, is thought to be the ancestral siglec, because no obvious orthologs of the mammalian siglecs have been found by genome analysis in fish [59].

Mammalian siglecs vary in their recognition of different sialic acids and can distinguish glycosidic linkages. For example, sialoadhesin and MAG only bind to α2,3-linked sialic acids, while CD22 recognizes α2,6-bonds. While N-acetyl hydroxylation (Neu5Gc) can modify the affinity, sialic acid-O-acetylation (Neu5,9Ac$_2$) abolishes it [27]. A shift in the affinity of CD22 for Neu5Ac or Neu5Gc has been observed during evolution. While mouse and ape CD22 preferentially bind to Neu5Gc, the human one prefers Neu5Ac. This mirrors the hominid development, resulting in a loss of the hydroxylase in humans that is responsible for the biosynthesis of Neu5Gc [1,2,4].

Another group of mammalian sialic acid-recognizing lectins are the selectins (see Chapter 16), which also possess repeating domains (for reviews see [27,34]). The E-, L-, and P-selectins occur on endothelial cells, leukocytes, and mainly blood platelets, respectively. They participate in the initial stage of adhesion of white blood cells to endothelia, along which they begin to roll and eventually penetrate into the tissue below, which may be damaged by the lack of oxygen, in, for example, transplants or infarcts, or if inflamed. Cytokines play an essential role here in selectin expression. Selectins preferably recognize α2,3-sialylated Lewis^x structures (Sia-Le^x), a fucosylated tetrasaccharide bound to glycoproteins or glycolipids. Since these oligosaccharide moieties also occur on tumor cells (e.g. on variant isoforms of CD44) [60], selectins can be involved in the formation of metastases.

Various cytokines possess carbohydrate-binding properties. Such lectin activity has also been shown for interleukin-4 which binds to the 1,7-intramolecular lactone of Neu5Ac [61]. This rare sialic acid species has been found in glycoproteins of human lymphocyte cell membranes.

6. CONCLUSION

The intention of this chapter was to show that the main role of sialic acids with their outstanding position on molecules and cells, their unique structure, and their chemical diversity seems to be the fine-tuning of a multitude of cellular events which may be explained by the capacity of these monosaccharides to regulate molecular interactions. We probably only see the tip of an iceberg at present, especially when it is considered that the ligand function of the sialic acid residues in mammalian cells was discovered only a few years ago. Only limited insight has also been obtained into the role of sialic acids in phylo- and ontogenesis, including aging. In pathology, sialic acids are increasingly recognized, not only in cancer and degenerative diseases, but impressively in the biology of infectious diseases. It is inevitable that more attention will be paid to sialic acids and their binding partners in pharmacology. In physiological and pathological processes it has mainly been Neu5Ac that has been described as involved, however, as was discussed here, other sialic acids, such as Neu5Gc and O-acetylated species are gaining importance and attention. With the availability of improved analytical tools it is expected that specific functions will be ascribed also to the other, often more common, neuraminic acid derivatives.

REFERENCES

1. Angata T & Varki A (2002) *Chem. Rev.* **102**, 439–469.
2. Schauer R & Kamerling JFG (1997) In: *Glycoproteins II*, pp. 243–402. Edited by J Montreuil, JFG Vliegenthart and H Schachter. Elsevier Science AV, Amsterdam.
3. Schauer R (2000) *Glycoconjugate J.* **17**, 485–499.
4. Schauer R (2004) *Zoology* **107**, 49–64.
5. Schwarzkopf M, Knobeloch K-P, Rohde E *et al.* (2002) *Proc. Natl Acad. Sci. USA* **99**, 5267–5270.
6. Malykh YN, Schauer R & Shaw L (2001) *Biochimie* **83**, 623–634.
7. Schauer R & Srinivasan GV (2005) *Glycoconj. J.* **22**, 288.
8. Zanetta J-P, Srinivasan V & Schauer R (2006) *Biochimie* **88**, 171–178.
9. Harduin-Lepers A, Mollicone R, Delannoy P & Oriol R (2005) *Glycobiology* **15**, 805–817.
10. Takashima S, Abe T, Yoshida S *et al.* (2006) *J. Biochem. (Tokyo)* **139**, 279–87
11. Iwersen M, Schmid H, Gasa S, Kohla G & Schauer R (2003) *Biol. Chem.* **384**, 1035–1047.
12. Srinivasan GV & Schauer R (2005) *Glycoconjugate J.* **22**, 305.
13. Higa HH, Butor C, Diaz S & Varki A (1989) *J. Biol. Chem.* **264**, 19427–19434.
14. Kamerling JP, Schauer R, Shukla A, Stoll S, van Halbeek H & Vliegenthart JFG (1987) *Eur. J. Biochem.* **162**, 601–607.
15. Tiralongo J & Schauer R (2004) *Trends Glycosci. Glycotechnol.* **16**, 1–15.
16. Lewis AL, Nizet V & Varki A (2004) *Proc. Natl Acad. Sci. USA* **101**, 11123–11128.
17. Deszo EL, Steenbergen SM, Freedberg DI & Vimr ER (2005) *Proc. Natl Acad. Sci. USA* **102**, 5564–5569.
18. Kooistra O, Lüneberg E, Lindner B, Knirel YA, Frosch M & Zähringer U (2001) *Biochemistry* **40**, 7630–7640.
19. Longworth E, Fernsten P, Miminni TL *et al.* (2002) *FEMS Immunol. Med. Microbiol.* **32**, 119–123.
20. Manzi AE, Sjoberg ER, Diaz S & Varki A (1990) *J. Biol. Chem.* **265**, 13091–13103.
21. Kelm A, Shaw L, Schauer R & Reuter G (1998) *Eur. J. Biochem.* **251**, 874–884.
22. Roggentin P, Schauer R, Hoyer LL & Vimr ER (1993) *Mol. Microbiol.* **9**, 117–129.

23. Miyagi T, Wada T, Yamaguchi K & Hata K (2004) *Glycoconj. J.* **20**, 189–198.
24. Traving C & Schauer R (1998) *Cell. Mol. Life Sci.* **54**, 1330–1349.
25. Strasser P, Unger U, Strobl B, Vilas U & Vlasak R (2004) *Glycoconj. J.* **20**, 551–561.
26. Schauer R (1985) *Trends Biochem. Sci.* **10**, 357–360.
27. Kelm S & Schauer R (1997) *Int. Rev. Cytol.* **175**, 137–240.
28. Ashwell G & Morell AG (1974) *Adv. Enzymol.* **41**, 99–128.
29. Jancik JM, Schauer R, Andres KH & von Düring M. (1978) *Cell Tissue Res.* **186**, 209–226.
30. Schauer R (2004) *Arch. Biochem. Biophys.* **426**, 132–141.
31. Müller E, Schröder C, Schauer R & Sharon N (1983) *Hoppe-Seyler's Z. Physiol. Chem.* **364**, 1419–1429.
32. Bratosin D, Mazurier J, Debray H *et al.* (1995) *Glycoconj. J.* **12**, 258–267.
33. Kelm S & Schauer R (1988) *Hoppe-Seyler's Z. Physiol. Chem.* **369**, 693–704.
34. Sharon N & Lis H (2003) *Lectins.* Kluwer Academic Publishers, Dordrecht.
35. Alper J (2003) *Science* **301**, 159–160.
36. Jelkmann W (2002) *Eur. J. Haematol.* **69**, 265–274.
37. Janas T & Janas T (2005) In: *Polysaccharides. Structural Diversity and Functional Versatility,* pp. 707–727. Edited by S Dumitriu. Marcel Dekker, New York.
38. Corfield T (1992) *Glycobiology* **6**, 509–521.
39. Ofek I & Doyle J (1994) *Bacterial Adhesion to Cells and Tissues.* Chapman & Hall, New York.
40. Hirst GK (1942) *J. Exp. Med.* **76**, 195–209.
41. von Itzstein M, Wu W-Y & Jin B (1994) *Carbohydr. Res.* **259**, 301–305.
42. Stevens J, Blixt O, Glaser L *et al.* (2006) *J. Mol. Biol.* **355**, 1143–1155.
43. Schauer R & Vlasak R (Eds) (2006) *Glycoconj. J.* **23**, 1–141.
44. Schwegmann-Weßels C, Zimmer G, Laude H, Enjuanes L & Herrler G (2002) *J. Virol.* **76**, 6037–6043.
45. Delorme C, Brüssow H, Sidoti J *et al.* (2001) *J. Virol.* **75**, 2276–2287.
46. Mahdavi J, Sondén B, Hurtig M *et al.* (2002) *Science* **297**, 573–578.
47. Johansson P, Nilsson J, Ångström J & Miller-Podraza H (2005) *Glycobiology* **15**, 625–636.
48. Gerhard M, Hirmo S, Wadström T *et al.* (2001) In: *Helicobacter pylori: Molecular and Cellular Biology,* pp. 185–206. Edited by S. Suerbaum. Horizon Scientific Press, Wymondham, UK.
49. Sumiyoshi W, Urashima T, Nakamura T *et al.* (2003) *Br. J. Nutr.* **89**, 61–69.
50. Urashima T, Saito T, Nakamura T & Messer M (2001) *Glycoconj. J.* **18**, 357–371.
51. Tsukamoto K, Kohda T, Mukamoto M *et al.* (2005) *J. Biol. Chem.* **280**, 35164–35171.
52. Martin MJ, Rayner JC, Gagneux P, Barnwell JW & Varki A (2005) *Proc. Natl Acad. Sci. USA* **102**, 12819–12824.
53. Pal S, Ghosh C, Mandal C *et al.* (2004) *Glycobiology* **14**, 859–870.
54. Kobayashi Y, Kobayashi K, Umehara K *et al.* (2004) *J. Biol. Chem.* **279**, 53048–53055.
55. Angata T & Brinkmann-Van der Linden ECM (2002) *Biochim. Biophys. Acta* **1572**, 294–316.
56. Crocker PR (2005) *Curr. Opin. Pharmacol.* **5**, 431–437.
57. Collins BE, Smith BA, Bengtson P & Paulson JC (2006) *Nat. Immunol.* **7**, 199–206.
58. Tedder TF, Poe JC & Haas KM (2005) *Adv. Immunol.* **88**, 1–50.
59. Lehmann F, Gäthje H, Kelm S & Dietz F (2004) *Glycobiology* **14**, 959–968.
60. Hanley WD, Napier SL, Burdick MM *et al.* (2006) **20**, 337–359.
61. Cebo C, Dambrouck T, Maes E *et al.* (2001) *J. Biol. Chem.* **276**, 5685–5691.

CHAPTER 13

Structure and carbohydrate specificity of β-prism I fold lectins

K. Sekar, A. Surolia and M. Vijayan

1. INTRODUCTION

The only feature common to lectins from different sources is their ability to specifically bind to different carbohydrates. They exhibit widely different sizes, oligomeric states, and folds [1–3]. Plant lectins themselves can be classified into five structural classes [3]. Of these five, the latest to be identified involves the β-prism I fold. The β-prism I fold was first characterized as a lectin fold in this laboratory in 1996 through the X-ray analysis of jacalin [4], one of the two lectins in jack fruit seeds. Since then, jacalin-like lectins have been studied extensively. Here we review these studies with particular reference to our own efforts.

2. EXAMPLES

2.1 Jacalin: new lectin fold and novel strategies for generating carbohydrate specificity

Jacalin is a tetrameric lectin (M_r 66 000) containing a long α-chain and a short β-chain in each subunit. It binds immunoglobulin A1 (IgA1) and other glycoproteins such as carcinoma-related mucins and is selectively mutagenic for human $CD4^+$ T-cells. The heavily glycosylated lectin is specific to galactose at the monosaccharide level. At the disaccharide level it binds the tumor-associated T-antigenic disaccharide Galβ13GalNAc. The two chains in each subunit are produced by posttranslational proteolysis. The short β-chain forms an integral part of the three-dimensional structure, specified essentially by the longer α-chain, although the two are not covalently linked.

Topologically, the structure of each subunit essentially consists of three Greek keys, one of them broken on account of the posttranslational proteolysis. The three keys form the sides of a triangular prism with approximate three-fold symmetry (Fig. 1). This symmetry is not, however, reflected in the sequence. The subunit structure is stabilized by a network of intra-key and inter-key hydrogen bonds and the presence of a hydrophobic core in the middle of the prism. The subunits assemble into tetramers with 222 symmetry.

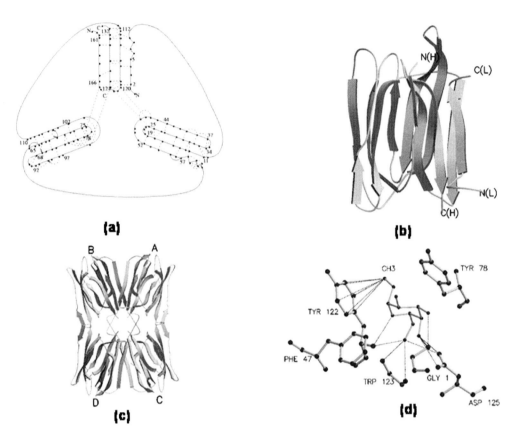

Figure 1. (a) The subunit topology of jacalin. (b) The tertiary structure of jacalin subunit as viewed perpendicular to the molecular pseudo three-fold axis. (c) The quaternary structure of jacalin. (d) Interactions between jacalin and methyl-α-galactose.

The structure of jacalin was initially solved as a complex with methyl-α-galactose (Me-α-Gal) [4]. Of the three Greek keys, only one carries a carbohydrate-binding site. The interaction between the lectin and the sugar is mediated by several hydrogen bonds and the stacking of an aromatic residues on the galactose ring (Fig. 1d). The interactions involving the N-terminus of the α-chain is of particular interest. This N-terminus is generated by the posttranslational proteolysis referred to earlier. It interacts with O-3 and O-4 of the sugar. O-4 is axial in galactose while it is equatorial in mannose and glucose. The position of the positively charged terminal amino group is such that it could not interact with O-4 if it were in the equatorial position. Therefore, the interaction is specific to galactose. This is the first observation of posttranslational modification as a strategy for generating ligand specificity.

Since then, the binding site of jacalin has been thoroughly characterized through further X-ray analysis of the complexes with Gal, Me-α-GalNAc, T-antigen, Me-α-T-antigen, GalNAcβ1-3Gal-α-O-Me and Galα1-6Glc (mellibiose) [5,6]. The binding mode of T-antigen in jacalin has also been compared with those in other relevant lectins. The structures of the complexes show that the carbohydrate-binding site of jacalin has three components: the primary binding site, secondary binding site A, and secondary binding site B (Fig. 2). Gal or GalNAc occupies the primary binding site with the aromeric carbon pointing towards secondary binding site A, which has variable geometry. α-Substituents interact, primarily hydrophobically, with this site. An interesting feature of this interaction is the presence of near-perfect O-H....π or C-H....π hydrogen bonds. The geometry of binding is such that β-substitution leads to severe steric clashes with this site. Therefore, when β- linked disaccharides bind to jacalin, the reducing sugar binds

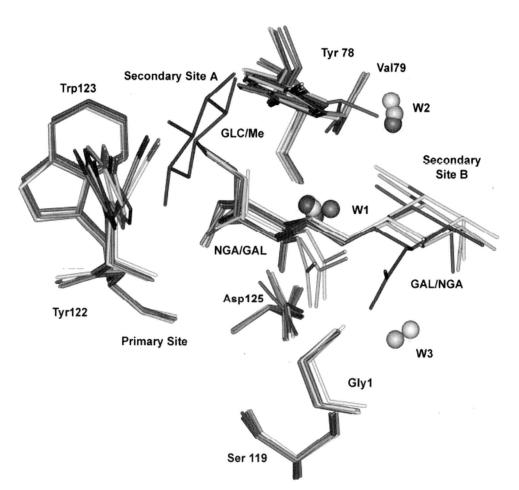

Figure 2. A composite view of the interactions at the carbohydrate binding site of a jacalin subunit with different sugars.

at the primary site with the non-reducing end located at secondary binding site B. The interactions at secondary site B are primarily through water-bridges.

Still further studies on jacalin were necessitated by a suggestion of promiscuity in the sugar-binding specificity of the lectin, on the basis of surface plasmon resonance and X-ray studies of a Me-α-Man complex of jacalin [7]. To further explore this suggestion, detailed isothermal titration calorimetric studies of Gal, Man, Glc, and their methyl derivatives and X-ray studies on jacalin-Me-Glc complex and a new form of the jacalin-Me-α-Man complex, were carried out [8]. Methyl substitution increases the affinity of Gal and Man by 27 times while the ratio between the affinities of Me-α-Gal and Me-α-Man is 20. Glc shows no measurable binding, while the binding affinity of Me-α-Glc is slightly less than that of Me-α-Man. These results could be readily explained in terms of interactions at the primary site and secondary site A observed in crystal structures. The affinity of Gal to jacalin is overwhelmingly larger than that of Man. Even the weak affinity of Me-α-Man to the lectin is mainly due to the interactions of the methyl group with the secondary binding site. Affinity and specificity are often treated separately. However, specificity loses much of its practical meaning when affinity is extremely weak. That appears to be the sitation in relation to the interaction of jacalin with sugars other than galactose and its derivatives.

The crystallographic studies outlined above and associated modeling have provided a structural explanation of the relative affinities of different subclasses of mucin-type *O*-glycans to jacalin [6]. Jacalin also binds many other *O*-linked glycoproteins. However, it does not bind major types of *N*-glycans which are based on mannopentose core. The same is true in relation to fucose-based *N*-glycans. Thus, jacalin is specific to *O*-glycans by virtue of the presence of galactose in them and its affinity to *N*-glycans is either extremely weak or non-existent.

2.2 Artocarpin: loop length as determinant of oligosaccharide specificity

The second lectin from jackfruit seeds, artocarpin, is a single-chain protein with considerable sequence similarity with jacalin. Unlike jacalin, it is specific to mannose. It has the highest affinity for the tetrasaccharide from horseradish peroxidase. The structure of the lectin and its binding site for the monosaccharide were characterized through the X-ray analysis of the crystals of the lectin and its complex with Me-α-Man [9]. The crystal structure and detailed modeling studies revealed the structural basis for the difference in the carbohydrate specificity of jacalin and artocarpin (Fig. 3a). First, unlike jacalin, artocarpin does not contain an N-terminus generated by posttranslational proteolysis. In jacalin, this N-terminus is important for generating specificity for galactose. Secondly, the carbohydrate-binding site of jacalin contains aromatic residues whereas that of artocarpin contains none. It turns out that stacking interaction with an aromatic residue is important for the binding of galactose; also the binding site is rich in aromatic residues, while such residues are usually absent in the carbohydrate binding sites of mannose specific lectins with the β-prism I fold. A comparison of the binding site of artocarpin with those of jacalin and heltuba [10], another mannose-binding β-prism I fold lectin, provides a striking example of this difference (Fig. 3).

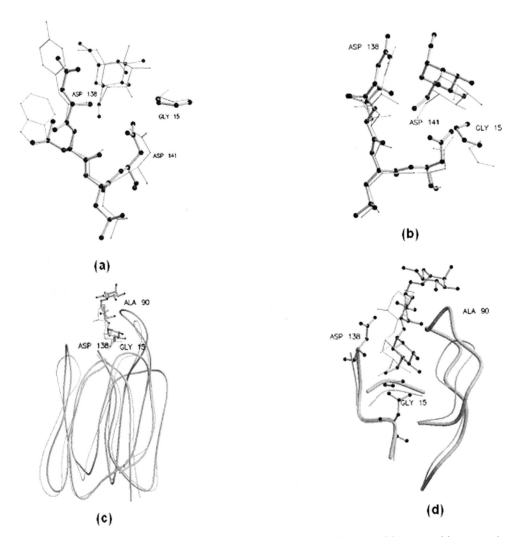

Figure 3. Superposition of the sugar-binding sites of artocarpin and (a) jacalin, (b) heltuba. (c) Structural superposition of artocarpin and heltuba along with the oligosaccharides bound to them. (d) A close-up of the sugar-binding sites of artocarpin and heltuba. Open bonds or bands correspond to artocarpin and the sugars bound to it.

While the specificity of, and the protein–carbohydrate interaction in, artocarpin and heltuba are the same at the monosaccharide level, they differ when di- or higher oligomers of mannose are considered. The structural basis for this difference could be clearly elucidated on the basis of the structures of the complexes of artocarpin with a mannotriose and a mannopentose determined in this laboratory [11], and a comparison of the structures with heltuba–carbohydrate complexes of known structure (Fig. 3). The carbohydrate-binding site of artocarpin (as indeed of heltuba) is made up of three loops 14–17, 86–95, and 137–141. Only loops 14–17 and 137–141 are involved in mannose recognition. They may be considered to constitute the primary binding site of the lectin, while loop 86–95 may be referred to as the secondary binding site.

In its complex with artocarpin (Fig. 3b,c), the first mannose residue of the trisaccharide (Manα1-3 Manα1-6 Man) interacts, through several hydrogen bonds, with the primary binding site. The second residue of the mannotriose interacts through hydrogen bonds and non-bonded contacts almost exclusively with the secondary binding site. The third residue has only van der Waals interactions with this site. Although mannopentose has two additional residues, they are not involved in additional interactions. Consequently, the mannotriose and the mannopentose have nearly the same affinity for artocarpin.

The crystal structure of the complex of the same mannotriose with heltuba is also available [10]. Only the first two residues are defined in the structure. A superposition of the binding sites in the complexes of artocarpin and heltuba (Fig. 3b,c) shows that the location, orientation, and the interactions of the first mannose residue are the same in the two complexes. This is related to the identity of the two loops constituting the primary binding site in the two lectins. The third loop constituting the secondary binding site is considerably shorter in heltuba. Consequently, the interactions of the two remaining residues in the trisaccharide are different in the two lectins. The orientations of the second mannose residue are different in them. This residue is involved in four hydrogen bonds in artocarpin as against one in heltuba. The third residue in the heltuba complex, not defined in the crystal structure, can be modeled in more than one way. However, none of the models lead to interactions with the lectin. On the contrary, the third residue has van der Waals interactions with loop 86–95 in the artocarpin complex. For all these reasons, the affinity of the mannotriose for heltuba is considerably weaker than that for artocarpin. The structural arguments presented above clearly demonstrate that this difference in affinity is caused by the difference in the length of a loop.

2.3 Banana lectin: two primary binding sites, specificity for branched oligosaccharides, evolutionary implications

Yet another β-prism I fold lectin X-ray analyzed in this laboratory [12] and elsewhere [13] is that from banana. The lectin interacts with branched mannans and glucans. The structure of the lectin from *Musa paradisiaca*, determined as a complex with Me-α-Man, has two binding sites, each located on a different Greek key (Fig. 4a). The two sites have nearly identical amino acid composition and geometry and are involved in exactly the same type of interactions with the sugar. We had earlier suggested that the β-prism I fold could have evolved through successive gene duplication and fusion of an ancestral Greek key motif [4]. In other lectins of known structure, all from dicots, divergent evolution appears to have obliterated the sequence similarities among the three motifs and the binding site got retained in only in one of them. The X-ray analysis of banana lectin constitutes the first structure determination of a β-prism I fold lectin from a monocot. In this lectin, the three motifs have not diverged enough to obliterate the sequence similarity among them. The binding site itself has been conserved on two motifs. The corresponding region on the third motif serves as a common secondary site.

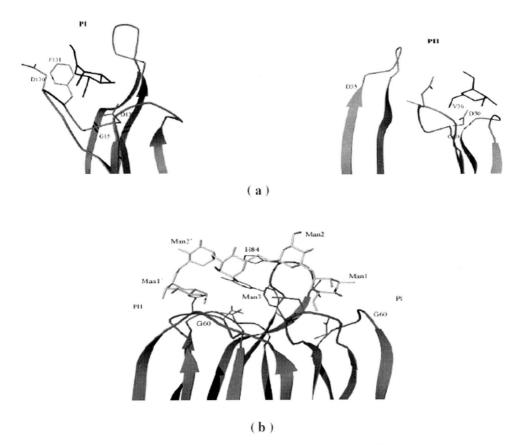

Figure 4. (a) The two primary binding sites in each subunit of banana lectin. (b) A model of the branched pentasacharide bound to a subunit of banana lectin.

Complexes of 1-2, 1-3 and 1-6 dimannosides with banana lectin could be separately modeled with non-reducing ends anchored at the primary sites in the same way as Me-α-Man is situated in the crystal structure. In almost all possible models, the second ring interacts with secondary binding site irrespective of which of the two primary sites the first ring is located in. An attempt was then made to link through a fifth sugar residue, the disaccharides anchored at the other primary site. This could be readily done in one instance, resulting in a branched pentasaccharide:

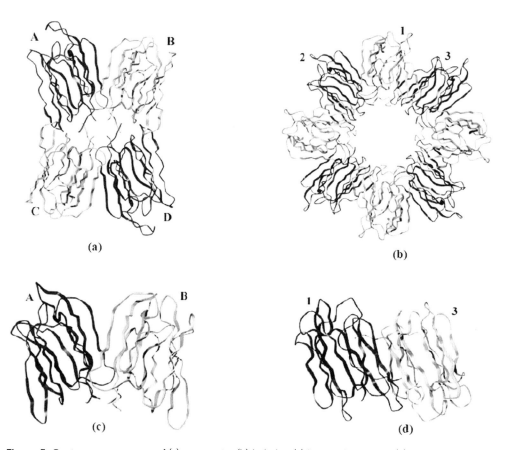

Figure 5. Quaternary structures of (a) artocarpin, (b) heltuba, (c) banana lectin, and (d) calsepa.

This result (Fig. 4b) explains the unique ability of the lectin to bind branched mannans. This ability arises directly as a consequence of the presence of two primary binding sites and a common secondary binding site. It is also interesting that all the branched oligomannosides known to bind banana lectin have branching involving 3 and 6 positions as in the model readily arrived at using the crystal structure.

3. CONCLUDING REMARKS

The structures of a number of β-prism I fold lectins (*Maclusa pamitera* lectin [14] and calsepa [15] in addition to those already mentioned) from different plant families are now available, all essentially with same tertiary structure. However, they exhibit a wide variety of quaternary association, as in the case of legume lectins. Thus, β-prism I fold lectins also constitute (see Fig. 5) a family in which small alterations in essentially the same tertiary structure lead to widely different quaternary structures. Recently a novel arrangement involving three tandem β-prism I domains in a single domain has been reported in Mimosoideae lectin [16]. Two such subunits dimerize to produce a circular arrangements of domains.

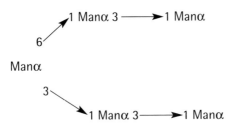

In addition to contributing to the elucidation of structural variety and carbohydrate specificity of lectins, studies on β-prism I fold lectins have provided further insights into strategies for generating ligand specificity. For instance, the structure of jacalin–sugar complexes revealed posttranslational modification as a novel strategy for generating specificity for galactose. A comparison of the binding sites of galactose-specific and mannose-specific jacalin-like lectins also brought out the importance of the presence of aromatic residues for galactose binding. The extended binding site of jacalin provides an interesting situation with the primary binding site flanked by two secondary binding sites, one hydrophobic with variable geometry and the other exposed to solvent. A comparison between artocarpin and heltuba, both mannose-specific, demonstrates how change in the length of a loop can be used to alter specificity for oligosaccharides. A similar observation was earlier made in the cases of legume lectins as well [17]. Finally the structure of banana lectin, containing two primary binding sites leading to specificity for branched oligosaccharides, provides a framework for exploring the evolution of β-prism I fold lectins.

Acknowledgments

The work is supported by the Department of Science and Technology, Government of India. MV is a Distinguished Biotechnologist Awardee of the Department of Biotechnology, Government of India.

REFERENCES

1. Lis H & Sharon N (1998) *Chem. Rev.* **98**, 227–246.
2. Vijayan M & Chandra NR (1999) *Curr. Opin. Struct. Biol.* **9**, 707–714.
3. **Bettler HM, Loris R & Imberty A.** Lectin database available online at http://www.cermav.cnrs.fr/lectins.

4. Sankaranarayanan R, Sekar K, Banerjee R, Sharma V, Surolia A, & Vijayan, M (1996) *Nat. Struct. Biol.* **3**, 596–603.
5. Jeyaprakash AA, Gettha Rani P, Bhanuprakash Reddy G *et al.* (2002) *J. Mol. Biol.* **321**, 637–645.
6. Jeyaprakash AA, Katiyar S, Swaminathan, CP, Sekar K, Surolia A & Vijayan M (2003) *J. Mol. Biol.* **332**, 217–228.
7. Bourne Y, Astoul CH, Zamboni V, Peumans WJ, Menu-Bonaoniche L, van Damme EJ, Barre A & Rougue P (2002) *Biochem. J.* **364**, 173–180.
8. Jeyaprakash AA, Jayashree G, Mahanta SK *et al.* (2005) *J. Mol. Biol.* **347**, 181–188.
9. Pratap JV, Jeyaprakash AA, Gettha Rani P, Sekar K, Surolia A & Vijayan M (2002) *J. Mol. Biol.* **317**, 237–247.
10. Bourne Y, Zamboni V, Barre A, Peumans WJ, van Damme EJ, & Rouge P. (1999) *Struct. Fold. Des.* **7**, 1473–1482.
11. Jeyaprakash AA, Srivastava A, K, Surolia A & Vijayan M (2004) *J. Mol. Biol.* **338**, 757–770.
12. Singh DD, Saikrishnan K, Kumar P, Surolia A, Sekar K & Vijayan M (2005) *Glycobiology* **15**, 1025–1032.
13. Meagher JL, Winter HC, Ezell P, Goldstein IJ & Stuckey JA (2005) *Glycobiology* **15**, 1033–1042.
14. Lee X, Thompson A, Zhang Z *et al.* (1998) *J. Biol. Chem.* **273**, 6312–6318.
15. Bourne Y, Roig-Zamboni V, Barre A *et al.* (2004) *J. Biol. Chem.* **279**, 527–533.
16. del Sol FG, Nagano C, Cavada BS & Calvete JJ (2006) *J. Mol. Biol.* **353**, 574–583.
17. Prabu MM, Suguna K & Vijayan M (1999) *Proteins Struct. Funct. Genet.* **35**, 58–69.

CHAPTER 14
Protein–glycan interactions in immunoregulation: galectins as tuners of the inflammatory response

Gabriel A. Rabinovich and Marta A. Toscano

1. INTRODUCTION

Protein–glycan interactions mediate a variety of biological processes including cell trafficking, activation, differentiation, and survival. Galectins, a growing family of animal lectins defined by shared consensus amino acid sequence and affinity for β-galactose-containing oligosaccharides, are found on various cells of the immune system, and their expression is associated with the differentiation and activation state of these cells. Recent studies indicate that galectins participate both in immune cell homeostasis and inflammatory processes by acting as modulators of innate and adaptive immune responses. In addition, research over the past decade has identified a key role for galectins in the regulation of inflammatory conditions such as bowel inflammation, autoimmune ocular disease, and articular degenerative diseases. Furthermore, recent evidence indicates that galectins may function as soluble mediators employed by tumor cells to evade the immune response. Given the potential use of these proteins as novel anti-inflammatory agents or targets for immunomodulatory drugs, we will summarize here recent advances on the role of galectin–carbohydrate interactions in different aspects of immune cell physiology and their impact in pathological conditions including chronic inflammation, autoimmunity, and cancer.

2. PROTEIN–GLYCAN INTERACTIONS IN HEALTH AND DISEASE

Changes in glycosylation are often a hallmark of physiological or pathological conditions [1,2]. The glycosylation pattern of a cell is a code for cellular information and the understanding of this code is starting to emerge. The responsibility of decoding this information is attributed in part to a growing number of carbohydrate-binding proteins or lectins [3,4]. Under physiological conditions, protein–glycan interactions control critical immunological processes (see Fig. 1), including pathogen recognition, cell–cell and cell–matrix interactions, lymphocyte migration, apoptosis, and cytokine secretion [5–7]. In addition, research over the past decade has enabled the association of specific glycan structures and carbohydrate-binding proteins with specific disease states including chronic inflammation, autoimmunity and cancer [2]. In this context, understanding the "sugar code" (see Chapter 3) is a major challenge for glycobiologists, particularly in the context of physiological processes. This information will support the design of novel therapeutic approaches for the manipulation of inflammatory and neoplastic disorders. These therapeutic strategies should be aimed at blocking specific glycan formation, controlling the expression of specific glycan-processing enzymes (glycosyltransferases and/or glycosidases) or interfering with protein–glycan interactions.

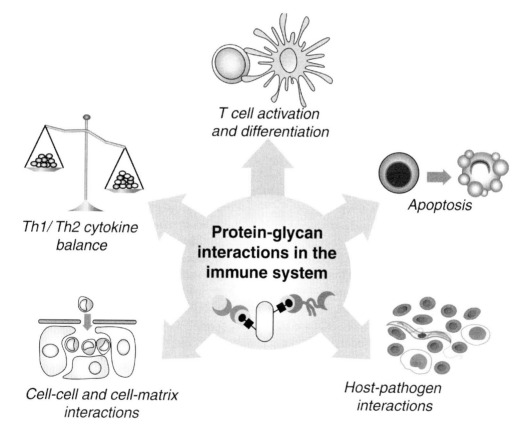

Figure 1. Protein–glycan interactions in immunoregulation. This scheme illustrates the influence of protein–glycan interactions in the control of critical immunological processes, including pathogen recognition, cell–cell and cell–matrix interactions, lymphocyte migration, apoptosis, activation, differentiation and cytokine secretion. Regulated glycosylation can affect immune functions by creating or masking ligands for endogenous glycan-binding proteins.

3. GALECTINS AS NOVEL IMMUNOMODULATORY AGENTS: BIOCHEMISTRY AND CELL BIOLOGY

Several families of carbohydrate-binding proteins or lectins have been implicated in a wide variety of immunological functions including first-line defense against pathogens, cell trafficking, and immune regulation [3,8]. These include, among others, the C-type lectins (collectins, selectins, mannose receptor, and others), S-type lectins (galectins), I-type lectins (siglecs and others), P-type lectins (phosphomannosyl receptors), pentraxins, and tachylectins [8]. This chapter will focus on the role of galectins, an evolutionarily conserved family of β-galactoside-binding proteins, in immunoregulation under physiological and pathological conditions.

Since the identification of discoidin-1 in the cellular slime mold *Dictyostelium discoideum* [9] and electrolectin in the electric organ tissue of the electric eel [10] early during the 1970s, the family of β-galactoside-binding lectins or galectins has received increasing attention. However it was only in the late 1990s that a growing body of experimental evidence emerged, illuminating a novel role for galectins in the regulation of physiological and pathological processes, particularly in the control of immune cell homeostasis and inflammation [11]. Members of the galectin family are defined by a conserved carbohydrate-recognition domain (CRD) with a canonical amino acid sequence and affinity for β-galactosides [12,13]. To date, 15 mammalian galectins have been identified, which can be subdivided into those that have one CRD and those that have two CRDs in tandem. In addition, galectin-3, a one-CRD galectin, is unique in that it contains unusual tandem repeats of short amino acid stretches fused onto the CRD [13,14].

Many galectins are either bivalent or multivalent with regards to their carbohydrate-binding activities. Some one-CRD galectins exist as dimers; two-CRD galectins have two carbohydrate-binding sites, and galectin-3 forms oligomers when it binds to multivalent carbohydrates [15]. Some galectins are distributed in a wide variety of tissues, where as others have a more restricted localization [14]. The expression of galectins is modulated during the activation and differentiation of immune cells and may be significantly altered under different physiological or pathological conditions [11].

Although galectins have the general properties of cytosolic proteins, they are secreted by non-classical (non-endoplasmic reticulum (ER)–Golgi) pathways [14]. Extracellularly, galectins can bind to cell surface glycoconjugates that contain suitable galactose-containing oligosaccharides. As galectins can bind either bivalently or multivalently, they can cross-link cell surface glycoconjugates which can trigger a cascade of transmembrane signaling events [15]. Through this mechanism, galectins can modulate processes that include apoptosis, cytokine secretion, cell adhesion, and migration [11]. Furthermore, intracellular galectins are engaged in processes that are essential for basic cellular functions such as pre-mRNA splicing, regulation of cell growth and cell cycle progression [14].

4. GALECTINS AS TUNERS OF THE ADAPTIVE IMMUNE RESPONSE

4.1 Galectins as regulators of T-cell survival

Immune cells are subject to cell death checkpoints at many stages during their lifespan to ensure proper development, maintain homeostasis, and prevent disease. Research over the past decade has demonstrated that galectins play a pivotal role in the regulation of T-cell survival [11] (see Fig. 2).

Galectin-1

A growing body of experimental evidence indicates that galectin-1, a 14.5 kDa member of the galectin family, induces cell growth inhibition and promotes apoptosis of developing thymocytes and peripheral T-lymphocytes [16-26]. Different glycoconjugates on the surface of activated T-cells appear to be primary

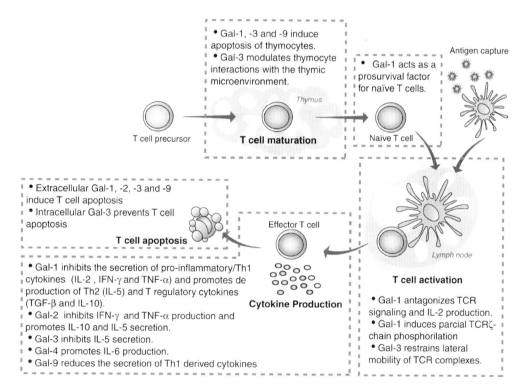

Figure 2. The role of galectins during the development of T-cell-mediated immune responses. T-lymphocytes are main effectors of the adaptive immune response. This scheme illustrates the influence of different members of the galectin family on different events of the dynamics of T-lymphocytes, including T-cell maturation, activation, differentiation, cytokine secretion, and apoptosis. Under distinct physiological or pathological conditions, different members of the galectin family may provide inhibitory or stimulatory signals to control immune cell homeostasis and regulate inflammation following an antigenic challenge.

receptors for galectin-1, including CD45, CD43, CD7, and CD2 [25,26]. Interestingly, galectin-1 binding to T-cells results in a marked redistribution of receptors into segregated membrane microdomains [30].

Susceptibility to galectin-1-induced cell death is tightly controlled by specific glycosyltransferases that are regulated during T-cell development, activation, and differentiation [27,28]. Galvan and colleagues [27] demonstrated that CD45-positive T-cells lacking the core-2-β-1,6-N-acetylglucosaminyltransferase (C2GnT) are resistant to galectin-1-induced cell death. This enzyme is responsible for creating branched structures on O-glycans of T-cell surface glycoproteins, such as CD45 [27]. In addition, other glycosyltransferases can also act to reduce galectin binding to indirectly or directly masking galectin saccharide ligands. In this sense, addition of α2,6-linked sialic acids to lactosamine units by the ST6Gal-I sialyltransferase has been shown to block galectin-1 binding and cell death [28]. Interestingly, recent findings suggest that altered glycosylation of T-cells during infection with human immunodeficiency virus type 1 (HIV-1) increases their susceptibility to galectin-1-induced cell death, suggesting that this apoptotic pathway may contribute to HIV-1-induced immunosuppression [29].

The signal transduction events triggered by galectin-1 are still controversial. It has been demonstrated that galectin-1 binding to primary T-cells triggers the activation of specific transcription factors (i.e. AP-1 and NFAT) [21,30], cytochrome c release [22], and modulation of caspases [22]. However, another study shows that apoptosis induced by galectin-1 in a T-cell line is not dependent on the activation of caspase-3 or on cytochrome c release [23]. Interestingly, a recent study proposed a model in which galectin-1 induces the release of ceramide, which in turn promotes downstream events including decrease of Bcl-2 protein, depolarization of the mitochondria, and activation of the caspase-9 and -3 [31]. All downstream events require the presence of two tyrosine kinases, p56^{lck} and ZAP-70 [32]. On the other hand, another study showed that galectin-1 promotes phosphatidylserine exposure in different cell types (including activated neutrophils and T-cell lines) with no apparent signs of apoptosis [33]. Therefore, it seems evident that galectin-1 may trigger different death pathways or different apoptosis end-points in different cell types.

In this regard, using a human allogeneic T-cell model, we have demonstrated that alternative mechanisms may operate to achieve T-cell immunosuppression including inhibition of proinflammatory cytokines in the non-apoptotic cell population [20]. Most recently Endharti and colleagues [34] demonstrated that, in contrast to the pro-apoptotic role of galectin-1 on activated T-cells, secretion of this protein by stromal cells is capable of supporting the survival of naïve T-cells without promoting proliferation [34]. Thus, galectin-1 might trigger different signals (i.e. apoptosis or survival) depending on a number of factors including the activation state of the cells, the spatiotemporal expression of specific glycosyltransferases and/or the nature of the target cell (primary cell cultures or immortalized T-cell lines).

While these findings suggest a pivotal role for galectin-1 in the maintenance and re-establishment of T-cell tolerance and homeostasis, targeted disruption of the galectin-1 gene in knockout mice resulted in the absence of major spontaneous abnormalities, suggesting potential redundancy between different members of the galectin family [13,14]. However, in contrast to these previous assumptions, recent work clearly indicates that galectin family members are not redundant and that there are subtle, but functionally relevant differences in the specificity and function of individual members of the galectin family in the regulation of inflammatory responses [13,14].

Galectin-2

Galectin-2, a member of the galectin family structurally related to galectin-1, also has the ability to promote T-cell apoptosis [35]. Investigation of the molecular mechanisms and intracellular pathways involved in the pro-apoptotic activity of this protein revealed binding to cell surface β-integrins, activation of caspases-3 and -9, enhanced cytochrome *c* release, disruption of the mitochondrial membrane potential and increase of the Bax/Bcl-2 ratio [35].

Galectin-3

Interestingly, galectin-3 has been shown to act in a dual manner either protecting cells from apoptosis or promoting apoptosis depending on whether the protein acts intracellularly [36] or is added exogenously [25,37]. Yang and colleagues [36] demonstrated that T-cell transfectants overexpressing galectin-3 display higher growth rates than control transfectants and are protected from apoptosis induced by a variety of agents including Fas ligation and staurosporine [36]. Interestingly, recent findings [37] showed that recombinant galectin-3 can signal apoptosis of human T-cells through binding to cell surface glycoconjugates resulting in activation of the mitochondrial pathway, cytochrome *c* release, and caspase-3, but not caspase-10 activation [37]. Furthermore, Hahn *et al.* [26] suggested a functional cross-talk between intracellular and extracellular galectins in the regulation of T-cell death; the authors demonstrated that galectin-1-induced cell death is inhibited by intracellular expression of galectin-3 [23].

Galectin-9

Similarly, galectin-9 induces apoptosis of developing thymocytes in a carbohydrate-dependent manner [38]. Furthermore, this two-CRD lectin also promotes death of peripheral $CD4^+$ and $CD8^+$ T-cells through a Ca^{2+}-calpain–caspase-1 signaling pathway [39]. Interestingly, recent findings highlight a novel role for galectin-9 as a binding partner of the Th1-specific molecule Tim-3 [40]. Through binding to Tim-3 on the surface of Th1 cells, galectin-9 can trigger apoptosis and immunosuppression.

Galectins-7, -8, and -12

Other members of the family, including galectins-7, -8, and -12 also regulate cell cycle progression, proliferation, and survival [41–43]; however their impact in the immune system has not yet been ascertained.

4.2 Galectins as modulators of T-cell activation, differentiation and the immunological synapse

In addition to their role in the modulation of T-cell survival and proliferation, galectins have been shown to modulate T-cell activation and differentiation and to influence the immunological synapse (see Fig. 2).

Galectin-1

Vespa and colleagues [44] demonstrated that galectin-1 can act as a potent modulator of T-cell receptor (TCR) signals and antagonize TCR-induced interleukin 2 (IL-2) production in a murine T-cell hybridoma clone [44]. Interestingly, the same group further demonstrated that galectin-1 induces partial TCR ζ-chain phosphorylation and is able to antagonize full TCR responses including the production of IL-2 [45].

Regarding the B-cell compartment, Schiff and colleagues [46] demonstrated that galectin-1 can act as a stromal ligand of the pre-B-cell receptor (pre-BCR) and contribute to synapse formation between pre-BCR and stromal cells [46]. Thus, by acting at early events of TCR or BCR signaling, galectin-1 can also control lymphocyte development and activation.

Galectin-3

In a very elegant study, Demetriou and colleagues [47] reported that galectin-3 may play a role in restricting TCR complex-initiated signal transduction. The authors hypothesized that galectin-3 might form multivalent complexes with N-glycans on the TCR,thereby restraining the lateral mobility of TCR complexes [47]. This effect was abrogated in mice deficient in β1,6 N-acetylglucosaminiltransferase (Mgat5), a crucial enzyme in the N-glycosylation pathway. These mice showed enhanced delayed-type hypersensitivity responses and increased susceptibility to autoimmunity [47]. Thus, galectin-3 may influence T-cell interactions with antigen-presenting cells and control T-cell activation by negatively regulating the immunological synapse [47].

Galectin-4

Hokama and colleagues [48] recently found that galectin-4 plays a key role in CD4$^+$ T-cell activation under specific inflammatory conditions [48]. Galectin-4-mediated stimulation of intestinal T-cells was reflected by a marked increase in IL-6 production, potentiating the severity of inflammatory bowel disease [48].

4.3 Galectins as regulators of cytokine secretion

Accumulating evidence indicates a pivotal role of galectins in the control of the cytokine balance, suggesting that these proteins may influence a variety of physiological processes including tissue remodeling, wound healing, and fetomaternal tolerance. In addition, the ability to control cytokine production endows these sugar-binding proteins with the capacity to positively or negatively interfere in pathological processes such as autoimmunity, chronic inflammation, allergy, and cancer (see Fig. 2).

Galectin-1

Galectin-1 has been shown to block secretion of Th1 and proinflammatory cytokines *in vitro*, including IL-2, interferon gamma (IFNγ) and tumor necrosis factor alpha (TNFα) [44,49]. In addition, studies in experimental models of chronic inflammation and autoimmunity showed the ability of galectin-1 to skew the balance towards a Th2-type cytokine profile, with decreased levels of IFNγ and increased secretion of IL-5 or IL-10 by pathogenic T-cells [50–53]. In addition, recent evidence indicates that, upon viral infection, galectin-1 can induce dendritic cell secretion of IL-6 [54]. Furthermore, van der Leij and colleagues [55] recently reported a marked increase in IL-10 mRNA and protein levels in non-activated and activated $CD4^+$ and $CD8^+$ T-cells following exposure to recombinant galectin-1 [55].

Galectin-2

Sturm and colleagues demonstrated that galectin-2 can modulate T-cell-derived cytokines *in vitro* and shift the balance towards a Th2 profile [25]. In addition it has been shown that this protein can regulate lymphotoxin-α secretion, which can in turn affect the degree of inflammation during myocardial infarction [56]. Therefore, galectin-2, as well as galectin-1, can modulate the balance of pro- and anti-inflammatory cytokines.

Galectin-3

Galectin-3 has been shown to act in most cases as a "proinflammatory cytokine," as has been clearly demonstrated by the attenuated inflammatory response in galectin-3 knockout mice [57]. In addition, this lectin suppresses Th2-mediated allergic inflammation by blocking IL-5 secretion by antigen-specific T-cell lines, suggesting their potential use in the treatment of allergic disorders [58]. Furthermore, we have recently demonstrated that galectin-3 is a critical intracellular mediator of IL-4-induced survival and differentiation of B cells into a memory phenotype [59].

Galectin-4

Recent evidence indicates that galectin-4 produced by epithelial cells induces IL-6 production by $CD4^+$ T-cells, an effect mediated by a protein kinase Cθ-associated pathway [48], suggesting that this two-CRD galectin functions as a T-cell activator by favoring secretion of proinflammatory cytokines.

Taken together, these data suggest that under distinct physiological or pathological conditions, different members of the galectin family may provide inhibitory or stimulatory signals to control immune cell homeostasis and regulate inflammation following an antigenic challenge. The mechanisms underlying galectin-mediated regulation of cytokine production and the association of this effect with the regulation of T-cell apoptosis and differentiation still remains to be elucidated.

5. GALECTINS AS MODULATORS OF INNATE IMMUNE RESPONSES

In addition to their effects in adaptive immunity, galectins have also been shown to influence the development of innate immune responses including leukocyte extravasation to inflamed tissues, chemotaxis, survival, phagocytosis, and respiratory burst.

5.1 Galectin-1

We have demonstrated, using the rat hind paw edema test, that galectin-1 suppresses the acute inflammatory response and inhibits neutrophil extravasation [60]. Furthermore, this lectin suppresses arachidonic acid release and modulates the arginine metabolism in activated macrophages [60,61]. Most recently, Perretti and colleagues showed that galectin-1 can inhibit neutrophil chemotaxis and transendothelial cell migration [62].

5.2 Galectin-3

In contrast to the anti-inflammatory effects of galectin-1, studies of acute peritonitis in galectin-3-deficient mice provided significant support for the proinflammatory role of this lectin during acute inflammatory responses [57,63]. After intraperitoneal injection of thyoglicolate, galectin-3-deficient mice had significantly reduced numbers of recoverable granulocytes compared to wild-type animals [57,63]. Interestingly, Karlsson and colleagues [64,65] showed that both galectin-1 and galectin-3 are able to induce activation of the superoxide-producing NADPH oxidase at similar levels in primed neutrophils [64,65]. Furthermore, galectin-3 induces neutrophil adhesion to laminin [66] and endothelial cells [67] *in vitro* following an immunological challenge.

In addition, Liu and colleagues highlighted a critical role for galectin-3 in phagocytosis by macrophages [68]. Compared with wild-type macrophages, galectin-3-deficient cells exhibited reduced phagocytic capacity [68]. In addition, Liu's work clearly demonstrated that galectin-3 promotes chemotaxis of human monocytes, through a Pertussis toxin-sensitive G-protein-mediated pathway [69]. In this context, we have recently shown that galectin-3 acts in concert with soluble fibrinogen to regulate neutrophil activation, degranulation and survival through alternative activation of mitogen-activated protein kinases (ERK1/2 or p38)-mediated pathways [70].

5.3 Galectins-8 and -9

Although not studied in detail as galectin-1 and galectin-3, other galectins also modulate innate immune responses. For example, galectin-8 can modulate neutrophil functions related to microbial killing [71]. On the other hand, galectin-

9 (so-called "ecalectin") can act as an eosinophil-specific T-cell-derived chemoattractant [72].

Finally, interesting findings revealed a critical role for galectin-9 and galectin-3 in host–pathogen interactions during *Leishmania major* infection [73,74].

Taken together, these observations indicate that galectins may act at different levels of the inflammatory cascade to regulate innate immune responses. In addition, these proteins may act as "danger" signals during host–pathogen interactions. Whether galectins might also modulate the cross-talk between innate and adaptive immune responses still remains to be elucidated.

6. GALECTINS IN IMMUNOPATHOLOGY

6.1. Autoimmunity and chronic inflammation

Galectins have been shown to play a key role in the regulation of T-cell-mediated inflammatory disorders mainly by targeting pathogenic effector T-cells [11]. We will summarize the clinical and immunological consequences of preventive or therapeutic administration of galectins in chronic inflammatory disorders and autoimmunity (see Table 1).

Table 1 Impact of galectins in experimental models of chronic inflammation and autoimmunity

Experimental models	Strategies used	Clinical outcome	Potential mechanisms involved
Experimental autoimmune myasthenia gravis (EAMG)	Injection of electrolectin to rabbits	Complete clinical recovery and delayed onset	No changes in circulating autoantibodies or modifications at the muscular level
Experimental autoimmune encephalomyelitis in (EAE)	Prophylactic administration of Gal-1 to MBP-immunized Lewis rats	Prevention of clinical and histopathological signs of the disease	N.D (Blockade of activation of pathogenic T-cells?)
Collagen-induced arthritis (CIA)	Gal-1 gene therapy and protein administration to DBA/1 mice	Suppression of clinical and histopathological manifestations	Increased IL-5 and decreased IFNγ production
			Increased T-cell susceptibility to activation induced cell death
Concanavalin A-induced hepatitis	Prophylactic administration of Gal-1 in BALB/c mice	Prevention of liver injury and T-helper cell liver infiltration	Suppressed TNFα and IFNγ production
			Increased apoptosis of activated T-cells
Inflammatory bowel disease (TNBS-induced colitis)	Prophylactic and therapeutic administration of Gal-1 in BALB/c mice	Suppression of clinical and histopathological manifestations	Reduced ability of mucosal T-cells to produce IFNγ
			Increased number of apoptotic T-cells within mucosal tissue

Interphotoreceptor retinoid-binding protein (IRBP)-induced experimental autoimmune uveitis (EAU)	Administration of Gal-1 during the afferent or efferent phase of EAU in B10.RIII mice	Suppression of ocular inflammatory disease	Ability to counteract Th1-mediated responses
			Promotion of a Th2 and T-regulatory (Tr1)-mediated anti-inflammatory response
Nephrotoxic nephritis (induced by anti-glomerular basement membrane serum)	Gal-1, Gal-3, Gal-9 administration to Wistar Kyoto rats	Clinical recovery	Gal-9 induces apoptosis of activated $CD8^+$ cells
			Gal-1 and Gal-3 block the accumulation of macrophages
Graft vs. host disease	Gal-1 administration to mice	Increased host survival following allogeneic hematopoietic stem cell transplant	Reduced production of IFNγ and IL-2
			Reduced alloreactivity
Experimental autoimmune encephalomyelitis	Gal-9 injection to MOG-immunized C57BL/6 mice	Reduced severity and mortality	Selective loss of IFNγ-producing cells. Apoptosis of $Tim-3^+$ Th1 cells
	siRNA *gal-9* to PLP-immunized SJL mice	Increased severity of the disease	
Inflammatory bowel disease (TNBS-colitis)	Epithelial-derived Gal-4	Exacerbates intestinal inflammation	Stimulates IL-6 production by $CD4^+$ T-cells

Gal, galectin; MBP, myelin-basic protein; MOG, myelin oligodendrocyte glycoprotein; PLP, myelin proteolipid protein; siRNA, small interfering RNA; TNBS, 2,4,6-trinitrobenzene sulfonic acid.

Experimental autoimmune myasthenia gravis (EAMG)

As early as 1983, Levi and colleagues [75] reported the preventive and therapeutic effects of electrolectin, an endogenous galectin from the fish *Electrophorus electricus* in EAMG induced by immunization of rabbits with purified acetylcholine receptors [83]. The administration of electrolectin to myasthenic rabbits led to clinical recovery; however this effect was not accompanied by any significant change in the level of circulating autoantibodies or modifications at the muscular level [75]. The authors suggested that electrolectin might play a role in the regulation of immune tolerance to self-antigens.

Experimental autoimmune encephalomyelitis (EAE)

Offner and colleagues demonstrated that galectin-1 prevented the development of clinical and histopathological signs of EAE in Lewis rats [76]. Although the mechanisms of action of galectin-1 were not investigated in this study, the authors proposed that galectin-1 might block the activation of encephalitogenic T-cells [76]. These pioneer studies in EAMG and EAE prompted our research to investigate *in vitro* and *in vivo* the molecular mechanisms involved in the immunoregulatory activity of galectin-1. Most recently, Kuchroo et al. demonstrated, using the EAE model in C57BL/6 and SJL mice, that galectin-9 binds and kills Th1 cells *in vivo* [40].

Collagen-induced arthritis (CIA)

We have demonstrated, using gene therapy strategies, that a single injection of syngeneic fibroblasts engineered to secrete galectin-1 at the day of the disease onset was able to abrogate clinical and histopathological manifestations of CIA, an experimental model of rheumatoid arthritis in DBA/1 mice [50]. Investigation of the mechanisms involved in this process revealed a shift from a Th1- to a Th2-polarized immune response [50]. This effect was manifested by reduced levels IFNγ and increased levels of IL-5 in draining lymph nodes from mice treated by gene therapy with galectin-1 [50]. In addition, sera from galectin-1-treated mice showed reduced levels of anti-collagen type II IgG2a and increased levels of anti-collagen type II IgG1 antibodies. In addition, lymph node cells from mice engaged in the galectin-1 gene therapy protocol had increased susceptibility to antigen-induced apoptosis. This study provided a strong correlation between the apoptotic properties of galectin-1 *in vitro* and its therapeutic potential *in vivo*.

Concanavalin A-induced hepatitis

Similarly, Santucci and colleagues [52] found that galectin-1 treatment prevented liver injury and T-helper cell liver infiltration in a model of concanavalin A-induced hepatitis [52]. Using *in vitro* and *in vivo* experiments the authors demonstrated the protective effects of galectin-1 in this model and confirmed that galectin-1 acts *in vivo* by promoting selective elimination of antigen-activated T-cells [61].

Inflammatory bowel disease

The preventive and therapeutic administration of galectin-1 has clearly been demonstrated in the hapten model of inflammatory bowel disease induced by intrarectal delivery of 2,4,6-trinitrobenzene sulfonic acid (TNBS) [51]. Galectin-1 treatment induced a reduction of the number of TNBS-activated mucosal T-cells and a decreased secretion of pro-inflammatory and Th1 cytokines [51].

Experimental autoimmune uveitis (EAU)

Given the potential role of galectin-1 in the maintenance of immune privilege in organs such as the eye, we have recently investigated the immunoregulatory effects of this protein in EAU, a Th1-mediated model of retinal disease [77]. Treatment with galectin-1 either early or late during the course of EAU was sufficient to suppress clinical ocular pathology, inhibit leukocyte infiltration and counteract pathogenic Th1 cells [77]. Administration of galectin-1 ameliorated retinal inflammation by skewing the uveitogenic response towards non-pathogenic Th2 or T regulatory (IL-10 and TGF-β)-mediated anti-inflammatory responses [77]. Interestingly, galectin-1 treatment generated an expansion of IL-10-producing regulatory T-cells (Tr1) in lymph nodes from treated mice. In addition, increased levels of apoptosis were detected in lymph nodes from mice treated with recombinant galectin-1 during the efferent phase of the disease [77]. These results highlight the ability of this endogenous lectin to counteract Th1-mediated responses through different, but potentially overlapping anti-inflammatory mechanisms. In addition, we found a striking correlation between the levels of anti-retinal galectin-1 autoantibodies in sera from uveitic patients and the severity of autoimmune retinal inflammation [78].

Immune-mediated renal disease

Regarding other members of the galectin family, it has been found that galectin-9 induces apoptosis of activated $CD8^+$ T-cells, while galectins-1 and -3 inhibit the accumulation of macrophages in the renal glomeruli in nephrotoxic serum nephritis, an immune-mediated renal disease in Kyoto rats [79].

6.2 Graft versus host disease (GVHD)

Baum and colleagues [53] investigated the efficacy of galectin-1 in a murine model of GVHD and found that 68% of galectin-1-treated mice survived, compared to 3% of vehicle-treated mice [53]. Galectin-1 treatment reduced inflammatory infiltrates in affected tissues and significantly improved reconstitution of normal splenic architecture following transplant. Similar to findings in autoimmune models, Th1 cytokines were markedly reduced, while production of Th2 cytokines was similar between galectin-1-treated and control animals [53]. These findings demonstrated that galectin-1 therapy is capable of suppressing GVHD without compromising engraftment or immune reconstitution following allogeneic hematopoietic stem cell transplant.

6.3 Galectins in cancer immunity

The association between galectin-1 expression in different tumor types and the aggressiveness of these tumors [14,80], prompted us to investigate the role of galectin-1 in tumor-immune escape. We hypothesized that tumor cells may impair T-cell effector functions through secretion of galectin-1 and that this mechanism may contribute in tilting the balance towards an immunosuppressive environment at the tumor site. By a combination of *in vitro* and *in vivo* experiments, we established a link between galectin-1-mediated immunoregulation and its contribution to tumor-immune escape (see Fig. 3). Blockade of the inhibitory

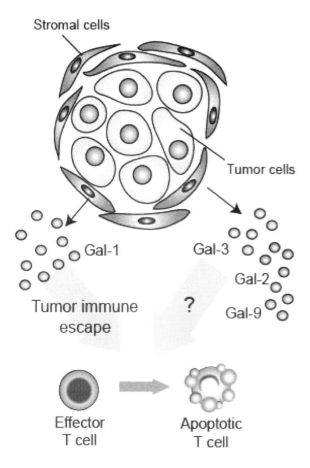

Figure 3. The role of galectin-1 in tumor-immune escape. Despite the existence of tumor-specific immune cells, most tumors have devised multiple strategies for evading the immune system. Tumors can evade the effector immune response by secreting immunosuppressive cytokines and soluble inhibitory factors including galectin-1. Galectin-1 contributes to the immune escape of tumors by modulating survival of effector T-cells and by skewing the balance of the immune response towards a Th2 anti-inflammatory cytokine profile. Other galectins, including galectin-2, galectin-3, and galectin-9, also induce T-cell apoptosis and modulate cytokine production; however their contributions to tumor immune escape *in vivo* have not yet been demonstrated.

effects of galectin-1 within tumor tissue resulted in reduced tumor mass and stimulated the generation of a tumor-specific T-cell response *in vivo* [24].

Supporting our results, Le and colleagues [81] have recently identified galectin-1 as a molecular link between tumour hypoxia and tumor-immune privilege. The authors found a strong inverse correlation between galectin-1 expression and CD3 staining in tumour sections corresponding to head and neck squamous cell carcinoma patients [81]. Taken together, these results support the concept that galectin-1 contributes to immune privilege of tumors by modulating survival and differentiation of effector T-cells.

6.4 Galectins in allergic disorders

Glycans and glycan-binding proteins may also play a role during allergic reactions and chronic airway inflammation [82]. The ability of galectin-3 to downregulate IL-5 secretion by eosinophils and T-cell lines [58] prompted Lahoz and colleagues to investigate the potential role of galectin-3 gene therapy in chronic airway inflammation [83]. The authors showed that intranasal instillation of a plasmid-encoding galectin-3 resulted in an improvement of clinical and immunological manifestations of chronic airway inflammation, including normalization of the eosinophil count, attenuation of hyperresponsiveness to metacholine, and reduction in lung collagen [83]. This result suggests an ability of galectin-3 to counteract pathogenic Th2 responses *in vivo*. Interestingly, however, a recent study indicates that galectin-3-deficient mice develop higher Th1 responses in a model of airway inflammation *in vivo* [84]. In fact, Liu and colleagues [84] showed that galectin-3 null mutant mice developed significantly less airway hyperresponsiveness and increased levels of Th1-mediated reactions following antigenic challenge in a murine model of asthma [84]. The discrepancies between these two studies might be possibly explained by the different strategies used by the authors to evaluate the role of galectin-3 *in vivo* (i.e. *gal-3* gene therapy versus induction of inflammation in $gal-3^{-/-}$ mice). Regarding other members of the galectin family, the ability of galectin-9 to specifically chemoattract eosinophils [72] suggests that this protein might also play a role in allergic reactions and chronic airway inflammation.

7. CONCLUSIONS AND FUTURE DIRECTIONS

Over the past decade galectins have emerged as novel regulators of immune cell homeostasis and inflammation. In addition to their role in regulating innate and adaptive responses, these glycan-binding proteins might be crucial in a diverse set of diseases, including autoimmunity, chronic inflammation, allergy, and cancer. In the present chapter we focused our attention on this fascinating family of animal lectins, exploring their possible functions under physiological and pathological conditions. The accumulating evidence for the multiple pleiotropic mechanisms regulated by galectins should indeed prompt further exploration of their potential use in the development of therapies for several autoimmune disorders, chronic inflammation, GVHD, and allergy. In addition, given the influence of galectins in different events of tumor escape and metastasis, a current challenge is the design of specific and potent galectin inhibitors as potential anticancer agents. The emergence of novel high-throughput screening platforms, including glycan arrays

[85], will facilitate the characterization and profiling of the specificity of a diverse range of galectins and the identification of more specific and potent inhibitors of galectin–carbohydrate interactions.

Acknowledgments

We apologize that we could not cite many excellent studies because of space limitations. Work in the authors' laboratory is supported by grants from Mizutani Foundation for Glycoscience (Japan), Agencia de Promoción Científica y Tecnológica (PICT 2003-05-13787), Fundación Sales, and University of Buenos Aires (UBACYT-M091). G.A.R. is a member of the National Council Research (CONICET). We thank J.M. Ilarregui, G. Bianco, L. Campagna, and D. Croci-Russo for continuous support.

REFERENCES

1. Lowe JB (2001) *Cell* **104**, 809–812.
2. Dube DH & Bertozzi CR (2005) *Nat. Rev. Drug. Discov.* **4**, 477–488.
3. Sharon N & Lis H (2004) *Glycobiology* **14**, 53R-62R.
4. Gabius HJ (2006) *Crit. Rev. Immunol.* **26**, 43–80.
5. Daniels MA, Hogquist KA & Jameson SC (2002) *Nat. Immunol.* **3**, 903–910.
6. Morgan R , Gao G, Pawling J, Dennis JW, Demetriou M & Li B (2004) *J. Immunol.* **173**, 7200–7208 .
7. Collins BE, Smith BA, Bengtson P & Paulson JC (2006) *Nat. Immunol.* **7**, 199–206.
8. Kilpatric DC (2002) *Biochem. Biophys. Acta* **1572**, 187–197.
9. Barondes SH (1986) In: *The Lectins*, pp. 467–491. Edited by IE Liener, N Sharon and IJ Goldstein. Academic Press, Orlando, FL.
10. Teichberg VI, Silman I, Beitsch DD & Resheff G (1975) *Proc. Natl Acad. Sci. USA* **72**, 1383–1387.
11. Rabinovich GA, Baum LG, Tinari N et al. (2002) *Trends Immunol.* **23**, 313–320.
12. Cooper DNW (2002) *Biochim. Biophys. Acta* **1572**, 209–231.
13. Leffler H, Carlsson S, Hedlund M, Qian Y & Poirier F (2004) *Glycoconj. J.* **19**, 433–440.
14. Liu FT & Rabinovich GA (2005) *Nat. Rev. Cancer* **5**, 29–41.
15. Brewer CF (2002) *Biochim. Biophys. Acta* **1572**, 255–262.
16. Blaser C, Kaufmann M, Muller C et al. (1998) *Eur. J. Immunol.* **28**, 2311–2319.
17. Perillo NL, Uittenbogaart CH, Nguyen JT & Baum LG (1997) *J. Exp. Med.* **97**, 1851–1858.
18. Perillo NL, Pace KE, Seilhamer JJ & Baum LG (1995) *Nature* **378**, 736–739.
19. Rabinovich GA, Iglesias MM, Modesti NM et al. (1998) *J. Immunol.* **160**, 4831–4840.
20. Rabinovich GA, Ramhorst RE, Rubinstein N et al. (2002) *Cell Death Diff.* **9**, 661–670.
21. Rabinovich GA, Alonso CR, Sotomayor CE, Durand S, Bocco JL & Riera CM (2000) *Cell Death Diff.* **7**, 747–753.
22. Matarrese P, Tinari A, Mormone E et al. (2005) *J. Biol. Chem.* **280**, 6969–6985.
23. Hahn HP, Pang M, He J et al. (2004) *Cell Death Diff.* **11**, 1277–1286.
24. Rubinstein N, Alvarez M, Zwirner NW et al. (2004) *Cancer Cell* **5**, 241–251.
25. Stillman BN, Hsu DK, Pang M et al. (2006) *J. Immunol.* **176**, 778–789.
26. Pace KE, Hahn HP, Pang M, Nguyen JY & Baum LG (2000) *J. Immunol.* **165**, 2331–2334.
27. Galvan M, Tsuboi S, Fukuda M & Baum LG (2000) *J. Biol. Chem.* **275**, 16730–16737.
28. Amano M, Galvan M, He J & Baum LG (2003) *J. Biol. Chem.* **278**, 7469–7475.
29. Lanteri M, Giordanengo V, Hiraoka N et al. (2003) *Glycobiology* **13**, 909–918.
30. Walzel H, Blach M, Hirabayashi J, Arata Y, Kasai KI & Brock J (2002) *Cell Signal.* **14**, 861–868.
31. Ion G, Fajka-Boja R, Kovacs F et al. (2006) *Cell Signal.* (2006) (in press).

32. Ion G, Fajka-Boja R, Toth GK et al. (2005) Cell Death Diff. **12**, 1145–1147.
33. Dias-Baruffi M, Zhu H, Cho M, Karmakar S, McEver RP & Cummings RD (2003) J. Biol. Chem. **278**, 41282–41293.
34. Endharti AT, Zhou YW, Nakashima I & Suzuki H (2005) Eur. J. Immunol. **35**, 86–97.
35. Sturm A, Lensch M, Andre S et al. (2004) J. Immunol. **173**, 3825–3837.
36. Yang RY, Hsu DK & Liu FT (1996) Proc. Natl Acad. Sci. USA **93**, 6737–6742.
37. Fukumori T, Takenaka Y, Yoshii T et al. (2003) Cancer Res. **63**, 8302–8311.
38. Wada J, Ota K, Kumar A, Wallner EI & Kanwar YS (1997) J. Clin. Invest. **99**, 2452–2461.
39. Kashio Y, Nakamura K, Abedin MJ et al. (2003) J. Immunol. **170**, 3631–3636.
40. Zhu C, Anderson AC, Schubart A et al. (2005) Nat. Immunol. **6**, 1245–1252
41. Kuwabara I, Kuwabara Y, Yang RY et al. (2002) J. Biol. Chem. **277**, 3487–3497.
42. Arbel-Goren R, Levy Y, Ronen D & Zick Y (2005) J. Biol. Chem. **280**, 19105–19114.
43. Yang RY, Hsu DK, Yu L, Ni J & Liu FT (2001) J. Biol. Chem. **276**, 20252–20260.
44. Vespa GN, Lewis LA, Kozak KR et al. (1999) J. Immunol. **162**, 799–806.
45. Chung CD, Patel VP, Moran M et al. (2000) J. Immunol. **165**, 3722–3729.
46. Gauthier L, Rossi B, Roux E, Termine E & Schiff C (2002) Proc. Natl Acad. Sci. USA **99**, 13014–13019.
47. Demetriou M, Granovsky M, Quaggin S & Dennis JW (2001) Nature **409**, 733–739.
48. Hokama A, Mizoguchi E, Sugimoto K et al. (2004) Immunity **20**, 681–693.
49. Rabinovich GA, Ariel A, Hershkoviz R, Hirabayashi J, Kasai KI & Lider O (1999) Immunology **97**, 100–106.
50. Rabinovich GA, Daly G, Dreja H et al. (1999) J. Exp. Med. **190**, 385–397.
51. Santucci L, Fiorucci S, Rubinstein N et al. (2003) Gastroenterology **124**, 1381–1394.
52. Santucci L, Fiorucci S, Cammilleri F, Servillo G, Federici B & Morelli A (2000) Hepatology **31**, 399–406.
53. Baum LG, Blackall DP, Arias-Magallano S et al. (2003) Clin. Immunol. **109**, 295–307.
54. Levroney EL, Aguilar HC, Fulcher JA et al. (2005) J. Immunol. **175**, 413–420.
55. van der Leij J, van den Berg A, Blokzijl T et al. (2004) J. Pathol. **204**, 511–518.
56. Ozaki K, Inoue K, Sato H et al. (2004) Nature **429**, 72–75.
57. Hsu DK, Yang RY, Pan Z et al. (2000) Am. J. Pathol. **156**, 1073–1083.
58. Cortegano I, del Pozo V, Cardaba B et al. (1998) J. Immunol. **161**, 385–389.
59. Acosta-Rodríguez EV, Montes CL, Motran CC et al. (2004) J. Immunol. **172**, 493–502.
60. Rabinovich GA, Sotomayor CE, Riera CM, Bianco I & Correa SG (2000) Eur. J. Immunol. **30**, 1331–1339.
61 Correa SG, Sotomayor CE, Aoki MP, Maldonado CA & Rabinovich GA (2003) Glycobiology **13**, 119–128.
62. La M, Cao TV, Cerchiaro G et al. (2003) Am. J. Pathol. **63**, 1505–1515.
63. Colnot C, Ripoche MA, Milon G, Montagutelli X, Crocker PR & Poirier F (1998) Immunology **94**, 290–296.
64. Karlsson A, Follin P, Leffler H & Dahlgren C (1998) Blood **91**, 3430–3438.
65. Almkvist J, Dahlgren C, Leffler H & Karlsson A (2002) J. Immunol. **168**, 4034–4041.
66. Kuwabara I & Liu FT (1996) J. Immunol. **156**, 3939–3944.
67. Sato S, Ouellet N, Pelletier I, Simard M, Rancourt A & Bergeron MG (2002) J. Immunol. **168**, 1813–1822.
68. Sano H, Hsu DK, Apgar JR et al. (2003) J. Clin. Invest. **112**, 389–397.
69. Sano H, Hsu DK, Yu L et al. (2000) J. Immunol. **165**, 2156–2164.
70. Fernández GC, Ilarregui JM, Rubel CJ et al. (2005) Glycobiology **15**, 519–527.
71. Nishi N, Shoji H, Seki M et al. (2003) Glycobiology **13**, 755–763.
72. Matsumoto R, Hirashima M, Kita H & Gleich GJ (2002) J. Immunol. **168**, 1961–1967.
73. Pelletier I, Hashidate T, Urashima T et al. (2003) J. Biol. Chem. **278**, 22223–22230.
74. Pelletier I & Sato S (2002) J. Biol. Chem. **277**, 17663–17670.
75. Levi G, Tarrab-Hazadai R & Teichberg VI (1983) Eur. J. Immunol. **13**, 500–507.
76. Offner H, Celnik B, Bringman TS, Casentini-Borocz D, Nedwin GE & Vandenbark AA (1990) J Neuroimmunol **28**, 177–184.
77. Toscano MA, Commodaro AG, Ilarregui JM et al. (2006) J. Immunol. **176**, 6323–6332.
78. Romero MD, Muino JC, Bianco GA et al. (2006) Invest. Ophthalmol. Vis. Sci. **47**, 1550–1556.

79. Tsuchiyama Y, Wada J, Zhang H *et al.* (2000) *Kidney Int.* **58**, 1941–1952.
80. Danguy A, Camby I & Kiss R (2002) *Biochem. Biophys. Acta* **1572**, 285–293.
81. Le QT, Shi G, Cao H *et al.* (2005) *J. Clin. Oncol.* **23**, 8932–8941.
82. van Die I & Cummings RD (2006) *Chem. Immunol. Allergy* **90**, 91–112.
83. Lopez E, del Pozo V, Miguel T *et al.* (2006) *J. Immunol.* **176**, 1943–1950.
84. Zuberi RI, Hsu DK, Kalayci O *et al.* (2004) *Am. J. Pathol.* **165**, 2045–2053.
85. Blixt O, Head S, Mondala T *et al.* (2004) *Proc. Natl Acad. Sci. USA* **101**, 17933–17938.

CHAPTER 15

Recognition of bacterial glycoepitopes by pulmonary C-type lectins as an arm of lung innate immunity

Itzhak Ofek, Hany Sahly, Yona Keisari and Erika Crouch

1. INTRODUCTION

Many bacterial species that cause severe infections are classified into serological types based on the immunological specificity of surface glycoconjugates (e.g. serotypic epitopes or glycoepitopes). The association of specific bacterial serotypes of a given genus with defined clinical infections is a common phenomenon that applies to a variety of Gram-positive and Gram-negative bacteria. There is no generally accepted explanation for the predominance of serotypes bearing a distinct glycoepitope. However, it is assumed that it reflects a random phenomenon that can change under appropriate immunological pressure.

In this review we will present evidence suggesting that the high frequency of isolation of serotypes bearing a specific glycoepitope emanates from the ability of these serotypes to evade the innate immune system operating at the site of infection. Conversely, lectins of the innate immune system that recognize serotype-specific glycoepitopes contribute to eradication of these serotypes, resulting in a lower frequency of isolation from clinical samples.

There are a number of surface glycoconjugates that are known to enable bacteria to evade host innate immunity, while at the same time bearing the glycoepitope responsible for their classification into various serotypes. For example, the polysaccharide capsule of *Streptococcus pneumoniae* enables this pathogen to resist phagocytosis, but there is still no satisfactory explanation as to why only 22 capsular serotypes out of more than 80 are responsible for most pneumococcal infections [1]. With this in mind, we have examined the relationship between the recognition of specific Gram-negative glycoepitopes by one or more lectins of the innate immune system and the capacity of the serotype-bearing organisms. As a model, we have studied the interactions of *Klebsiella pneumoniae* capsular and O-serotypes with soluble and cellular lectins of the pulmonary innate immune system.

Glycobiology (C. Sansom and O. Markman, eds.)
© Scion Publishing Limited, 2007

The following review will focus on the interactions of *K. pneumoniae* glycoconjugates with lung C-type-lectins (see Chapters 12 and 20) as a paradigm of glycoepitope–innate immune interactions in the lung. We will summarize studies that have examined the biological consequences of these interactions and their role in innate immunity. We will first briefly describe the various components of the lung innate immune system. Second, we will describe the specific *Klebsiella* glycoepitopes that interact with the innate immune components and the role of each interaction in the infectious process. Lastly, we will summarize strong circumstantial evidence supporting the hypothesis that poor recognition of *Klebsiella* glycoepitopes by one or more lectin components of the innate immune system contributes to the predominance of these strains as causative agents of pneumonia and bacterial sepsis.

1.1 C-type lectins and pulmonary host defense

Numerous studies described the role of C-type lectins in the innate, natural defence system of the lung [2–4]. These lectins include the mannose/N-acetylglucosamine-receptor (MR) of the alveolar macrophages and the collagenous carbohydrate binding surfactant proteins A and D (SP-A and SP-D). SP-A, SP-D, and the MR are members of a family of calcium-dependent (C-type) carbohydrate-binding proteins.

The MR of macrophages, which mediates binding and subsequent uptake of glycoconjugates terminating in mannose, fucose or N-acetylglucosamine, is a major player in carbohydrate recognition within the immune system [5,6]. The MR (also known as CD220) has been studied extensively, especially with respect to its role in host defense [7,8]. It is a member of a family of receptors comprising of four molecular species that share significant homology but differ in their sugar specificity, function and cell type expression. The MR is expressed by a number of cell types including macrophages, lymphatic and hepatic endothelium, renal mesangial cells, tracheal smooth muscle cells, and retinal pigment epithelium. It is likely, therefore, that MR has diverse physiological functions, one of which is related to its innate immune functions associated with its expression on macrophages, including alveolar macrophages. On macrophages, it is expressed as a transmembrane glycoprotein of 220kDa with eight sequential carbohydrate-binding domains [5]. At least *in vitro*, MR can mediate non-opsonic (lectino-) phagocytosis of bacteria by recognizing corresponding sugar residues on the bacterial surface [8].

SP-A and SP-D are collagenous C-type lectins, also known as collectins [9]. The molecules assembled as oligomers of trimeric subunits stabilized by N-terminal disulfide cross-links. Although SP-D most commonly occurs as a tetramer of trimers (dodecamer), SP-A often occurs as a hexamers of trimers (octadecamers)

The C-type lectins of the lung show specific interactions with a wide variety of microorganisms, as well as phagocytic cells [3,7,9–11]. In addition, the levels of lung collectins can increase following microbial challenge [12]. Mice lacking a functional SP-A gene show increased bacterial proliferation, more intense lung inflammation, and increased dissemination following challenge with a variety of bacteria [13,14]. SP-D-deficient mice have a more complex structural phenotype

[15,16]. However, they also show decreased uptake of bacteria by macrophages and an altered inflammatory response to bacterial challenge [15]. Although SP-A and SP-D are synthesized by type II and bronchiolar epithelial cells, SP-D is found in many other tissues, including the salivary glands, where it could interact with potential pulmonary pathogens that initially colonize the upper respiratory tract [17].

2. GLYCOEPITOPES OF *KLEBSIELLA* AND THEIR INTERRELATIONSHIP WITH THE INNATE IMMUNE RESPONSE

K. pneumoniae strains possess two types of glycoconjugate structures that show distinct affinities for C-type lectins: outer-membrane lipopolysaccharides (LPS) and capsular polysaccharides (CPS). The former are predominantly recognized by SP-D, while the latter are recognized by SP-A and MR. Relevant data regarding the roles of these glycoconjugates as virulence factors, and their role in evading C-type lectin-mediated innate immunity is summarized.

2.1 Interaction of *Klebsiella* capsular polysaccharide with the mannose receptor

K. pneumoniae is able to produce a prominent capsule composed of complex acidic polysaccharides. The capsular repeating subunits consist of four to six sugars with negatively charged uronic acids. Based on the structural variability of the CPS subunits, K. pneumoniae has been classified into 77 serotypes [18,19]. In an attempt to verify the significance of the distinct composition and sequence of repeating units of the 77 capsular polysaccharides, a number of strains belonging to different serotypes were tested for their ability to bind to rat alveolar macrophages in a serum-free system [20]. K. pneumoniae serotypes (e.g. K21a) that express capsular polysaccharides that contain Manα2/3Man or Rhaα2/3l-Rha sequences bound avidly to alveolar macrophages [20–23]. This recognition by the MR can result in the ingestion and killing of the organisms. In contrast, serotypes that lack such sequences (e.g. K2) did not bind to the MR and were not internalized. Consistent with these observations, purified CPSs containing Manα2/3Man or Rhaα2/3lRha bound to guinea-pig alveolar macrophages, while CPSs of those lacking these disaccharide units did not. The serotype binding specificity of the macrophage lectin was confirmed by inter-serotype switching of the CPS genes by reciprocal recombination. In this manner it was possible to generate capsule switched recombinant strains K2(K21a) and K21a(K2), which retained their respective recipient K2 and K21 strain backgrounds, but inherited genes encoding for CPSs of the parental donor strain [21,22]. Specifically, the capsule-switched derivative K2(K21a) retained the CPSs of the K21a donor recognized by macrophage MR, and the K21a(K2) derivative bound poorly to macrophages because it inherited the capsule genes of the donor K2 strains, which are not recognized by the macrophage MR.

Mannose receptor is not expressed by blood monocytes and when the latter mature to macrophages they do express the MR [24]. To see if *Klebsiella* binding is associated with MR expression, the binding of the bacteria to fresh blood monocytes and to cells after maturation *in vitro* was examined [20]. The expression of MR only on mature monocyte-derived macrophages was confirmed with anti-MR antibodies. The results show that *Klebsiella* bound only to mature monocyte-derived macrophages in a calcium-dependent manner. Consistent with the suggested binding specificity of the macrophage MR, the binding was inhibited by mannan.

The contribution of glycoepitopes recognized by MR to the virulence of *K. pneumoniae* was examined in mice using serotype K2 and K21a and their respective capsule-switched derivatives [22]. The results suggest that switching of *cps* genes in *K. pneumoniae* serotypes affects interactions of the bacteria with alveolar macrophages and blood clearance, and thus their virulence. Moreover, *Klebsiella* serotypes that express CPSs recognized by the MR were significantly less virulent than serotypes expressing CPSs not recognized by the MR. Capsule types such as K21a, which are recognized by the macrophage lectin, have a lower virulence, probably because of effects on phagocytic uptake and killing. Although the K2 serotype was highly virulent, the capsule-switched derivative K21a(K2) expressing K2 capsule was more virulent than the parent K21a strains, but less virulent than the K2 donor *Klebsiella* strain. Together the data suggest that the chemical structure of the capsule partially determines the virulence of *K. pneumoniae* in mice

2.2 Interaction of *Klebsiella* capsular polysaccharides with SP-A

The interaction of SP-A with *Klebsiella* was examined employing the K21a and K2 serotypes and their capsule-switched derivatives [25,26]. SP-A specifically agglutinated the K21a serotype, which expresses the Manα2Man sequence in its CPS. SP-A also bound to immobilized parent K21a strain and to a recombinant strain of K2 that expresses the K21a capsule. In contrast, there was no significant binding of SP-A to the K2 parent strain, which lacks Manα2Man sequences. Furthermore, the CPS of K21a bound to immobilized SP-A, and the binding was inhibited by mannan but not by *Klebsiella* LPS nor by K2 CPS [26]. Taken together, the data suggest that SP-A recognizes the same capsular structure as recognized by macrophage MR. SP-A did not agglutinate a non-capsulated phase variant of K21a, suggesting that, as for MR, structures underneath the capsule are not recognized by SP-A [25].

Macrophages express at least one SP-A receptor [11]. Because SP-A binds to *Klebsiella* capsule and to macrophages in a lectin-dependent and lectin-independent manner its ability to opsonize the K21a serotype was tested. Pretreatment of the bacteria with SP-A followed by washing off excess unbound SP-A caused a significant increase in the number of bacteria associated with alveolar macrophages. Other experiments showed that the increase of *Klebsiella* association with macrophages was followed by ingestion and killing of the bacteria, suggesting that SP-A acts as an opsonin in bridging between the capsulated K21a and the alveolar macrophages [26].

Previous studies have shown that SP-A bound to macrophages is rapidly internalized [27], consistent with recognition by specific receptors. A marked increase in the association of *Klebsiella* with alveolar macrophages was also observed when the macrophages were pretreated with SP-A [26]. The SP-A-induced association of K21a with alveolar macrophages was inhibited by mannan and did not occur to a significant extent with K2 or the capsule-switched derivative K21a(K2) that expresses the K2 CPS. Subsequent experiments revealed that SP-A-treated alveolar macrophages also bound increased amounts of mannan, the ligand of MR. Moreover, SP-A-induced enhancement of *Klebsiella* and mannan binding gradually decreased over a period of 5hours after washing off excess SP-A [26]. The data collectively indicate that SP-A upregulates MR expression, resulting in increased association of *Klebsiella* with macrophages. This conclusion is supported by the findings showing that SP-A upregulated MR expression in human monocyte-derived macrophages plated on SP-A matrix, by using mannan as ligand and anti-human MR to monitor the receptor activity [28].

2.3 Relationship between *Klebsiella* O-polysaccharide structure and binding to SP-D

Early studies identified the core region of Gram-negative LPS as the glycoepitope that is preferentially recognized by SP-D [29]. This interaction was subsequently shown to involve the lectin domain of the collectin which bound to LPS of *K. pneumoniae* and other Gram-negative bacteria [30]. SP-D interacts with non-capsulated strains of *Klebsiella* but not with their corresponding capsulated variants [31]. Subsequent studies with *K. pneumoniae* showed that SP-D reacts differently with non-capsulated strains of the bacterium, which results in bacterial agglutination as well as enhanced internalization and killing by macrophages. For example, SP-D was more effective in the agglutination of a non-capsulated phase variant of the O3 serotype than a non-capsulated phase variant of the O1 serotype. The minimal SP-D concentration needed to induce agglutination of the O1 serotype was approximately 10-fold higher than that needed for the O3 serotype [31,32]. These finding indicate that non-capsular factors, such as the molecular structure of the O-antigen, influence the binding affinity of SP-D to *Klebsiella* and that SP-D contributes to the clearance of *Klebsiella* by alveolar macrophages.

In order to examine the effects of O-antigen structure on SP-D binding, LPS was extracted from O1, O3, O4, and O5 serotypes, resolved by SDS-PAGE, blotted and overlaid with SP-D [32]. SP-D bound to LPS molecules of all serotypes when the LPS lacked an O-antigen or contained only small numbers of attached oligosaccharide repeating subunits. In contrast, LPS molecules containing longer chain of oligosaccharides (e.g. higher number of repeating units) were recognized by SP-D only if derived from O3 or O5 serotypes. The glycoepitope of the O-antigen of these SP-D reactive serotypes contains alternating alpha-2 and -3 mannose residues [18], which seem to facilitate SP-D binding.

Various studies have shown that non-capsulated phase variants are required for efficient epithelial attachment and invasion [33,34], presumably through the enhanced exposure of bacterial adhesins. SP-D showed dose-dependent inhibition of bacterial adhesion to the alveolar epithelial cells [32]. Significantly, SP-D decreased bacterial adhesion to alveolar epithelial cells *in vitro*, and the potency of this inhibitory effect was greater for the SP-D "reactive" than for the "non-reactive" O-serotypes (e.g. the minimal concentration of SP-D needed to inhibit the adhesion of non-capsulated *Klebsiella* to the epithelial cells was four times higher for the O1 serotype than that required for the O3 serotype). Recently it was shown that SP-D impaired the ability of *Pseudomonas aeruginosa* to adhere to corneal epithelial cells, suggesting that the anti-adhesion effect of this collectin is not limited to Klebsiella [35].

2.4 Relationship between SP-D binding to *Klebsiella* and macrophage stimulation

In an *in vitro* survey it was shown than coating of non-capsulated phase variants of *Klebsiella* strains with SP-D resulted in marked stimulation of cytokine mRNA accumulation in both macrophages and peripheral blood monocytes. This is consistent with the hypothesis that innate immunity against *Klebsiella* involves SP-D, which can act as an opsonin to enhance the interaction of macrophages with non-capsulated phase variants originating from the upper respiratory tract [31,25]. This notion is consistent with evidence for at least two classes of SP-D receptors on phagocytic cells. SP-D can bind to cells through interactions of the carbohydrate-recognition domain with cell surface glycoconjugates (lectin-dependent) or via a lectin-independent mechanism, presumably involving one or more protein receptors [9].

Interactions of SP-D-coated O3-bearing non-capsulated *Klebsiella* with human monocyte-derived macrophages induced significant cytokine release (see Fig. 1), whereas the SP-D-treated O1-bearing bacteria were without effect. In recent mouse studies cytokine production elicited by SP-D-coated bacteria *in vivo* and intrapulmonary growth of the same bacterial serotypes was examined. Non-capsulated mannose-containing O3 serotypes, which bind to SP-D *in vitro*, triggered higher IL-1β and IL-6 production, and were more efficiently cleared from the lungs of mice. By contrast, galactose-containing O1 serotypes, which interact poorly with SP-D, showed a lower cytokine response and were inefficiently cleared. The stimulation caused by O3 serotypes did not occur in macrophage-depleted mice. These findings are consistent with *in vitro* results showing that production of IL-1β and IL-6 mRNA and IL-6 protein by human macrophages exposed to mannose bearing *Klebsiella* O-serotypes is significantly increased by SP-D [36].

Coating of non-capsulated variants of O3 bacteria with SP-D stimulated nitrous oxide (NO) production [31], suggesting that complexes of *Klebsiella* with SP-D interact efficiently with receptors on macrophages to trigger cytokine and NO production. In order to see how the engagement of SP-D with its ligand on the bacterial surface affects the ability of the bound SP-D to interact with its cognate receptor on macrophage and activate the cells, we employed a *Klebsiella*-free model system to avoid any possible effect of potential other bacterial

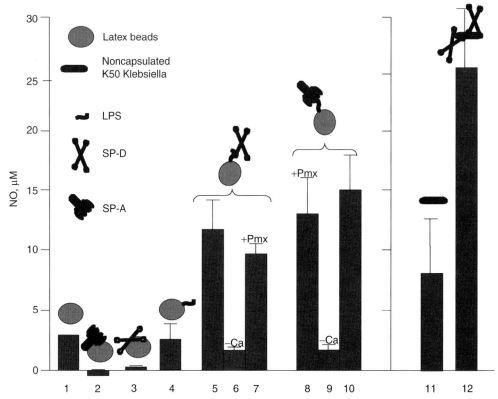

Figure 1. Stimulation of nitrous oxide (NO) production by macrophages exposed to latex beads coated with lipopolysaccharide (LPS) and collectins. Rat alveolar macrophages cell line (N-8383) were exposed to (1) uncoated latex beads or to beads coated with (2) SP-D, (3) SP-A, (4) LPS, (5) LPS followed by SP-D, (6) LPS followed by SP-D in calcium (Ca)-free medium, (7) LPS followed by SP-D in medium supplemented with polymyxin B (Pmx), (8) LPS followed by SP-A, (9) LPS followed by SP-A in calcium-free medium, (10) LPS followed by SP-A in medium supplemented with polymyxin B. For comparison, NO production induced by non-capsulated O3 K50 Klebsiella uncoated (11) and SP-D coated (12) is reproduced from elsewhere (ref. 31). The results with the latex beads were obtained in the laboratory of the late Moshe Kalina, at Sackler Faculty of Medicine, Tel Aviv University in collaboration with I Ofek and E Crouch.

constituents. Specifically, latex particles were coated either with SP-D, *Klebsiella* LPS, or sequentially with LPS and SP-D and delivered to macrophage monolayers. Increases in NO production were only observed when the beads were coated sequentially with LPS and SP-D [31]. The addition of soluble LPS and/or SP-D at amounts 100-fold greater than present on the washed beads did not cause detectable stimulation of macrophage NO production. These effects were calcium dependent and did not require ingestion of the particles (see Fig. 1, Kalina, Crouch, and Ofek, unpublished observations). Taken together, the data suggest that the engagement of SP-D with its LPS ligand on bacteria or particles enable the collectin to better recognize its cognate receptors on macrophages to subsequently stimulate the cells.

3. GENERAL OVERVIEW AND CONCLUDING REMARKS

From the foregoing it appears that two of the C-type lectins, SP-A and MR, recognize capsular polysaccharides sharing the same di-mannose or di-rhamnose sequence. By contrast, SP-D preferentially recognizes the LPS core region. It can also interact with O-antigen-bearing LPS if the O-antigen contains repeating mannose units. Enhancement of *Klebsiella* phagocytosis appears to be a major mechanism through which these C-type lectins provide protection against *Klebsiella* infections. In the case of MR and SP-A, uptake by macrophages is enhanced either directly by MR, or following opsonization and receptor-mediated uptake of the SP-A-coated organisms. A clue to this seemingly redundant function might come from the studies of Chroneos and Shepherd [37]. It was found that agents that suppress either MR or SP-A receptor *in vitro* or *in vivo*, upregulate the other receptor. Thus, it seems that the defense mechanisms provided by MR and SP-A are directed mainly against the di-mannose and di-rhamnose-expressing capsular serotypes. Other factors are undoubtedly involved, but the data seem to indicate that there is a role for these C-type lectins in protecting against at least half of the capsular serotypes of *K. pneumoniae*.

Based on the available information from various studies, we suggest the following roles of C-type lectins in pulmonary innate immunity. Colonization of the upper respiratory tract by Gram-negative bacteria precedes entry of the organisms into the lung. Because capsule interferes with the expression of adhesins required for colonization of epithelial cells by the organisms, it is likely that most of the bacteria colonizing the upper respiratory tract (or other mucosal surfaces) are in the non-capsulated phase. *Klebsiella* opsonization and agglutination by SP-D might provide early protection against all strains of non-capsulated phenotypes because the LPS core region, which reacts with SP-D, is conserved. However, as discussed above, productive binding of SP-D to the core region is influenced by the molecular structure of the O-antigen. Mannose-containing O-serotypes bind avidly to SP-D, while those that lack mannose do not. *In vitro* studies have shown that *Klebsiella* can phase vary between capsulated and non-capsulated phenotypes [40]. Thus, capsulated bacteria that emerge during the infection as a result of the phase variation phenomenon are prone to escape SP-D recognition. Mannose receptor-equipped macrophages in conjunction with SP-A may provide additional protection by eliminating specific capsulated *Klebsiella* through recognition of the di-mannose and di-rhamnose sequences in the capsular polysaccharide. SP-A, which opsonizes and agglutinates the di-mannose-containing *Klebsiella*, may also augment expression of MR, which in turn mediates phagocytosis of the organisms. Thus, *Klebsiella* capsular serotypes that are not recognized by SP-A and MR may become the predominant infective capsular serotypes (Table 1).

Epidemiological data are consistent with this prediction as discussed above. First, sero-epidemiological studies have shown that all capsular serotypes containing the di-mannose or di-rhamnose residues in their capsular polysaccharide are isolated at a significantly lower frequency as compared to those capsular serotypes lacking those sequences [8,19]. Second, the O-serotypes (e.g. O3 and O5) that contain poly-mannose and bind with higher affinity to SP-D are isolated at significantly higher frequency than those O-serotypes (e.g. O1 and O4) lacking such SP-D-reactive residues [32,38,39]. Finally, studies on the survival of the various *Klebsiella* serotypes in the lung of infected mice have shown that survival of inhaled bacteria in the lung depends partially on their LPS structure and this correlates with the ability of SP-D and other lung C-type lectins to recognize the LPS and alter phagocyte responses to the organism *in vitro* [36]. In spite of the relationship between the epidemiological data and the *in vitro* interactions implicating the MR, and the lung collectins SP-D and SP-A as major molecules of innate host defense, there are certain serotypes isolated at high frequency but their capsular structure is recognised by the MR. Conversely, other serotypes express surface glycoconjugates not recognized by the MR but are isolated at low frequency. The field of innate immunity is rapidly growing and new molecules that recognize specific structures are revealed and it is likely that they play a role in clearance of pulmonary pathogens. For example, the C-type lectins DECTIN-2 and SIGNR1 are expressed on some macrophages or dendritic cells and can bind to mannan and other microbial glycoconjugates [41]. Further studies are required to examine the role of these C-type lectins in innate immunity against *Klebsiella* infections in relation to the frequency of isolation of the various serotypes.

Although this review focused on *K. pneumoniae*, the lung C-type lectins were found to interact with corresponding glycoconjugates on the surface of other bacterial species including *Pseudomonas aeruginosa* [42], *Escherichia coli* [29], *Haemophilus influenzae* [43], *Streptococcus pneumoniae* [44] *Bacillus subtilis*, and *Staphylococcus aureus* [45], group B streptococci [14], *Mycobacterium tuberculosis* [28,46,47], *Mycobacterium avium* [48], and *Mycoplasma pneumoniae* [49].

Recent studies confirm the role of SP-D and SP-A in providing innate immunity employing mice deficient in either one of the collectins or in both of them [13–16]. Significantly, it was found that stimulation of phagocytosis and cytokine secretions were the major mechanism providing innate immunity against non-mucoid strain of *P. aeruginosa*. Thus, the studies summarized above for *K. pneumoniae* may represent a paradigm for other bacterial lung infections.

In summary, the studies indicate important interactions between *K. pneumoniae* and at least three innate immune effectors of the C-type lectin family. These molecules interact with distinct glycoepitopes expressed by specific bacterial serotypes. We anticipate these observations as relevant for host interactions with a variety of other lung pathogens, and suggest that future studies should focus on those serotypes that escape recognition by the innate immune system of the hospitalized patient (e.g. K2/O1 serotype, see Table 1) but are eradicated by the innate immune system of otherwise healthy individuals. Such studies could assist with the development of new therapeutic strategies to manage nosocomial pneumonia.

Acknowledgments

We wish to acknowledge the excellent assistance of Dr Ariella Matityahou and Ms Shoshana Riklis in the unpublished data on the latex particles generated in the laboratory of the late Dr Kalina Moshe. This work was partially supported by grant 2001055 from the United States-Israel Binational Science Foundation.

Table 1 Relationship between the recognition of glycoepitope structure, by C-type lectin, mouse virulence and frequency of isolation

Lung C-type lectin	Reacting variant[a]	Serotype and glycoepitope structure	Lectin recognition[b]	Mouse virulence[c]	Frequency of isolation[d]
SP-A MR	Capsulated	K21a like 3GlcAα3<u>Manα2-Man</u>3αGalβ[d] α3Gal	+	Low	Low
		K21 like 4Glcβ4Manβ4Glcβ[d] 4αGlcA	–	High	High
SP-D	Non-capsulated	O3 LPS like 3 <u>Manα3 Manα2</u> Manα2[e]	++++	Low	Low
		O1 LPS like 3Galα3Galα	+	High	High

[a]Irrespective of the capsular or LPS serotypes, the capsulated variants do not interact with SP-D [31] and non-capsulated variants do not interact with MR or SP-A [8,9,25].
[b]Recognition by lectin results in either agglutination of the bacteria or opsonization (SP-A and SP-D). This is associated with cytokine release (SP-D) or lectinophagocytosis via the mannose receptor (MR) of alveolar macrophages or inhibition of adhesion to epithelial cells (SP-D) (adapted from [20,36,31,32,51]).
[c]Mouse virulence was determined on the number of live bacteria surviving in the lung after 3 days of inoculation [36] and on blood clearance in mice [22].
[d]Shown are the repeating units of the K21 a capsular polysaccharide. The underlined sequence is the glycoepitope recognized by the lectins. Other serotypes with same glycoepitope or with glycoepitopes containing rhamnose instead of of mannose (Man) with alpha 3 instead of 2 linkage also react same way with the indicated lectin [20,26]. For other capsular serotypes expressing these SP-A and MR-reactive glycoepitopes see [8,9,25].
[e]Shown are the repeating units of the O-antigen of the LPS. On blots of LPS gel electrophoresis SP-D reacts with LPS molecules of both serotypes containing core region or core region with one or more repeating units whereas binding of SP-D to LPS molecules with a high number of repeating units occurred only with LPS with O-antigen containing the underlined structure in the O3 antigen. Other O-antigens of LPS serotypes containing such SP-D-reactive glycoepitopes (see [32]).
[f]Frequency of isolation of capsular serotypes containing the glycoepitopes reactive with the SP-A and MR lectins is based on serotype distribution [19] and on CPS structure of the serotypes (see [20] for references).

REFERENCES

1. Hausdorff WP, Feikin DR & Klugman KP (2005) *Lancet Infect. Dis.* **5**, 83–93.
2. Wright JR (2005) *Nat. Rev. Immunol.* **5**, 58–68.
3. Whitsett JA (2005) *Biol. Neonate* **88**, 175–80.
4. Stahl PD & Ezekowitz RA (1998) *Curr. Opin. Immunol.* **10**, 50–55.
5. McGreal EP, Martinez-Pomares L & Gordon S (2004) *Mol. Immunol.* **41**, 1109–1121.
6. Weis WI, Taylor ME & Drickamer K (1998) *Immunol. Rev.* **163**, 19–34.
7. East L & Isacke CM (2002) *Biochim. Biophys. Acta* **1572**, 364–386.
8. Ofek I, Goldhar J, Keisari Y & Sharon N (1995) *Annu. Rev. Microbiol.* **49**, 239–276.
9. Crouch EC, Hartshorn K & Ofek I (2000) *Immunol. Rev.* **173**, 52–65.
10. Sano H & Kuroki Y (2005) *Mol. Immunol.* **42**, 279–287.
11. Van de Wetering JK, van Golde LM & Batenburg JJ (2004) *Eur. J. Biochem.* **271**, 1229–1249.

12. Reading PC, Allison J, Crouch EC & Anders EM (1998) *J. Virol.* **72**, 6884–6887.
13. Korfhagen TR, Bruno MD, Ross GF *et al.* (1996) *Proc. Natl Acad. Sci. USA* **93**, 9594–9599.
14. LeVine AM, Kurak KE, Wright JR *et al.* (1999) *Am. J. Respir. Cell Mol. Biol.* **20**, 279–286.
15. Korfhagen TR, Sheftelyevich V, Burhans MS *et al.* (1998) *J. Biol. Chem.* **273**, 28438–28443.
16. Botas C, Poulain F, Akiyama J *et al.* (1998) *Proc. Natl Acad. Sci. USA* **95**, 11869–11874.
17. Madsen J, Kliem A, Tornoe I, Skjodt K, Koch C & Holmskov U (2000) *J. Immunol.* **164**, 5866–5870.
18. Ørskov I & Ørskov F (1984) In: *Methods in Microbiology*, pp. 143–164. Edited by T Bergan. Academic Press, London.
19. Cryz S J, Fürer E & Germanier R (1984) *Infect. Immun.* **43**, 440–441.
20. Athamna A, Ofek I, Keisari Y, Markowitz S, Dutton GS & Sharon N (1991) *Infect. Immun.* **59**, 1673–1682.
21. Ofek, I, Kabha K, Athamna A *et al.* (1993) *Infect. Immun.* **61**, 4208–4216.
22. Kabha K, Nissimov L, Athamna A *et al.* (1995) *Infect. Immun.* **63**, 847–852.
23. Keisari Y, Kabha K, Nissimov L, Schlepper-Schafer J & Ofek I (1997) *Adv. Dent. Res.* **11**, 43–49.
24. Shepherd VL, Campbell EJ, Senior RM & Stahl PD (1982) *J. Reticuloendothel. Soc.* **32**, 423–431.
25. Ofek I, Crouch E & Keisari Y (2000) In: *The Biology and Pathology of Innate Immunity Mechanisms*, pp. 27–36. Edited by Y Keisari and I Ofek. Kluwer Academic/Plenum, London.
26. Kabha K, Schmegner J, Keisari Y, Parolis H, Schlepper-Schaefer J & Ofek I (1997) *Am. J. Physiol.* **272**, 344–352.
27. Manz-Keinke H, Egenhofer C, Plattner H & Schlepper-Schafer J (1991) *Exp. Cell Res.* **192**, 597–603.
28. Gaynor CD, McCormack FX, Voelker DR, McGowan SE & Schlesinger LS (1995) *J. Immunol.* **155**, 5343–5351.
29. Kuan SF, Rust K & Crouch E (1992) *J. Clin. Invest.* **90**, 97–106.
30. Lim BL, Wang JY, Holmskov U, Hoppe HJ & Reid KB (1994) *Biochem. Biophys. Res. Commun.* **202**, 1674–1680.
31. Ofek I, Mesika A, Kalina M *et al.* (2001) *Infect. Immun.* **69**, 24–33.
32. Sahly H, Ofek I, Podschun R Brade H, He Y & Crouch E (2002) *J. Immunol.* **169**, 3267–3274.
33. Sahly H, Podschun R, Kekow J *et al.* (2000) *Infect. Immun.* **68**, 6744–6749.
34. Favre-Bonte S, Joly B & Forestier C (1999) *Infect. Immun.* **67**, 554–561.
35. Ni M, Evans DJ, Hawgood S, Anders EM, Sack RA & Fleiszig SM (2005) *Infect. Immun.* **73**, 2147–2156.
36. Kostina E, Ofek I, Crouch E *et al.* (2005) *Infect. Immun.* **73**, 8282–8290.
37. Chroneos Z & Shepherd VL (1995) *Am. J. Physiol.* **269**, 721–726.
38. Sahly H, Aucken H, Benedi VJ *et al.* (2004) *Eur. J. Clin. Microbiol. Infect. Dis.* **23**, 20–26.
39. Sahly H, Aucken H, Benedi VJ *et al.* (2004) *Antimicrob. Agents Chemother.* **48**, 3477–3482.
40. Matatov R, Goldhar J, Skutelsky E *et al.* (1999) *FEMS Microbiol. Lett.* **179**, 123–130.
41. Cambi A, Koopman M & Figdor CG (2005) *Cell. Microbiol.* **7**, 481–488.
42. Bufler P, Schmidt B, Schikor D, Bauernfeind A, Crouch EC & Griese M (2003) *Am. J. Respir. Cell Mol. Biol.* **28**, 249–256.
43. McNeely TB & Coonord JD. (1994) *Am. J. Respir. Cell Mol. Biol.* **11**, 114–122.
44. McNeely TB & Coonrod JD (1993) *J. Infect. Dis.* **167**, 91–97.
45. Van de Wetering JK, van Eijk M, van Golde LM, Hartung T, van Strijp JA & Batenburg JJ (2001) *J. Infect. Dis.* **184**, 1143–1151.
46. Ferguson JS, Voelker DR, McCormack FX & Schlesinger LS (1999) *J. Immunol.* **163**, 312–321.
47. Nigou J, Gilleron M, Rojas M, Garcia LF, Thurnher M & Puzo G (2002) *Microbes Infect.* **4**, 945–953.

48. Kudo K, Sano H, Takahashi H et al. (2004) *J. Immunol.* **172**, 7592–7602.
49. Chiba H, Pattanajitvilai S, Evans AJ, Harbeck RJ & Voelker DR (2002) *J. Biol. Chem.* **277**, 20379–20385.
50. Giannoni E, Sawa T, Allen L, Wiener-Kronish J & Hawgood S (2006) *Am. J. Respir. Cell Mol. Biol.* **34**, 704–710.
51. Keisari Y, Wang H, Mesika A, Matatov R, Nissimov L, Crouch EC & Ofek I (2001) *J. Leuk. Biol.* **70**, 135–141.

CHAPTER 16
Selectin and integrin recognition of ligands under shear flow: affinity and beyond

Ronen Alon, Revital Shamri and Sara Feigelson

1. INTRODUCTION

Immune cells (leukocytes) and hematopoietic progenitor cells circulating the body must exit blood vessels near specific target sites of injury, infection, inflammation, or proliferation [1–3]. Recruitment of different subsets of leukocytes, as well as the trafficking of some malignant cells to these sites and to lymphoid organs, is tightly regulated by sequential adhesive interactions between specific protein receptors on their surface and their respective ligands on the blood vessel endothelial wall [4] (Fig. 1). Accumulated data from *in vivo* and *in vitro* studies

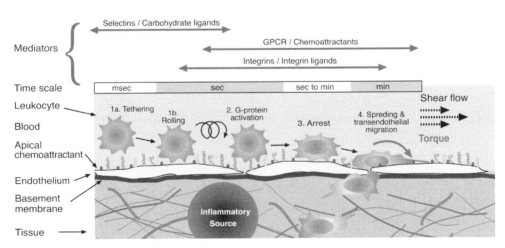

Figure 1. The multi-step model of leukocyte recruitment at specialized sites of emigration to inflamed or lymphoid tissues. Selectins (leukocyte or endothelial expressed), endothelial-displayed chemoattractants (primarily chemokines) and integrins sequentially coordinate in capturing a circulating leukocyte to the vessel wall at specific sites and in generating cell resistance to detachment by the blood shear forces.

Glycobiology (C. Sansom and O. Markman, eds.)
© Scion Publishing Limited, 2007

suggest that the major players in this multistep process are members of two adhesive families, the C-type lectins, called selectins (see Chapters 16 and 22), and integrins, which are structurally and functionally adapted to operate under disruptive shear forces exerted on leukocytes at the vessel wall by the blood flow.

The primary attachment or tethering of circulating leukocytes to the vessel wall is labile, mediated primarily by selectins and their glycoprotein counterreceptors [5]. When present at sufficient and uniform density on the endothelial target, these labile adhesive interactions give rise to leukocyte rolling in the direction of flow, which allows the transiently attached immune cells to survey the endothelial lining and bring them into proximity to activating chemoattractants (mostly of the chemokine family) presented on the endothelial surface [6]. These vessel wall-displayed chemotactic cytokines bind specific G protein-coupled receptors (GPCRs) on recruited leukocytes and trigger, within subseconds, the activation on the leukocyte surface of a second class of adhesion receptors, integrins, which can then firmly bind to their respective endothelial ligands, causing the immune cell to arrest on the blood vessel [7].

A remarkable feature of integrins is that their affinity state and adhesive activity are regulated *in situ* by local GPCR signals without accompanied increase of their surface expression [8,9]. Selectins and their glycoproteins ligands regulate their activity via membranal and cytoskeletal associations [10–12], but their intrinsic affinity is not subjected to *in situ* modulation by endothelial-displayed chemoattractant signals. Thus, immune cell integrins, but not selectins or selectin ligands, can rapidly adapt their adhesive behavior at specific endothelial sites within target tissues according to tissue- and context-restricted patterns of chemokine or chemoattractant expression at these sites.

2. SELECTIN RECOGNITION OF LIGANDS UNDER SHEAR STRESS: AFFINITY, AVIDITY AND BEYOND

Selectins are the major adhesive receptors in the vasculature that mediate the initial tethering of flowing leukocytes to the vessel wall and the propagation of these tethers into continuous rolling adhesions [13]. They comprise three family members, two inducibly expressed on endothelial surfaces or platelets and one, L-selectin, expressed by most circulating leukocytes. The binding of selectins to their carbohydrate ligands is the fastest cell–cell recognition event known in nature. With the exception of the P-selectin–P-selectin glycoprotein ligand-1 (PSGL-1) pair [14,15], the majority of selectin interactions are specialized to capture (tether) circulating cells from the bloodstream and support rolling adhesions on their respective endothelial ligands with extremely low affinity [16,17] (Table 1). Using flow chamber assays simulating blood flow at sites of leukocyte emigration [18] and high resolution video-microscopic analysis of cellular motions over surfaces reconstituted with known densities of purified endothelial-derived selectins or selectin ligands, several labs, including ours, have gained key insights into the molecular basis of selectin adhesiveness [12,19–28].

Table 1 Affinity and kinetic constants of key selectin and integrin–ligand interactions measured by surface plasmon resonance (Biacore)

Immobilized molecule	Soluble molecule	On-rate (k_{on}, M^{-1}s^{-1})	Off-rate (k_{off}, s^{-1})	Dissociation constant (K_D, μM)	Reference
PSGL-1 (h)[a]	P-selectin (h)	4.4×10^6	1.4 ± 0.1	0.32 ± 0.020	15
ESL-1 (m)	E-selectin (m)	7.4×10^4	4.6	62	56
GlyCAM-1 (m)	L-selectin (m)	10^5	10	108	16
ICAM-1 (h)	Wild-type (h) sLFA-1	2.2×10^5	0.12	0.5	57
ICAM-1 (h)	LFA-1 I-domain[b] (h, open conformation)	$1.05–1.39 \times 10^5$	0.014–0.045	0.15–0.36	58,59
ICAM-1 (h)	LFA-1 I-domain (h, intermediate conformation)	$0.89–1.33 \times 10^5$	0.43–0.76	3–9.4	58,59
	LFA-1 I-domain (h, closed conformation)	$0.02–0.034 \times 10^5$	1.2–4.6	450–1760	58,59
LFA-1 (h), high affinity conformation[c]	ICAM-1 (h)	1.2×10^5	0.029	0.24	60

[a](h, m) = human, murine.
[b]The affinity of the I-domain to ICAM-1 is assumed comparable to that of intact LFA-1 to ICAM-1 [58,59].
[c]Determined by flow cytometry-based staining.

A powerful approach to elucidate the basis for the extremely efficient recognition between low affinity selectin-ligand interactions like L-selectin with its endothelial ligands under shear flow has been to measure the kinetics of individual reversible tethers mediated by selectin-ligand interactions at subsecond adhesive contacts [29]. These transient tethers are the smallest unit of adhesive interaction observable in shear flow. Site density measurements of selectins or ligands indicate that these tethers are mediated by singular adhesive pairs with durations ranging from 20 ms to 1 s [12,29]. Our studies suggest that the high efficiency by which low-affinity L-selectin-ligand pairs interact under shear flow is due to fast kinetics of bond formation. Although monomeric recognition between L-selectin and its carbohydrate ligand has a moderate on-rate in the range of 10^5s^{-1}M^{-1} (Table 1 [16]), this recognition must be facilitated by local rebinding of the selectin to neighboring carbohydrate ligands on its mucin glycoprotein ligand. Indeed mucin L-selectin ligands such as GlyCAM-1, CD34, and MadCAM-1 [30] are decorated with tandem O-glycans, some of which carry two functional L-selectin carbohydrate-binding units. Indeed the affinity of multivalent L-selectin association with GlyCAM-1 is enhanced by orders of magnitude due to rebinding of GlyCAM-1 to surface-presented L-selectin [16]. Likewise, the majority of E-selectin counterreceptor glycoproteins are decorated with multiantennary N-linked glycans or with closely spaced biantennary O-glycans carrying numerous closely spaced E-selectin ligands.

Since the duration of initial contact between selectin and its ligand falls in the millisecond range [31], selectin rebinding must be facilitated by unique cellular properties in addition to the close spacing between selectin carbohydrate ligands. For instance, presentation of L-selectin on leukocyte microvilli [32] and selectin dimerization [27] may both contribute to the exceptional cell-capturing activities of this selectin, compensating for its low intrinsic bond association rate (Table 1). L-Selectins are indeed often found clustered on these cell surface projections [32,33]. Another role for these projections predicted by recent computer simulations [34] is that they facilitate the ability of L-selectins to rebind their endothelial ligands at confined adhesive contacts. In addition, cellular rotation over adhesive substrates drives the generation of multimicrovillar contacts which help distribute the shear forces applied on the captured leukocyte between multiple receptor–ligand pairs.

L-selectin, as well as other selectin interactions taking place under shear stress, share unique biophysical properties, in particular, high tensile strength (i.e. low dependence of their dissociation rate constants on applied force) [29]. This tensile strength arises primarily from intrinsic mechanical properties (i.e. unique structural motifs shared by the three selectins but not by other structurally related C type lectins). For instance, mannose-binding proteins whose lectin–carbohydrate interface resembles that of all three selectins [35] fail to mediate cell capture under shear flow, even if mutated to recognize the selectin recognition carbohydrate motif sLex (Dwir, Drickamer, and Alon, unpublished results). It is thus possible that the EGF domains of selectins contribute to the exceptional mechanical stability of their lectin–carbohydrate bonds. Indeed we could show that the EGF domain of L-selectin can modulate the binding of its lectin domain to ligands under shear flow without affecting the equilibrium binding properties of the selectin towards the same ligands in shear-free conditions [24]. Thus, biochemical, mechanical and topographical properties seem to determine the ability of weak L-selectin interactions (Table 1) to efficiently translate into functional adhesive tethers and rolling adhesions under physiological conditions of shear flow.

3. CYTOSKELETAL ASSOCIATIONS OF L-SELECTIN REGULATE MECHANICAL STABILIZATION OF ADHESIVE BONDS UNDER SHEAR FLOW

Our studies suggest that for cell surface microvillar projections to facilitate leukocyte encounter of endothelial ligands, L-selectin should be properly anchored to the cytoskeletal network underlying the microvilli surface. Stabilization of L-selectin tethers under flow is highly susceptible to mild cell treatment with actin disrupting reagents [10,12,36]. It has also become increasingly evident that associations of L-selectins and of the P-selectin ligand, PSGL-1, with the cell cytoskeleton contribute to adhesiveness under shear flow and do so without altering selectin affinity to soluble ligand [10–12,37]. Notably, the cytoskeletal anchorage of L-selectin does not regulate selectin clustering or presentation on tips of microvilli [38] and does not protect L-selectin from enzymatic shedding [12]. A specific and apparently preformed L-selectin association with the actin cytoskeleton through its cytoplasmic domain [38] was found by us to increase selectin tether duration under flow and to enhance the mechanical stability of these tethers [12]. Physical anchorage of a cell-free tail truncated L-selectin to a solid surface was found both necessary and sufficient to rescue its defective adhesion, suggesting that L-selectin anchorage to the cytoskeleton is obligatory for the selectin tethers to acquire high stability under shear flow. The enhanced stability of tethers mediated by cytoskeletally anchored L-selectin is unlikely to arise from the selectin protection from membrane uprooting, a process much slower than cytoskeletally mediated tether stabilization [39]. The cell surface microvillar projections to which L-selectin is cytoskeletally anchored may thus function as shock absorbers, reducing the effective shear forces applied on the selectin or its ligand [40]. Thus, cytoskeletal anchorage of L-selectin stabilizes a microvillar complex mechanically adapted to tolerate shear forces and thereby further increase the intrinsically high tensile strength of L-selectin–ligand bonds [12].

4. VLA-4 AND LFA-1 INTEGRINS ALSO UTILIZE CYTOSKELETAL ASSOCIATIONS TO STABILIZE THEIR ADHESIVE BONDS UNDER SHEAR FLOW

Circulating leukocytes captured or rolling through selectin-mediated adhesions arrest on target endothelium conditional to their ability to develop firm adhesion to vessel wall ligands through their various integrin receptors $\alpha_4\beta_1$ (VLA-4), $\alpha_4\beta_7$, $\alpha_L\beta_2$ (LFA-1), and $\alpha_M\beta_2$ (Mac-1) [41]. Recent structural and biophysical studies predict that integrins transition between inactive bent conformers and variable extended conformers with intermediate and high affinity to ligand (Fig. 2). Integrins bind their respective endothelial ligands under shear flow at much lower efficiency than selectins [1] since the majority of integrins on circulating leukocytes are kept at low-affinity bent state, in which the ligand-binding headpiece is unavailable for ligand recognition (Fig. 2). In contrast, unfolded LFA-1 is predicted to extend 25 nm over the cell surface and can readily bind immobilized ligand on a countersurface albeit at low binding affinity (Table 1).

LFA-1 as well as other leukocyte integrins can undergo an instantaneous conformational switch upon exposure to endothelial chemokine signals (Fig. 2) [42]. Dissecting this transition in a flow chamber using monoclonal antibody (mAb) reporters for distinct LFA-1 conformations, we found that immobilized chemokines stabilized *in situ*, within a fraction of a second, an extended LFA-1 conformer without inducing a high-affinity integrin conformation [43]. Chemokine extended integrins can then rearrange their headpiece upon ligand binding subject to a critical duration of contact. Our studies also suggest that this subsecond chemokine signaling to integrin requires GPCR proximity to the

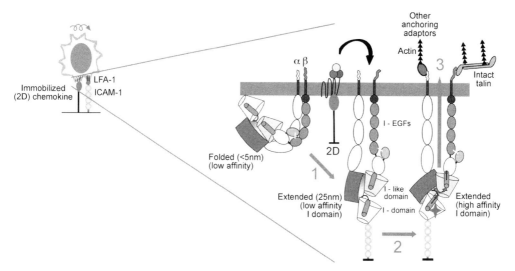

Figure 2. Chemokine signaling to integrins under shear flow, a rapid localized activation within a leukocyte microvillus tethered to endothelial displayed integrin ligand and juxtaposed chemokine. A rolling leukocyte tethered to integrin ligands must encounter juxtaposed chemokine at the site of integrin activation. A quarternary complex between integrin, ligand, chemokine, and G protein-coupled receptor (GPCR) must form within milliseconds to locally activate the integrin–ligand complex via a bidirectional signaling event. Only a fully activated integrin can arrest the rolling leukocyte on the vessel wall. Multistep bidirectional activation switches the LFA-1 integrin from an inactive bent state to a high-affinity extended state by proximal chemokine and integrin–ligand signals. Upon initial encounter, endothelial-bound (immobilized, two-dimensional) chemokines transduce, within 0.1 s, leukocyte GPCRs signals converting inactive (folded) integrin to its extended conformation (step 1). GTP-bound RhoA is involved in this step. This critical chemokine-driven inside-out event primes the integrin to transiently bind endothelial ligands on the counter endothelial surface. The I-domain containing integrin undergoes a further conformational shift upon ligand binding (step 2), resulting in full integrin activation with an open I domain (depicted by the star) and lymphocyte arrest. The ligand-driven step is predicted to result in further separation of the integrin subunit tails (step 3), increasing the binding strength. Our studies suggest the involvement of the Rap-1 and the RhoA GTPases in the GPCR-transduced inside-out integrin activation event. The integrin adaptors talin and paxillin (not shown) regulate the outside-in activation events.

integrin at the adhesive contact, although the precise distance between the integrin activating GPCR and its target integrin is still unknown. We refer to this mode of chemokine signaling as 2D-GPCR signaling, since the rapid integrin stimulatory GPCR signals are confined to a 2D interface and cannot be reconstituted with soluble chemokine. A support for this model came from studies on LFA-1 locked in an intermediate affinity state in which the integrin is extended but its headpiece is inactive. This LFA mediates leukocyte rolling on high-density ICAM-1 but cannot arrest on this ligand regardless of density [44,45]. The VLA-4 integrin can also mediate either rolling or firm arrest on its endothelial ligand VCAM-1 depending on both the activation states of its ectodomain and the duration of the contact, which is tightly regulated by ligand density [46,47]. As with selectins, preformed integrin microclustering (e.g. on microvillar projections) may also facilitate ligand rebinding and thereby increase the probability of integrin headpiece rearrangement and activation leading to arrest [48].

Cytoskeletal constraints of integrins have been suggested to restrict integrin patching and thereby reduce integrin adhesiveness at intercellular contacts [9,49,50]. Our earlier studies suggested that integrins must properly anchor to the cytoskeleton to support adhesive interactions under shear flow [51]. In a most recent study we characterized the effect of disrupting the association between the α_4 tail and the cytoskeletal adaptor paxillin [52] and found that blocking this interaction markedly impaired the integrin's ability to anchor to the cytoskeleton. While not essential for $\alpha_4\beta_1$ affinity (i.e. extension state), ligand-induced activation (i.e. rearrangement of its headpiece), surface clustering and topography on microvilli, paxillin recognition by $\alpha_4\beta_1$ and the related integrin $\alpha_4\beta_7$ was crucial for the ability of α_4 integrins to mediate capture and adhesion strengthening on their respective ligands, VCAM-1 and MadCAM-1, under shear stress [48]. Paxillin associations with $\alpha_4\beta_1$–VCAM-1 bonds thus allow these bonds to better resist externally applied mechanical strains [48]. These results suggest that, in analogy to L-selectin bonds, subsecond stabilization of integrin tethers depends on the ability of ligand-occupied integrins to properly anchor to the cytoskeleton via a bridge formed by paxillin binding to the $\alpha_4\beta_1$ cytoplasmic domains. Although a similar cytoskeletal partner of the LFA-1 integrin has not been yet defined, disruption of the actin cytoskeleton, which does not affect LFA-1 binding of immobilized ICAM-1 under shear-free conditions, strongly abrogates the integrin adhesiveness to its endothelial ligand ICAM-1 under shear flow [43].

5. WHY DO SELECTIN-MEDIATED ADHESIONS FAIL TO ARREST ROLLING LEUKOCYTES?

The unique capacity of fast-forming selectin–carbohydrate bonds to withstand mechanical strain may account for their exceptional ability to capture rapidly flowing leukocytes on endothelial target sites [53]. The transient nature of most selectin-mediated adhesions reflected in their high dissociation rate constants (Table 1) restricts these receptors, however, from arresting rolling leukocytes on their endothelial targets, even when some leukocyte endothelial contacts are stabilized by as many as 20–30 adhesive bonds during slow rolling processes mediated by L-selectin and E-selectin [54]. Notably, the P-selectin interaction with its major ligand PSGL-1 is much stronger than most L-selectin or E-selectin bonds and has comparable affinity to that between activated LFA-1 and ICAM-1 (Table 1). An additional stabilization of P-selectin-mediated tethers is provided by covalent dimerization of PSGL-1 and P-selectin dimers [26]. Nevertheless, the lifetime of singular P-selectin PSGL-1 bonds is as short as that of E-selectin ligand bonds, 30- and 100-fold shorter than that of high-affinity LFA-1 ICAM-1 bonds (Table 1). Thus, chemokine-activated ligand-rearranged LFA-1 bound to ICAM-1 has a predicted lifetime of 20–70 s, sufficient to arrest cells and recruit more

LFA-1 to the nascent arrest site and thereby further strengthen adhesion. P-selectin–PSGL-1 bonds and the weaker L-selectin and E-selectin bonds dissociate within second time-frames and so would fail to arrest rolling leukocytes. A key point to consider, however, is that all the kinetic information discussed above is based on Biacore measurements of unstressed bonds. Both integrin and selectin bonds are destabilized by shear stress and so the effective lifetime of their respective bonds subjected to shear stress is considerably shorter. Yet, the relative degree of this destabilization is not fully understood and can vary for individual molecular pairs [55]. Nevertheless, these kinetic Biacore-based estimates provide a reasonable prediction for why individual high-affinity integrin ligand bonds experiencing shear stress are more stable than the highest affinity selectin bonds, namely, the P-selectin–PSGL-1 interaction.

What makes integrins so versatile in their adhesive properties is the large heterogeneity of their conformational states and their unique ability to undergo ligand-induced rearrangement of their headpiece, a property unmatched by any known selectin–ligand pair. The ability of integrins to modulate their adhesive activities at endothelial contacts within a fraction of a second in response to local chemokine signals altering their cytoplasmic tails [42] is also unshared by any selectin–ligand interactions.

6. CONCLUSIONS

It has become increasingly evident that productive ligand recognition under shear flow is not only a uniquely fast adhesive event completed within subsecond time frames, but a complex adhesive event which involves specialized mechanical features of the interacting counterreceptors. Integrin and selectin bonds may share intrinsic mechanical properties such as high tensile strength of bonds which depend on the ability of these adhesion molecules to couple to the actin cytoskeleton upon ligand occupancy, and thereby stabilize adhesion specifically under stress [12,48]. A deeper understanding of how specific cytoskeletal assemblies of integrins and selectins vary among different subsets of leukocytes and cancer cells should help define new therapeutic targets to interfere with the function of these key adhesion receptors in leukocyte trafficking to tissues in numerous pathologies ranging from autoimmune diseases, allergic responses, graft rejection, and mestastasis (see Chapter 22).

Acknowledgments

R. Alon is the incumbent of the Linda Jacobs Chair in Immune and Stem Cell Research. The research discussed here has been supported by the Israel Science Foundation, MAIN, the EU6 Program for Migration and Inflammation, and by GIF, the German-Israeli Foundation for Scientific Research and Development.

REFERENCES

1. **Springer TA** (1994) *Cell* **76**, 301–314.
2. **Butcher EC & Picker LJ** (1996) *Science* 2 **72**, 60–66.

3. Mazo IB & von Andrian UH (1999) *J. Leukoc. Biol.* **66**, 25–32.
4. Muller WA (2003) *Trends Immunol.* **24**, 327–334.
5. McEver RP (2002) *Curr. Opin. Cell Biol.* **14**, 581–586.
6. Mackay CR (2001) *Nat. Immunol.* 2, 95–101.
7. Campbell JJ & Butcher EC (2000) *Curr. Opin. Immunol.* **12**, 336–341.
8. Hynes RO (2002) *Cell* **110**, 673–687.
9. Carman CV & Springer TA (2003) *Curr. Opin. Cell Biol.* **15**, 547–556.
10. Kansas GS, Ley K, Munro JM & Tedder TF (1993) *J. Exp. Med.* **177**, 833–838.
11. Setiadi H, Sedgewick G, Erlandsen SL & McEver RP (1998) *J. Cell Biol.* **142**, 859–871.
12. Dwir O, Kansas GS & Alon R (2001) *J. Cell Biol.* **155**, 145–156.
13. Kansas GS (1996) *Blood* **88**, 3259–3287.
14. Moore KL, Eaton SF, Lyons DE, Lichenstein HS, Cummings RD & McEver RP (1994) *J. Biol. Chem.* **269**, 23318–23327.
15. Mehta P, Cummings RD & McEver RP (1998) *J. Biol. Chem.* **273**, 32506–32513.
16. Nicholson MW, Barclay AN, Singer MS, Rosen SD & van der Merwe PA (1998) *J. Biol. Chem.* **273**, 763–770.
17. van der Merwe PA (1999) *Curr. Biol.* 9, R419–R422.
18. Lawrence MB & Springer TA (1991) *Cell* **65**, 859–873.
19. Alon R, Chen S, Puri KD, Finger EB & Springer TA (1997) *J. Cell Biol.* **138**, 1169–1180.
20. Lawrence MB, Kansas GS, Kunkel EJ & Ley K (1997) *J. Cell Biol.* **136**, 717–727.
21. Puri KD, Finger EB & Springer TA (1997) *J. Immunol.* **158**, 405–413.
22. Alon R, Chen S, Fuhlbrigge R, Puri KD & Springer TA (1998) *Proc. Natl Acad. Sci. USA* **95**, 11631–11636.
23. Ramachandran V, Nollert MU, Qiu H *et al.* (1999) *Proc. Natl Acad. Sci. USA* **96**, 13771–13776.
24. Dwir O, Kansas GS & Alon R (2000) *J. Biol. Chem.* **275**, 18682–18691.
25. Chen S & Springer TA (2001) *Proc. Natl Acad. Sci. USA* **98**, 950–955.
26. Ramachandran V, Yago T, Epperson TK *et al.* (2001) *Proc. Natl Acad. Sci. USA* **98**, 10166–10171.
27. Dwir O, Steeber DA, Schwarz US *et al.* (2002) *J. Biol. Chem.* **277**, 21130–21139.
28. Yago T, Leppanen A, Qiu H *et al.* (2002) *J. Cell Biol.* **158**, 787–799.
29. Alon R, Hammer DA & Springer TA (1995) *Nature* **374**, 539–542.
30. Rosen SD (2004) *Annu. Rev. Immunol.* **22**, 129–156.
31. Hammer DA & Lauffenburger DA (1989) *Cell Biophys.* **14**, 139–173.
32. von Andrian UH, Hasslen SR, Nelson RD, Erlandsen SL & Butcher EC (1995) *Cell* **82**, 989–999.
33. Picker LJ, Warnock RA, Burns AR, Doerschuk CM, Berg EL & Butcher EC (1991) *Cell* **66**, 921–933.
34. Schwarz US & Alon R (2004) *Proc. Natl Acad. Sci. USA* **101**, 6940–6945.
35. Somers WS, Tang J, Shaw GD & Camphausen RT (2000) *Cell* **103**, 467–479.
36. Finger EB, Puri KD, Alon R, Lawrence MB, von Andrian UH & Springer TA (1996) *Nature* **379**, 266–269.
37. Snapp KR, Heitzig CE & Kansas GS (2002) *Blood* **99**, 4494–4502.
38. Pavalko FM, Walker DM, Graham L, Goheen M, Doerschuk CM & Kansas GS (1995) *J. Cell Biol.* **129**, 1155–1164.
39. Shao JY & Hochmuth RM (1999) *Biophys. J.* **77**, 587–596.
40. Shao JY, Ting-Beall HP & Hochmuth RM (1998) *Proc. Natl Acad. Sci. USA* **95**, 6797–6802.
41. Alon R & Feigelson S (2002) *Semin. Immunol.* **14**, 93–104.
42. Laudanna C (2005) *Nat. Immunol.* **6**, 429–430.
43. Shamri R, Grabovsky V, Gauguet JM *et al.* (2005) *Nat. Immunol.* **6**, 497–506.
44. Salas A, Shimaoka M, Chen S, Carman CV & Springer T (2002) *J. Biol. Chem.* **277**, 50255–50562.
45. Salas A, Shimaoka M, Kogan AN, Harwood C, von Andrian UH & Springer TA (2004) *Immunity* **20**, 393–406.
46. Feigelson SW, Grabovsky V, Winter E *et al.* (2001) *J. Biol. Chem.* **276**, 13891–13901.
47. Kinashi T, Aker M, Sokolovsky-Eisenberg M *et al.* (2004) *Blood* **103**, 1033–1036.
48. Alon R, Feigelson SW, Rose DM *et al.* (2005) *J. Cell Biol.* **171**, 1073–1084.

49. Constantin G, Majeed M, Giagulli C et al. (2000) Immunity 13, 759-769.
50. Kim M, Carman CV, Yang W, Salas A & Springer TA (2004) J. Cell Biol. 167, 1241-1253.
51. Chen C, Mobley JL, Dwir O et al. (1999) J. Immunol. 162, 1084-1095.
52. Liu S, Thomas SM, Woodside DG et al. (1999) Nature 402, 676-681.
53. Greenberg AW, Brunk DK & Hammer DA (2000) Biophys. J. 79, 2391-2402.
54. Chen S & Springer TA (1999) J. Cell Biol. 144, 185-200.
55. de Chateau M, Chen S, Salas A & Springer TA (2001) Biochemistry 40, 13972-13979.
56. Wild MK, Huang MC, Schulze-Horsel U, van der Merwe PA & Vestweber D (2001) J. Biol. Chem. 276, 31602-31612.
57. Lupher ML, Jr., Harris EA, Beals CR, Sui LM, Liddington RC & Staunton DE (2001) J. Immunol. 167, 1431-1439.
58. Shimaoka M, Lu C, Palframan RT et al. (2001) Proc. Natl Acad. Sci. USA 98, 6009-6014.
59. Shimaoka M, Xiao T, Liu JH, et al. (2003) Cell 112, 99-111.
60. Sarantos MR, Raychaudhuri S, Lum AF, Staunton DE & Simon SI (2005) J. Biol. Chem. 280, 28290-28298.

SECTION 6
Carbohydrate-modifying enzymes

CHAPTER 17
Complex carbohydrate-modifying enzymes

Karen J. Loft and Spencer J. Williams

1. INTRODUCTION

Carbohydrate-based polymers are the most complex biological polymers known. The capacity for structural diversity in simple carbohydrate chains far exceeds that possible from chains of nucleotides (DNA/RNA) or amino acids (proteins). The complexity of carbohydrate structures arises from ring stereochemistry (D-glucose, D-galactose, etc.), linkage patterns (1→2, 1→3, 1→4, etc), alternate ring forms (furanose, pyranose, acyclic, etc.), and anomeric stereochemistry (α or β). Further complexity results from the presence of carbohydrate ring modifications, including deoxygenation, and the presence of amino and carboxy groups. A group of modifications that provides an additional, unique level of complexity to carbohydrates are ester functions with inorganic and organic acids. These modifications include the presence of sulfate, phosphate, and carboxylic acid esters.

Unlike stereochemical, linkage, and ring isomers, the modification of carbohydrates with esters is a potentially reversible process. Thus, the installation and removal of esters provides a simple and powerful mechanism by which the bioactivity of a carbohydrate entity can be modulated, potentially in a dynamic fashion. For example, modification of a specific lipidated tetrasaccharide (a nodulation (Nod) factor) with a sulfate ester allows the symbiotic nitrogen-fixing bacterium *Sinorhizobium meliloti* to elicit the formation of specialized root nodules on alfalfa, which it colonizes, resulting in a beneficial supply of ammonia to the plant host. Loss of the sulfate ester from this tetrasaccharide results in a change of the host specificity of *S. meliloti* from alfalfa to vetch. In this article we will present an overview of the occurrence of sulfate, phosphate, and carboxylic acid esters in carbohydrate structures, the enzymes responsible for their installation and removal, and examples of the biological functions mediated by the presence or absence of these groups.

2. CARBOHYDRATE SULFATION

Sulfated carbohydrates play indispensable roles in biology. Sulfate esters can be introduced to or removed from various sugars in order to modulate their bioactivity. In bacteria sulfated carbohydrates have been identified as osmoprotectants in halophilic bacteria, as modulators of interspecies interactions, and have been implicated in several pathogenic processes (Fig. 1). In mammals, modification by sulfation is a major pathway for the detoxification and excretion of xenobiotics [1]. In higher organisms sulfated carbohydrates are commonly found on cell surfaces or in the extracellular matrix. Here, in addition to their important structural roles, sulfated carbohydrates have been implicated in cellular signaling, pathogenicity, and normal cellular development. Consequently, the enzymes that catalyze the installation (sulfotransferases) and removal (sulfatases) of sulfate esters on biomolecules are important for regulation of the corresponding biological processes.

2.1 Carbohydrate sulfotransferases

Generally, biological sulfation is achieved through the action of sulfotransferases, which catalyze the transfer of a sulfuryl group (SO_3^-) from the activated sulfate donor 3'-phosphoadenosine-5'-phosphosulfate (PAPS; Fig. 1b) onto either a free alcohol or amine, producing an O-sulfate or N-sulfate, respectively (Fig. 1).

Eukaryotic sulfotransferases can be classified according to whether they are soluble or membrane-associated. The soluble or cytosolic sulfotransferases are responsible for the sulfation of small molecules such as hormones, neurotransmitters, drugs, and xenobiotics, and their broad specificities allow modification of a wide variety of substrates. Membrane-associated sulfotransferases are localized within the Golgi network and are typically more specific in the substrates that they recognize. These sulfotransferases are responsible for the posttranslational modification of peptides, proteins, and carbohydrates. The importance of sulfotransferases in normal biological function is underscored by the observation that defects in sulfotransferase activity result in a variety of developmental diseases including: macular corneal dystrophy, a disease characterized by opacity of the cornea leading to blindness; and spondyloepiphyseal dysplasia, which results in severe skeletal developmental defects [2,3].

The X-ray crystal structures of two membrane-associated [4,5], six cytosolic [6,7], and one bacterial sulfotransferase [8] have been determined. Although the cytosolic and membrane-associated enzymes share little sequence similarity they are all approximately spherical in shape and share similar α/β-folds. This fold consists of a central four- or five-stranded β-sheet surrounded by α-helices containing two well-conserved regions: an N-terminal region and a central sequence. Both of these regions are involved in binding the sulfate donor PAPS [9]. The similarity of all three-dimensional structures and the presence of these two conserved regions, both separated by a similar number of residues, has led to the suggestion that all sulfotransferases evolved from a common ancestral gene [9].

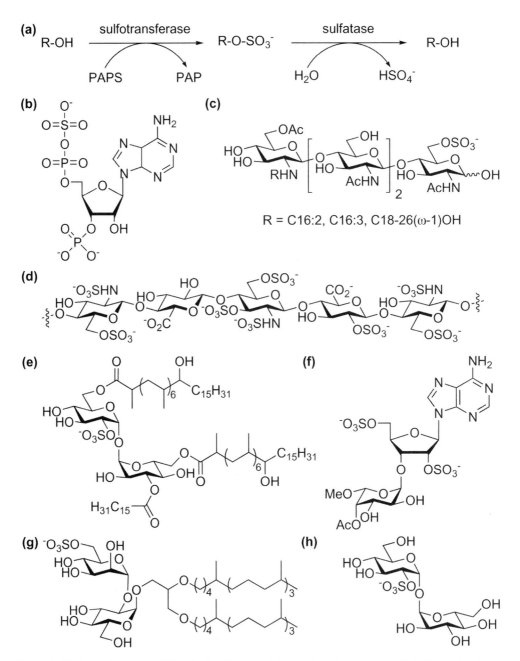

Figure 1. (a) General scheme of biological sulfation and desulfation. Examples of sulfated biomolecules: (b) sulfate donor 3′-phosphoadenosine-5′-phosphosulfate (PAPS); (c) Nod factor from *Sinorhizobium meliloti*; (d) heparin pentasaccharide antithrombin III-binding sequence; (e) sulfolipid 1 (SL-1) from *Mycobacterium tuberculosis*; (f) spider venom component HF-7 from *Holena curta*; (g) sulfated glycolipid from the halophile *Halofax* S-DGA-1; and (h) trehalose 2-sulfate from *Natronococcus* sp.

Sulfuryl group transfer by sulfotransferases that utilize PAPS proceeds by an in-line attack of the nucleophile (ROH or RNH$_2$) on the sulfuryl group of PAPS (Fig. 2a), assisted by a conserved active-site histidine residue [7]. In all examples studied to date, the mechanism of sulfuryl transfer proceeds by a sequential Bi-Bi mechanism, in which both substrates must bind to the enzyme to form a ternary complex prior to sulfuryl transfer.

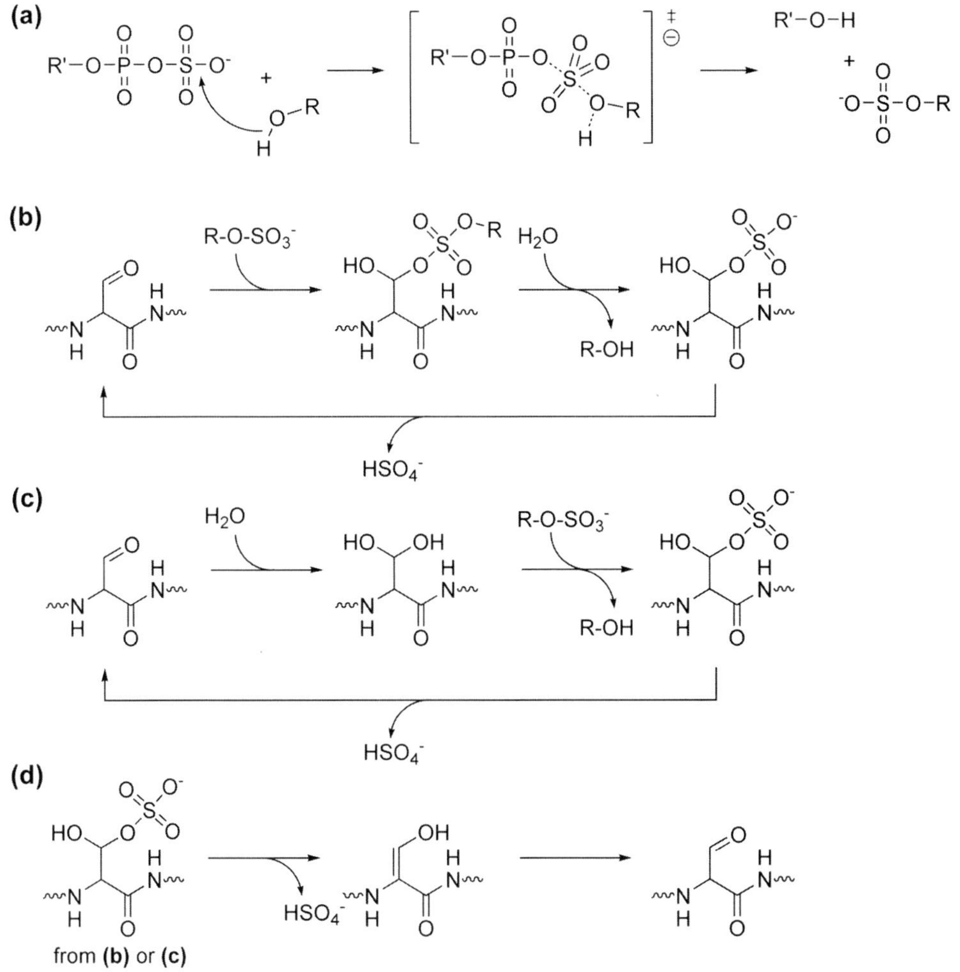

Figure 2. (a) Proposed mechanism of sulfotransferase-catalyzed sulfate ester formation. (b–d) Proposed mechanisms for sulfatase-catalyzed sulfate ester hydrolysis.

2.2 Carbohydrate sulfatases

Whilst sulfotransferases have received intense investigation for many years, sulfatases, which catalyze the hydrolysis of sulfate esters, have received considerably less attention owing largely to the misconception that their function was restricted to the degradation of organic sulfate esters in soil [10]. Since the discovery of inherited lysosomal storage disorders (see Chapters 2, 20, and 25) caused by sulfatase deficiencies in the 1970s sulfatases have emerged as important players in cellular and biological processes [11].

Human carbohydrate sulfatases can be grouped into two classes depending on whether they are found in the lysosome or the extracellular space. The lysosomal enzymes are the best characterized and are involved in the catabolism of sulfatides and glycosaminoglycans. The extracellular sulfatases, termed Sulfs, have only recently been reported and have been attributed crucial roles in signaling and embryonic development; their dysfunction is linked to tumor metastasis [10,12,13]. In contrast to the lysosomal carbohydrate sulfatases, which are *exo*-acting and remove sulfate esters from the ends of oligosaccharide chains, the extracellular Sulfs act in an *endo*-fashion, hydrolyzing sulfate esters on sugar residues found within the chain [14].

Sulfatases share a unique posttranslational modification, where an active-site cysteine (or serine in some prokaryotic sulfatases) is oxidized to a 2-amino-3-oxopropanoic acid or C_α-formylglycine (FGly), residue [15]. Multiple sulfatase deficiency (MSD), a rare autosomal recessive disorder, results in severely decreased activity of all sulfatases, causing severe developmental defects [11]. Studies of sulfatase peptides in fibroblasts derived from MSD patients revealed that the synthesis of sulfatase polypeptides occurs without modification to FGly, an observation that ultimately led to the discovery of this unique posttranslational modification [16]. The FGly modification is found exclusively in sulfatases and the enzyme that performs this modification in higher organisms has recently been identified and cloned [17,18].

Ongoing studies have revealed that sulfatases are members of an evolutionarily highly conserved gene family, with substantial sequence homology between bacterial, lower eukaryotic, and human enzymes [19]. The sulfatases are also structurally similar; of the four sulfatases whose crystal structures have been solved, all show similar mixed α/β topology [10]. Additionally, all four contain a highly conserved N-terminal region that is critical for directing the unique and unusual posttranslational modification that all functional sulfatases possess.

Currently, there are two proposed mechanisms of sulfatase action that involve the FGly residue. The first mechanism proposes that hydrolysis is initiated by the addition of the sulfated substrate to the aldehyde of FGly; subsequent attack of this intermediate by water at sulfur liberates the desulfated substrate leaving the sulfate bound to the enzyme; finally, elimination of sulfate to regenerate FGly completes the cycle (Fig. 2b) [20]. The second proposed mechanism bears similarity to that of alkaline phosphatase. It is proposed that FGly is first hydrated to form a geminal diol; next, one geminal hydroxyl group attacks the sulfur of the sulfate ester, expelling the parent alcohol or amine and leading to a covalently sulfated enzyme intermediate; finally, elimination of sulfate again completes the catalytic cycle (Fig. 2c) [21]. While some have argued for the latter mechanism (Fig. 2c) on the basis of X-ray structures of *Pseudomonas aeruginosa* arylsulfatase, where the geminal hydrate was directly observed [22], no definitive evidence has been presented that allows a clear-cut distinction between these two mechanisms.

A third mechanistic alternative is possible. In this mechanism, the formation of a covalent enzyme–sulfate intermediate can occur by either of the two proposed mechanisms above; next, elimination occurs by loss of the α-proton on the FGly residue to form an enol; finally, this enol tautomerizes to regenerate the active site aldehyde (Fig. 2d).

2.3 Involvement of sulfotransferases and sulfatases in heparin metabolism

Heparin is a heavily sulfated linear polymer composed of repeating units of 1→4 linked uronic acid and D-glucosamine residues. Approximately 90% of the uronic acid residues are L-iduronic acid, while the remaining 10% are D-glucuronic acid residues [23]. Sulfation can occur at the 2-position of the uronic acid residue, or the 3- and 6-positions of the D-glucosamine residue. Additionally, the amino group of D-glucosamine may be substituted with an acetyl or sulfuryl group, or can remain unsubstituted, leading to considerable structural complexity. Due to the high degree of sulfation, heparin has the highest negative charge density of any biological macromolecule, with the average heparin disaccharide unit possessing 2.7 sulfate monoesters.

Heparin, and the biosynthetically related heparan sulfate (HS), are ubiquitously associated with mammalian cells. Essentially all cells produce HS proteoglycans as part of basement membranes, on the cell surface, and within the extracellular matrix [24]. HS is a major component of the cellular environment (see Chapter 9) and has been implicated in a wide variety of biological processes such as cell proliferation, differentiation, adhesion, migration, morphogenesis, inflammation, blood coagulation, and tumor cell invasion [2,25,26]. Additionally, some pathogens such as *Pseudomonas aeruginosa*, *Plasmodium falciparum* (malaria), *Mycobacterium tuberculosis* [27], and HIV-1 can bind to cell surface HS proteoglycans to effect entry and virulence [23, 28].

Heparin is used widely for its anticoagulant properties. Heparin binds to antithrombin (ATIII), a serine protease inhibitor, resulting in a conformational change of ATIII that allows it to act as an inhibitor of thrombin and other proteases in the blood coagulation cascade. Inhibition is dependent on the presence of the ATIII-binding pentasaccharide (see Fig. 1d), and of signal importance is the 3-*O*-sulfate group found on the central glucosamine residue [29]. Synthetic heparin pentasaccharide is produced commercially and is in clinical use as an antithrombotic marketed as fondaparinux (Arixtra).

The biosynthesis of heparin begins with the synthesis of the unsulfated polysaccharide chain, composed of repeating β-1,4-GlcNAc-β-1,4-D-GlcA units, in the Golgi apparatus. Modification of this chain then proceeds in a tightly orchestrated manner. The bifunctional enzyme glucosamine *N*-deacetylase/*N*-sulfotransferase (NDST) starts the process by deacetylating some of the *N*-acetylglucosamine residues along the chain to generate glucosamine residues. The

N-sulfotransferase activity of the dimer then catalyzes the installation of a sulfate ester from PAPS, although this process is frequently incomplete. Following the action of NDST, a C-5 epimerase converts D-glucuronic acid residues to L-iduronic acid. Sulfation then occurs at the 2-position of the L-iduronic acid (catalyzed by a 2-*O*-sulfotransferase) succeeded by the sequential sulfation at the 6- and then 3-positions of the glucosamine residues (catalyzed by a 6-*O*-sulfotransferase and a 3-*O*-sulfotransferase, respectively). Further modification is effected by the action of the Sulf sulfatases; these *endo*-acting sulfatases hydrolyze 6-*O*-sulfate esters on heparin within the extracellular matrix as well as during heparin biosynthesis in the Golgi apparatus [14].

Degradation of heparin chains occurs within the lysosome and requires the complementary action of two different types of enzymes: sulfatases and glycosidases. Within the lysosome of mammalian cells there are sulfatases specific for each substitution found in the heparin chain, namely a 2-*O*-, 3-*O*-, and 6-*O*-sulfatase, and an *N*-sulfatase (sulfamidase). These enzymes are *exo*-acting and degrade the heparin chain from the non-reducing terminus in a highly ordered fashion. Generally, removal of the exposed sulfate(s) on the non-reducing terminus is required before the terminal carbohydrate residue can be cleaved by the action of a glycosidase.

The mucopolysaccharidoses (MPS) are a family of heritable disorders caused by deficiency of individual lysosomal enzymes needed to degrade glycosaminoglycans. These disorders are chronic and progressive, and often show a wide spectrum of clinical severity, even within a single enzyme deficiency [11]. Deficiencies in the heparin-degrading enzymes iduronate 2-*O*-sulfatase, *N*-sulfatase, and *N*-acetylglucosamine 6-sulfatase result in MPS II (Hunter syndrome), and MPS III (Sanfilippo syndrome) types A and D, respectively. Symptoms can range from skeletal abnormalities and organomegaly (enlarged organs), to mental retardation and central nervous system dysfunction.

2.4 Sulfated Nod factors: lipooligosaccharides that modulate interspecies symbiosis

Sulfated carbohydrates are relatively rare in bacteria, having only been reported in various mycobacteria [27], halophiles [30], and in a few members of the rhizobium family (genera *Sinorhizobia*, *Rhizobia*, *Azorhizobia*, and *Bradyrhizobia*) [31].

Rhizobia are bacteria capable of forming a symbiotic relationship with legumes. They infect specific leguminous hosts resulting in the formation of root nodules – specialized organs that are occupied by the bacteria. Within nodules the bacteria fix atmospheric nitrogen in the form of ammonia that the plant host uses, thereby promoting plant growth [32]. Nodulation (Nod) factors are molecules that stimulate the formation of root nodules and are released from the bacteria in response to signaling compounds (usually flavonoids) secreted in turn by the legume. Most Nod factors possess a backbone of 3, 4, or 5 β-1,4-linked *N*-acetylglucosamine residues, modified at the non-reducing end by *N*-acylation with a long-chain fatty acid (see Fig. 1c). Additionally, at the non-reducing end they may be *N*-methylated, *O*-acetylated, or *O*-carbamoylated, while the reducing end carbohydrate may be *O*-acetylated, *O*-sulfated, or substituted with L-fucosyl, 2-*O*-methyl-L-fucosyl, or 4-*O*-acetyl-L-fucosyl residues.

All Nod factors released by *Sinorhizobium meliloti* are sulfated at the reducing end, enabling the bacteria to infect its native host alfalfa [33]. Sulfation of the glucosamine backbone has been found to be dependent upon three genes; *nodH*, *nodP*, and *nodQ* [33]. The products of *nodP* and *nodQ* are ATP sulfurylase and APS kinase, respectively, enzymes involved in the synthesis of the universal activated sulfate donor PAPS, whereas the product of *nodH* is a sulfotransferase that catalyzes the installation of the sulfate ester. Deletion of the NodH sulfotransferase results in synthesis of unsulfated Nod factors that are unable to induce nodulation in the native host alfalfa. Remarkably, these desulfo-Nod factors can now induce nodulation and infection of the non-native host vetch [33]. This shift in host range demonstrates the importance of a sulfate group for the species specificity of nodulation by the bacterium.

Nod factors are hydrolyzed by chitinases, glycosidases that are specific for β-1,4-linkages within chitooligosaccharide chains. Certain variations within the oligosaccharide chain such as chain length, acetylation, and sulfation have been shown to confer stability towards hydrolysis by chitinases [34]. Thus, the presence of a sulfate ester may result in longer residence times for Nod factors within soil.

3. CARBOHYDRATE PHOSPHORYLATION

Phosphorylated carbohydrates play a major role in central cellular metabolism. Phosphate esters are found on a variety of hexoses and pentoses; for example, within metabolic pathways such as the glycolysis pathway and those of amino acids, in DNA and RNA as diesters, and on nucleotides such as AMP. This section will focus on phosphorylated carbohydrates that are not involved in nucleic acids, cellular signal transduction, or metabolism, and which have a more permanent and functional role. In comparison to the large number of phosphorylated carbohydrates that are metabolic intermediates there are relatively few non-intermediary phosphorylated carbohydrates. In eukaryotes it has been found that a phosphorylated mannose residue, mannose-6-phosphate, plays a critical role in the trafficking of lysosomal enzymes within the cell. There are also a few examples of phosphorylated carbohydrates present in the cell walls of microorganisms such as yeast [35], *Mycobacteria*, and *Leishmania* [36].

3.1 Mannose 6-phosphate: a lysosomal targeting signal

Lysosomal enzymes, secretory proteins, and plasma membrane proteins are all synthesized on membrane-bound polysomes in the rough endoplasmic reticulum (ER). They are then transferred into the lumen of the ER where they undergo glycosylation and further modification. The newly modified proteins migrate to the Golgi apparatus where they can be further modified and are sorted for targeting to their ultimate destination. Proteins destined for the lysosome undergo phosphorylation at the 6-position to generate mannose-6-phosphate

(M6P) (Fig. 3a). M6P binds to the M6P receptor, which ensures specific targeting to the lysosome. I-cell disease is an inherited lysosomal deficiency that results from loss of M6P. I-cell disease is characterized by a loss of lysosomal function that arises not from a lack of enzyme activity, but rather incorrect targeting of the lysosomal enzymes to their required subcellular destination [37].

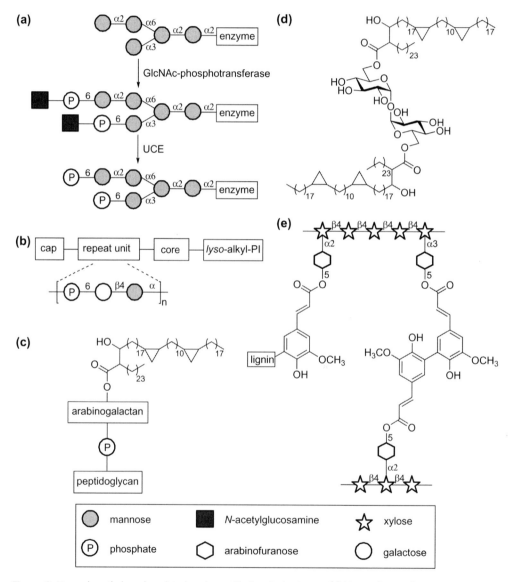

Figure 3. Examples of phosphorylated and esterified carbohydrates: (a) biosynthesis of mannose-6-phosphate; (b) repeating unit of leishmanial lipophosphoglycan (LPG); (c) mycobacterial mycolylarabinogalactan–peptidoglycan complex; (d) trehalose dimycolate (cord factor); and (e) hemicellulosic ferulic acid esters.

The biosynthetic approach used by the cell for the installation of the phosphate ester of M6P differs considerably from the biosynthesis of sulfate esters. Rather than transferring phosphate onto mannose using an activated phosphate donor, the phosphate ester is installed in two steps (Fig. 3a). Within the Golgi, a phosphate diester is first formed by the *en bloc* transfer of a phosphorylated carbohydrate onto the 6-hydroxyl of the mannosyl group to be phosphorylated. Second, a glycosidase hydrolyzes the transferred carbohydrate, leaving behind a phosphate monoester.

The first enzyme involved in the biosynthesis of the M6P group is UDP-*N*-acetylglucosamine:lysosomal-enzyme *N*-acetylglucosamine-1-phosphotransferase (*N*-acetylglucosamine-phosphotransferase). This enzyme transfers an *N*-acetylglucosamine-1-phosphate from UDP-*N*-acetylglucosamine to the C-6 position of a mannosyl residue on a glycoprotein, resulting in an intermediate phosphodiester-modified glycoprotein [38,39]. Suitable mannosyl residues for *N*-acetylglucosamine-1-phosphate transfer by *N*-acetylglucosamine-phosphotransferase are those found on α-1,3- and α-1,6-branches of high-mannose containing oligosaccharide chains on the surface of the lysosomal enzyme. While the signal that leads to *N*-acetylglucosamine-1-phosphate transfer is poorly understood, it has been found that lysosomal enzymes are phosphorylated at least 100 times more efficiently than non-lysosomal enzymes or glycoproteins that contain the same type of high-mannose oligosaccharides on their surface [40]. It is postulated that the signal for this modification is determined by the tertiary structure of the lysosomal proteins, due to the lack of sequence similarity among the lysosomal proteins characterized to date. Furthermore, heat denaturation of lysosomal proteins leads to reduced or inhibited *N*-acetylglucosamine-1-phosphate transfer [40]. Recent studies on the structurally unrelated lysosomal enzymes cathepsins A, B, and D indicate that there are several residues and loop structures that are involved in enzyme recognition leading to the installation of the *N*-acetylglucosamine mannose phosphodiester [41].

N-Acetylglucosamine-1-phosphodiester α-*N*-acetylglucosamidase (uncovering enzyme, UCE) is the second enzyme involved in the synthesis of M6P residues. This enzyme is also located in the Golgi apparatus and is a glycosidase that removes the terminal *N*-acetylglucosaminyl residue to reveal the phosphate monoester. After removal of the *N*-acetylglucosamine residue, and formation of the M6P moiety, the phosphorylated enzyme is transported to the lysosome through interaction with the M6P receptor. Once within the lysosome the phosphate ester is removed by one or more of numerous phosphatases found within the lysosomal compartment. To date it is unclear which enzyme hydrolyzes the M6P ester, although it has been suggested that purple acid phosphatase (tartrate-resistant phosphatase type V, uteroferrin), one of the major phosphatases present in lysosomes, is responsible [42].

3.2 Lipophosphoglycans of *Leishmania* parasites

Protozoa of the genus *Leishmania* are the sandfly-transmitted parasites that are the causative agents of leishmaniasis, a disease that affects millions of people annually. In order for the leishmania parasite to survive throughout its life-cycle it must be resistant to a variety of degradative enzymes within the gut of the sandfly, the bloodstream of the host animal, and within the macrophage. Leishmania parasites synthesize a cell surface phosphorylated carbohydrate, lipophosphoglycan (LPG), which plays a central role in enabling the survival of the parasite against these harsh conditions [43]. LPG also plays an important role in the colonization of the sandfly vector by the parasite. Upon ingestion the parasite attaches to the gut wall of the sandfly preventing excretion along with the rest of the digested bloodmeal, whilst offering some protection from digestive enzymes. Mutants lacking the ability to synthesize LPG are unable to colonize the sandfly vector, and cannot be transmitted to the mammalian host [44].

Leishmania LPG can be divided into four major regions (Fig. 3b): a phosphatidylinositol lipid anchor, a glycan core, a repeating phosphorylated disaccharide unit, and a small oligosaccharide cap. The repeating phosphorylated disaccharide unit of LPG is conserved, consisting of Gal-β-1,4-Man-α-1-PO_4-6- repeats, but can be substituted with different branching oligosaccharide chains depending on the species.

Synthesis of the unique disaccharide repeating units of LPG occurs by the stepwise addition of galactose and mannose-1-phosphate residues from their respective nucleotide derivatives, UDP-galactose and GDP-mannose. Two enzymes, α-D-mannosylphosphate transferase (MPT) and β-D-galactosyl-transferase (β-GalT), are responsible for the synthesis of the disaccharide [45,46]. Mutant *L. donovani* strains defective in LPG biosynthesis have demonstrated that the addition of the first mannose-1-phosphate residue is carried out by a mannosylphosphotransferase specific for the LPG glycan core-phosphatidylinositol substrate, and which has therefore been termed the LPG "initiating" phase MPT (iMPT) [47]. The second enzyme responsible for mannose-1-phosphate addition is the "elongation" phase MPT (eMPT), which is responsible for the extension of the disaccharide repeat unit to generate the full-length phosphoglycan. This enzyme recognizes the Gal-β-1,4-Man-α-1-PO_4 disaccharide unit, with the negatively charged phosphodiester linkage playing a major role in substrate specificity [48]. A β-linkage between the two sugar moieties is also essential, whilst substitutions to the backbone of the previous repeat unit can either enhance or reduce the rate of catalysis [49].

Transfer of the galactosyl residues are catalyzed by a β-1,3-galactosyltransferase; initial studies have shown that the minimum structural element required for binding of the enzyme is the terminal phosphate moiety [50].

4. CARBOHYDRATE CARBOXYLIC ACID ESTERS

Many carboxylic acid esters are found on carbohydrates with the most common types being esters of acetic acid and longer chain fatty acids. These modifications are particularly common in cell wall carbohydrates found in bacteria. A complex system of carbohydrate carboxylic acid esters is involved in biosynthesis and structure of the cell wall of mycobacteria. Within plants cell wall carbohydrates are esterified with hydroxycinnamic acids. Ultimately, through cross-linking reactions, these esters provide one of the major structural elements that endow woody plants with much of their strength.

4.1 Carbohydrate esters in the mycobacterial cell wall

Tuberculosis is one of the world's leading causes of death and over one-third of the global population is infected. The causative agent of tuberculosis is *Mycobacterium tuberculosis*. Members of the genus *Mycobacteria* possess a unique cell wall that is essential for growth and survival of the organism. The mycobacterial cell wall is composed of a superpolymer of covalently linked peptidoglycan, arabinogalactan, and mycolic acid moieties, which together form the mycolylarabinogalactan–peptidoglycan complex (mAGP complex). The arabinan of the mAGP complex is modified by esterification with mycolic acids, resulting in the formation of a second membrane-like structure that parallels the plasma membrane (Fig. 3c).

Carbohydrate carboxylic acid esters play a central role in the transfer of mycolic acids from the interior of the cell where they are biosynthesized to their eventual destination in the mAGP complex. Mycolic acids are long-chain (C_{60}–C_{90}) fatty acids that are β-hydroxylated and substituted with a long β-alkyl side-chain. The biosynthesis of mycolic acids occurs within the cytoplasm through the action of the multi-enzyme complex, type I and II fatty acid synthetase. Mycolic acids must then be transported from the cytoplasm to their eventual destination in the cell wall. Strong evidence has accumulated that movement of mycolic acids is facilitated by a carrier molecule, the disaccharide trehalose. Mycolic acids are transferred onto the primary hydroxyls of trehalose to generate trehalose dimycolate (TDM, also termed cord factor, see Fig. 3d). TDM is transported out of the cell where it forms a significant non-covalently associated portion of the cell wall. Transfer of mycolic acid to the arabinan of the mAGP complex is catalyzed by a set of transesterases, the antigen 85 proteins (85A, 85B, and 85C) [51]. The central role of trehalose as a carrier molecule for mycolic acid is illustrated by the effect of insertionally inactivating trehalose biosynthesis in the fast-growing mycobacterium, *M. smegmatis*. A mutant incapable of synthesizing its own trehalose could only be maintained in media containing trehalose or related trehalose derivatives [52].

The role of the antigen 85 transesterases is complex; aside from catalyzing the transfer of mycolic acid from TDM to the mAGP complex these enzymes can also reversibly catalyze the transfer of mycolic acids between TDM and trehalose to form two molecules of trehalose monomycolate, a reaction that presumably provides a ready flux of mycolyl groups for cell wall biosynthesis. The antigen 85 proteins are major secreted proteins of *M. tuberculosis*, consistent with their role in cell wall manufacture. They share a high degree of sequence homology (68–80%), a common α/β hydrolase fold, and possess a catalytic triad, Ser-Glu-His, which they share with lipases. X-ray crystal structures have revealed the presence of two structurally conserved regions; an active site carbohydrate-binding pocket, and a large surface region that is postulated to be the site of interaction with fibronectin [53,54].

The α/β hydrolase fold is common to esterases, lipases, and proteases, and consists of a mostly parallel, eight-stranded β-sheet surrounded by α-helices. This structure serves as a scaffold for the conserved active-site catalytic residues: a nucleophile (serine, cysteine, or aspartate), an acid (aspartate or glutamate), and a histidine residue. Ester cleavage occurs through the nucleophilic attack of the serine (or other residue) on the carbonyl carbon to form a covalently modified enzyme intermediate. Hydrolysis of this acyl-enzyme intermediate regenerates the active-site nucleophile.

4.2 Plant lignin–carbohydrate complexes

Plant cell wall polysaccharides are the most abundant organic materials on earth. After cellulose, the hemicelluloses are the next most common type of polysaccharides found within plants, constituting 20–35% by weight. Their major roles are to confer structural rigidity to plant cell walls and provide protection against microbial attack. Hemicelluloses are heterogeneous branched polymers consisting of pentoses (xylose, arabinose), hexoses (mannose, glucose, galactose), and uronic acids (glucuronic, 4-O-methylglucuronic acids); these polymer chains may be further methylated, or esterified with acetic, ferulic, or p-coumaric acids (see Fig. 3e). The major function of ferulic acid esters is to enhance the strength of the cell wall by forming diferuloyl cross-links between polymer chains or with lignin, a complex polyphenolic compound found in the extracellular matrix.

Enzymatic degradation of hemicelluloses has attracted considerable interest because of practical applications in various industrial processes, such as fuel and chemical production, delignification of paper pulp, enhancement of animal feedstocks, clarification of juices, and improvement of the consistency of beer [55]. Degradation of hemicelluloses occurs through the synergistic action of main-chain degrading glycosidases and a series of carbohydrate esterases, including acetylesterases, methylesterases, p-coumaroyl esterases and feruloyl esterases. Carbohydrate esterases are largely found within bacteria and fungi that catabolize hemicelluloses, but are also present in plant genomes, and presumably function to allow remodeling of cell wall hemicellulose during development.

All of the X-ray crystal structures determined for feruloyl esterases, cinnamoyl esterases and acetylxylan esterases reveal an α/β hydrolase fold in addition to the conserved catalytic triad, Ser-His-Asp, common to lipases [56,57]. These two features are common to carboxylesterases and the feruloyl esterases therefore constitute a subclass of this group [58].

5. CONCLUSION

The role of carbohydrate esters in biology has been slow to emerge as the problems inherent in the study of carbohydrates, including the difficulty of obtaining homogeneous samples, of structural characterization, and of chemical synthesis, are compounded for ester derivatives. As the examples herein show, modifications to carbohydrate structures by various organic and inorganic esters provides a unique mechanism with which to control biological function in an easily modifiable manner. The enzymes that mediate these modifications are in most cases unified by common mechanisms and this provides a base through which to integrate apparently unconnected observations and studies in the area. While a number of cases are now known where a detailed description of biological function for carbohydrate esters is well described, further investigations are needed to understand the biological rationale for the presence of carbohydrate esters in a wide range of other biomolecules.

Acknowledgments

Financial support from the Australian Research Council, the ANZ Medical Research and Technology in Victoria, and the University of Melbourne are gratefully acknowledged.

REFERENCES

1. Kertesz MA (1999) *FEMS Microbiol. Rev.* **24**, 135–175.
2. Honke K & Taniguchi N (2002) *Med. Res. Rev.* **22**, 637–654.
3. Thiele H, Sakano M, Kitagawa H, *et al.* (2004) *Proc. Natl Acad. Sci. USA* **101**, 10155–10160.
4. Kakuta Y, Sueyoshi T, Negishi M & Pedersen LC (1999) *J. Biol. Chem.* **274**, 10673–10676.
5. Edavettal SC, Lee KA, Negishi M, Linhardt RJ, Liu J & Pedersen LC (2004) *J. Biol. Chem.* **279**, 25789–25797.
6. Lee KA, Fuda H, Lee YC, Negishi M, Strott CA & Pedersen LC (2003) *J. Biol. Chem.* **278**, 44593–44599.
7. Chapman E, Best MD, Hanson SR & Wong C-H (2004) *Angew. Chem. Int. Ed.* **43**, 3526–3548.
8. Mougous JD, Petzold CJ, Senaratne RH *et al.* (2004) *Nat. Struct. Mol. Biol.* **11**, 721–729.
9. Kakuta Y, Pedersen LG, Pedersen LC & Negishi M (1998) *Trends Biochem. Sci.* **123**, 129–130.
10. Hanson SR, Best MD & Wong C-H (2004) *Angew. Chem. Int. Ed.* **43**, 5736–5763.
11. Neufeld EF & Muenzer J (1995) In: *The Metabolic and Molecular Basis of Inherited Disease*, 7th edn, pp. 2465–2494. Edited by CR Scriver, AL Beaudet, WS Sly and D Valle. McGraw-Hill, New York, USA.
12. Dhoot GK, Gustafsson MK, Ai X, Sun W, Standiford DM & Emerson CP (2001) *Science* **293**, 1663–1666.
13. Morimoto-Tomita M, Uchimura K, Werb Z & Hemmerich S (2002) *J. Biol. Chem.* **277**, 49175–49185.
14. Morimoto-Tomita M, Uchimura K, Werb Z, Hemmerich S & Rosen SD (2002) *J. Biol. Chem.* **277**, 49175–49185.
15. Dierks T, Miech C, Hummerjohann J, Schmidt B, Kertesz MA & von Figura K (1998) *J. Biol. Chem.* **273**, 25560–25564.
16. Rommerskirch W & von Figura K (1992) *Proc. Natl Acad. Sci. USA* **89**, 2561–2565.
17. Fang Q, Peng J & Dierks T (2004) *J. Biol. Chem.* **279**, 14570–14578.
18. Sardinello M, Annunziata I, Roma G & Ballabio A (2005) *Hum. Mol. Genet.* **14**, 3203–3217.
19. Schmidt B, Selmer T, Ingendoh A & von Figura K (1995) *Cell* **82**, 271–278.
20. Bond CS, Clements PR, Ashby SA *et al.* (1997) *Structure* **5**, 277–289.
21. von Bulow R, Schmidt B, Dierks T, von Figura K & Uson I (2001) *J. Mol. Biol.* **305**, 269–277.
22. Boltes I, Czapinska H, Kahnert A *et al.* (2001) *Structure* **9**, 483–491.
23. Caplia I & Linhardt RJ (2002) *Angew. Chem. Int. Ed.* **41**, 390–412.

24. Strott CA (2002) *Endocr. Rev.* **23**, 703–732.
25. Sasisekharan R, Shriver Z, Venkataraman G & Narayanasami U (2002) *Nat. Rev. Cancer* **2**, 521–528.
26. Perrimon N & Bernfield M (2000) *Nature* **404**, 725–728.
27. Mougous JD, Green RE, Williams SJ, Brenner SE & Bertozzi CR (2002) *Chem. Biol.* **9**, 767–776.
28. Park PW, Pier GB, Hinkes MT & Bernfield M (2001) *Nature* **411**, 98–102.
29. Petitou M, Casu B & Lindahl U (2003) *Biochimie* **85**, 83–89.
30. Kates M (1993) *Experientia* **49**, 1027–1036.
31. Denarie J, Debelle F & Prome J-C (1996) *Annu. Rev. Biochem.* **65**, 503–535.
32. D'Haeze W & Holsters M (2002) *Glycobiology* **12**, 79R–105R.
33. Roche P, Debelle F, Maillet F et al. (1991) *Cell* **67**, 1131–1143.
34. Staehelin C, Schultze M, Kondorosi E, Mellor RB, Boller T & Kondorosi A (1994) *Plant J.* **5**, 319–330.
35. Jigami Y & Odani T (1999) *Biochim. Biophys. Acta* **1426**, 335–345.
36. Turco SJ & Descoteaux A (1992) *Annu. Rev. Microbiol.* **46**, 65–94.
37. von Figura K & Hasilik A (1986) *Annu. Rev. Biochem.* **55**, 167–193.
38. Tabas I & Kornfeld S (1980) *J. Biol. Chem.* **255**, 6633–6639.
39. Hasilik A, Klein U, Waheed A, Strecker G & von Figura K (1980) *Proc. Natl Acad. Sci. USA* **77**, 7074–7078.
40. Reitman ML & Kornfeld S (1981) *J. Biol. Chem.* **256**, 11977–11980.
41. Lukong KE, Elsliger M-A, Mort JS, Potier M & Pshezhetsky AV (1999) *Biochemistry* **38**, 73–80.
42. Bresciani R & von Figura K (1996) *Eur. J. Biochem.* **238**, 669–674.
43. McConville MJ. (1996) In: *Molecular Biology of Parasitic Protozoa*, pp. 205–228. Edited by DF Smith and M Parsons. Oxford University Press, New York, USA.
44. Sacks DL, Modi G, Rowton E et al. (2000) *Proc. Natl Acad. Sci. USA* **97**, 406–411.
45. Carver MA & Turco SJ (1991) *J. Biol. Chem.* **266**, 10974–10981.
46. Carver MA & Turco SJ (1992) *Arch. Biochem. Biophys.* **295**, 309–317.
47. Descoteaux A, Mengeling BJ, Beverley SM & Turco SJ (1998) *Mol. Biochem. Parasitol.* **94**, 27–40.
48. Brown GM, Millar AR, Masterson C, Brimacombe JS & Nikolaev AV (1996) *Eur. J. Biochem.* **242**, 410–416.
49. Routier FH, Higson AP, Ivanova IA et al. (2000) *Biochemistry* **39**, 8017–8025.
50. Ng K, Handman E & Bacic A (1996) *Biochem. J.* **317**, 247–255.
51. Belisle JT, Vissa VD, Sievert T, Takayama K, Brennan PJ & Besra GS (1997) *Science* **276**, 1420–1422.
52. Woodruff PJ, Carlson BL, Siridechadilok B et al. (2004) *J. Biol. Chem.* **279**, 28835–28843.
53. Ronning DR, Klabunde T, Besra GS, Vissa VD, Belisle JT & Sacchettini JC (2000) *Nat. Struct. Biol.* **7**, 141–146.
54. Ronning DR, Vissa VD, Besra GS, Belisle JT & Sacchettini JC (2004) *J. Biol. Chem.* **279**, 36771–36777.
55. Saha BC (2003) *J. Ind. Microbiol. Biotechnol.* **30**, 279–291.
56. Kroon PA, Williamson G, Fish NM, Archer DB & Belshaw NJ (2000) *Eur. J. Biochem.* **267**, 6741–6752.
57. Faulds CB, Molina R, Gonzalez R et al. (2005) *FEBS J.* **272**, 4362–4371.
58. Matthew S & Abeaham TE (2004) *Crit. Rev. Biotechnol.* **24**, 59–83.

CHAPTER 18

Glycosyltransferases specific for the synthesis of mucin-type *O*-glycans

Inka Brockhausen

1. INTRODUCTION

Oligosaccharides linked through GalNAcα-Ser/Thr to peptide are called mucin-type *O*-glycans. These *O*-glycans are common in mucin glycoproteins where they have protective and receptor functions. *O*-Glycans may also control protein secretion and stability, cell adhesion, fertilization, and immune functions. The glycosyltransferases that assemble the core structures of *O*-glycans, as well as several sialyltransferases, are specific for this type of glycosylation, and their activities can be altered by hormones, cytokines, and other external factors. Important factors controlling the biosynthesis of *O*-glycans include the composition of the Golgi assembly line, the distinct substrate specificities, and the relative activities of transferases.

During development, apoptosis, and in many diseases such as cancer and cystic fibrosis, significant alterations have been observed in the expression and activities of transferases which predict new ranges of structures and altered functions of *O*-glycans.

Glycosyltransferases can be classified into more than 80 families having similar amino acid sequences and activities within a family. All of the enzymes involved in *O*-glycan biosynthesis are Golgi membrane-bound. To date, seven inverting and retaining glycosyltransferases involved in *O*-glycosylation have been crystallized and shown to have the overall GT-A structure with similar nucleotide sugar-binding domains. Knowledge of the enzyme active sites will be useful in designing specific glycosyltransferase inhibitors that have potential as therapeutic drugs.

1.1 Structures and functions of mucin-type *O*-glycans

Several types of carbohydrate-Ser/Thr *O*-linkages are found in glycoproteins, including Manα-*O*-Ser/Thr in yeast and mammalian glycoproteins, GlcNAcβ-*O*-Ser/Thr in nuclear and cytoplasmic proteins, Glcβ-*O*-Ser/Thr and Fucα-*O*-Ser/Thr in blood clotting factors, Xylβ-*O*-Ser-linked *O*-glycans in proteoglycans, and Galβ-*O*-hydroxy-Lys linkages in collagens. This review deals with the mucin-type

O-glycans that are O-linked through GalNAcα to Ser or Thr, and are most commonly found in mucins.

Mucins are high molecular weight monomeric or polymeric secreted glycoproteins in the viscous mucous gel that protects the mucous epithelium. In addition, mucins are found as cell surface-bound molecules that play important roles in cell adhesion. Mucins exhibit a high degree of heterogeneity at the levels of DNA, mRNA, protein, and carbohydrate. Their characteristic feature is the occurrence of variable number of tandem repeat (VNTR) regions that are highly O-glycosylated. O-Glycans can comprise 80% of the mass of the molecule and dictate its viscous, gel-forming, and adhesive properties [1]. The extensively O-glycosylated domains of mucins render the protein in a "bottlebrush" conformation that is resistant to protease degradation. These mucin-type O-glycans can also be found both in heavily O-glycosylated mucin-like or in sparsely glycosylated non-mucin glycoproteins.

Unmodified GalNAc residues attached to Ser/Thr of the peptide are rarely found in glycoproteins from normal tissues but this structure is common in mucins from cancer cells and has been designated Tn antigen (Table 1) [2]. Additional sugars are found in core structures 1–4 (Table 1), which are common in mucins from goblet or epithelial cells dedicated to the production of the mucous gel, although minor core structures also occur [3]. The core 1 structure is normally substituted with sulfate, sialic acid, or other sugars, but in cancer remains often unsubstituted as the T antigen [2,4]. Core structures can be extended by repeats of Gal-GlcNAc units, which are usually further substituted with sialic acid, with other terminal sugar structures, and sulfate esters. Many of the terminal structures have been recognized as antigens, and can represent blood group antigens and those recognized by lectins or antibodies. Mucins are particularly rich in blood group and Lewis antigens. O-Glycan structures have been shown to be significantly altered in cancer, leukemia, cystic fibrosis, and several other diseases [2–5].

Table 1 O-Glycan structures found in mammalian mucins

Tn antigen	GalNAcα-Thr/Ser
Sialyl-Tn (STn)	Sialylα2-6GalNAcα-Thr/Ser
Core 1, T antigen	Galβ1–3 GalNAcα-Ser/Thr
Sialyl-T	Sialylα2-3Galβ1-3GalNAcα-Thr/Ser
	Sialylα2-6(Galβ1-3)GalNAcα-Thr/Ser
Disialyl-T	Sialylα2-6(Sialylα2-3Galβ1-3)GalNAcα-Thr/Ser
Core 2	GlcNAcβ1-6(Galβ1-3)GalNAcα-
Core 3	GlcNAcβ1-3 GalNAcα-
Core 4	GlcNAcβ1-6(GlcNAcβ1-3)GalNAcα-
Core 6	GlcNAcβ1-6 GalNAcα-
I antigen	Galβ1-4GlcNAcβ1-6(Galβ1-4GlcNAcβ1-3)Galβ-
Lewis^x	Galβ1-4(Fucα1-3)GlcNAcβ1-3Gal-
Sialyl-Lewis^x (SLe^x)	Sialylα2-3Galβ1-4(Fucα1-3)GlcNAcβ1-3Gal-
Blood group A	GalNAcα1-3(Fucα1-2)Gal-
Blood group B	Galα1-3(Fucα1-2)Gal-
Linear B determinant	Galα1-3Gal-

O-Glycan functions have been determined using cultured cells with deficiencies in O-glycosylation, with the use of inhibitors of O-glycosylation, and by deleting glycosyltransferases (GTs) and sulfotransferases in mice and cultured cells [3,5–7]. O-Glycans control protein conformation and cell surface expression, and protect glycoproteins from protease action. O-Glycans also bind to carbohydrate-binding proteins from bacteria and viruses, and to mammalian lectins and immunoglobulins having carbohydrate-binding domains. Blood group substances recognized by antibodies are often found at the termini of O-glycans.

Mucins and mucin-like glycoproteins are important components of the immune system and bind to lectins on natural killer (NK) cells, lymphocytes, dendritic cells, and other leukocytes [4,8,9]. For example, sialyl-Lewis[x] (see Table 1) displayed preferentially by O-glycans is a ligand for selectins, and plays a critical role in the rolling of leukocytes along endothelial surfaces, the homing of lymphocytes, and the extravasation of leukocytes in the inflammatory process. Cancer cells may also bind via their O-glycans to selectins and endothelial cells *in vitro*, thus suggesting that this process contributes to the invasive and metastatic properties of tumor cells [6]. Other important lectins involved in cell adhesion include DC-SIGN of dendritic cells, sialic acid-binding siglecs, and galactose (Gal)-binding galectins. Some of the receptors for these lectins are found on O-glycans.

O-Glycans are also involved in sperm–egg interactions and several other aspects of fertilization [10,11]. Although O-glycans contain carbohydrate epitopes recognized by antibodies, complex oligosaccharides have the ability to mask underlying peptide antigens [9]. Some O-glycans are characteristic of cancer and are useful for the development of anticancer vaccines [12].

1.2 Biosynthesis of O-glycans

Mucin-type O-glycans are synthesized by families of Golgi membrane-bound, highly specific GTs that transfer sugar residues from nucleotide sugar donor to peptide or carbohydrate acceptor substrates [3] (Fig. 1). In addition, Golgi-bound specific sulfotransferases transfer sulfate from 3′-phosphoadenosine-5′phosphosulfate (PAPS) to either the 3- or the 6-hydroxyl of Gal, or the 6-hydroxyl of GlcNAc [13]. In the first step of the biosynthesis of O-glycans, GalNAc is transferred from UDP-αGalNAc to either Ser or Thr of the peptide backbone by a member of the polypeptide GalNAc-transferase family with retention of α-anomeric configuration of GalNAc. Subsequently, several different core structures can be synthesized, extended and terminated.

The mechanisms regulating the assembly of O-glycan chains include transcriptional and translational controls, transport of substrates through the Golgi, and concentrations of cofactors. Protein–protein interactions are also important and can influence GT activities, stability, and specificity. Several GTs are known to form homodimers or protein heterocomplexes [14]. β4Gal-transferase binds to α-lactalbumin in mammary tissues which changes its acceptor substrate specificity, while the chaperone Cosmc binds to and stabilizes core 1 β3Gal-

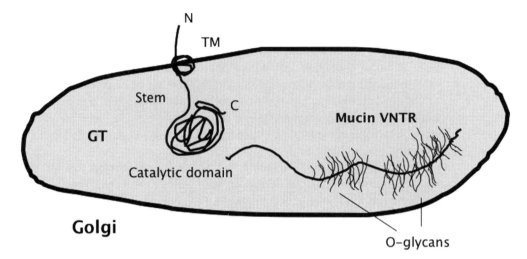

Figure 1. Schematic representation of a Golgi membrane-bound glycosyltransferase. Glycosyltransferases (GT) involved in *O*-glycosylation are bound to Golgi membranes, with the catalytic region near the C-terminus extending into the Golgi lumen. The N-terminus in the cytosolic compartment, the transmembrane domain (TM), the stem region (stem), and possibly other determinants contribute to the appropriate Golgi localization of the enzymes. The mucin molecule having a heavily *O*-glycosylated VNTR region, is efficiently glycosylated in the Golgi lumen by GT.

transferase (see Section 4). More commonly, GTs may associate less specifically with other proteins forming heterocomplexes that can affect activities.

The synthesis of *O*-glycans is initiated and completed in various compartments of the Golgi that consist of an efficient assembly line, with the flow of processed glycoproteins from the *cis*- to the *trans*-Golgi. GTs involved in *O*-glycosylation are all type II membrane glycoproteins anchored in the Golgi membrane (Fig. 1).

With some exceptions, early acting enzymes are concentrated in the *cis*-Golgi, and enzymes catalyzing the addition of extensions and terminal sugars are concentrated in the *trans*-Golgi and *trans*-Golgi network. The transmembrane domain of GT as well as adjacent domains determine the localization in specific Golgi compartments. For example, in many cell types, the α6-sialyltransferase ST6GalNAc-I, which synthesizes the STn antigen, sialylα2–6GalNAc-, localizes to the *trans*-Golgi. When the membrane anchor of the enzyme synthesizing core 2, C2GnT1, which normally resides in the *cis*/medial Golgi, was exchanged with that of ST6GalNAc-I [15], C2GnT1 relocalized to the *trans*-Golgi in CHO cells, resulting in inefficient core 2 synthesis (Fig. 2).

ST3Gal-I, which is active in the monomeric form [14], localizes mainly to the medial and *trans*-Golgi compartments in mammary cells, where it partially overlaps with C2GnT1 [9]. This allows the two enzymes to effectively compete for their common core 1 substrate, and the relative activities of these two enzymes

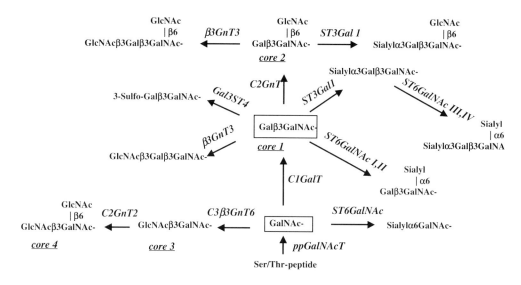

Figure 2. Pathways specific for the assembly of O-glycans. The scheme shows the biosynthesis of the common O-glycan core structures 1-4. The names of enzymes catalyzing the conversions (above the arrows) are explained in Table 2. After the synthesis of cores 1-4, further extension and termination reactions can occur.

control sialylation, branching patterns and size of O-glycans of MUC1 mucin. An overexpression of ST3Gal-I relative to C2GnT1 results in sialylated and short O-glycans with a suppression of complex core 2 structures [16]. Several GTs have been shown to form dimers and heterocomplexes with other transferases [14]. The role of these protein complexes in enzyme localization and colocalization remains to be further elucidated.

Mucins carry heterogeneous arrays of O-glycans, and intermediate structures in the biosynthetic pathways are often found on a glycoprotein. This variety of O-glycan structures arises from the fact that some structures can serve as substrates for several different competing GT reactions, and that enzyme products may serve as substrates for more than one subsequent reaction. The specificities of transferases, as well as their relative activities at the site of the Golgi where the appropriate substrate is available, will determine the range of individual O-glycan structures produced. Some structures contain a terminal residue that blocks further conversions. For example, sialylα2–6GalNAc (the STn antigen) is no longer substrate for the addition of sugars. The biosynthetic pathways (see Fig. 2) illustrate the complexity potentially leading to hundreds of different O-glycan structures. Pathological conditions, growth, differentation, and apoptosis as well as cytokines can influence the expression and activities of GTs and sulfotransferases and thus alter the glycan structures of glycoproteins [5,17]. This review discusses the synthesis of the four common O-glycan core structures 1 to

4, and some of the extension and termination reactions, with a focus on the structures and functions of GTs.

2. GLYCOSYLTRANSFERASE STRUCTURES

Each step in the biosynthesis of O-glycans (see Fig. 2) is catalyzed by a specific member of a GT family of enzymes with related amino acid sequences. More than 80 GT families have been identified (http://afmb.cnrs-mrs.fr/CAZy). Each GT is specific for both the sugar donor and the acceptor substrate, and all of these enzymes, as well as specific sulfotransferases involved in O-glycan biosynthesis (Table 2), are localized in the Golgi with a common domain structure of type II membrane proteins. Many of these transferases are glycosylated [14,18]. Members of any one GT family are predicted to have the same overall fold structure and similar catalytic mechanism, but can have diverse substrate specificities. In some examples, GTs have the same substrate specificity but have little sequence similarity and are not necessarily within the same GT family [19].

Studies of a limited number of GT crystal structures revealed the existence of three major structural superfamilies, GT-A, GT-B, and GstII. There are also enzymes

Table 2 Glycosyltransferases and sulfotransferases specific for the synthesis of O-glycans

Enzyme	Short name	EC	CAZy	GT acceptor	References
Enzymes specific for O-glycans:					
Polypeptide GalNAc-T	ppGalNAcT	2.4.1.41	27	Ser/Thr-peptide	31,32
				Ser/Thr-glycopeptide	
Core 1 β3Gal-T	C1GalT	2.4.1.122	31	GalNAcα-R	40,43
Core 3 β3GlcNAc-T	C3β3GnT6	2.4.1.147	31	GalNAcα-R	46,47
Core 2 β6GlcNAc-T (L)	C2GnT1	2.4.1.102	14	Galβ3GalNAcα-R	16,50
Core 2/core 4 β6GlcNAc-T (M)	C2GnT2	2.4.1.148	14	Galβ3GalNAcα-R	50,58,59
				GlcNAcβ3GalNAcα-R	
				GlcNAcβ3Galβ-R	
Core 2 β6GlcNAc-T (L)	C2GnT3	2.4.1.102	14	Galβ3GalNAcα-R	51
α6-Sialyl-T	ST6GalNAc-I	2.4.99.3	29	GalNAcα-peptide	64
				Galβ3GalNAcα-peptide	
				Sialylα3Galβ3GalNAcα-peptide	
α6-Sialyl-T	ST6GalNAc-II	2.4.99.3	29	Galβ3GalNAcα-peptide	64
α6-Sialyl-T	ST6GalNAc-III	2.4.99.7	29	Sialylα3Galβ3GalNAcα-R	64
α6-Sialyl-T	ST6GalNAc-IV	2.4.99.7	29	Sialylα3Galβ3GalNAcα-R	64
Enzymes selective for O-glycans:					
Elongation β3GlcNAc-T	β3GnT3	2.4.1.146	31	Galβ3GalNAcα-R	62,63
				GlcNAcβ6(Galβ3)GalNAcα-R	
α3-Sialyl-T	ST3Gal-I, IV	2.4.99.3	29	Galβ3GalNAcα-R	64
Gal-3-O-sulfo-T	Gal3ST4	2.8.2.11		Galβ3GalNAcα-R	13

-T, -transferase; -R can be peptide or a hydrophobic group; CAZy, carbohydrate-active enzymes.

that share properties between these groups. In contrast, glycosidases, which bind and cleave specific carbohydrates, exhibit many different types of folds. The structures of seven enzymes involved in O-glycosylation (Table 3) are known, and all of them have the GT-A fold characterized by α/β/α sandwich, with several different strands in the central β-sheet. These transferases often have a Asp-x-Asp (DxD) motif that plays a role in catalysis through the action of the acidic amino acids. The proteins have a short cytoplasmic tail at the N-terminus, a transmembrane domain (TM), a stem region that allows the catalytic domain to protrude into the lumen of the Golgi, and a large globular catalytic domain. The nucleotide sugar and the acceptor substrate binding domains are separate but tightly associated.

Table 3 Glycosyltransferases with known crystal structures involved in O-glycosylation

Enzyme name	CAZy GT	Fold	DxD	Complex	References
Mouse ppGalNAcT1	27 retaining	A	+	Mn^{2+}	24
Human ppGalNAcT2	27 retaining	A	+	Mn^{2+}	37
Mouse C2GnT1	14 inverting	A	+	–	22
Bovine β4GalT	7 inverting	A	+	Mn^{2+}	23,25
Human A α3GalNAcT	6 retaining	A	+	Mn^{2+}	26
Human B α3GalT	6 retaining	A	+	Mn^{2+}	26
Bovine linear B α3GalT	6 retaining	A	+	Mn^{2+}	27

GT-B folds have two α/β/α sandwich domains with the catalytic site between these domains [19]. An example of this type is the prokaryotic β-glucosyltransferase [20]. GT-B enzymes so far have not revealed bound metal ion, or the DxD motif often found in GT-A enzymes. A third type of fold, GstII, found in a sialyltransferase from *Campylobacter jejuni* [21], also consists of α/β/α sandwich but with different topology of the β-strands.

Transferases that retain the anomeric configuration of the sugar in the donor substrate are retaining enzymes, while inverting transferases invert this configuration. Inverting GTs act with a sequential ordered mechanism; the acceptor hydroxyl to be glycosylated appears to link to the carbon 1 of the sugar in the donor substrate through a nucleophilic attack and this results in a single displacement reaction with inversion of configuration. Thus, an acidic amino acid appears to assist the deprotonation of the acceptor hydroxyl that is to be glycosylated. Inverting GTs depend on the presence of a divalent cation such as Mn^{2+} or another cationic group [19]. In many of the enzymes, the Mn^{2+} ion interacts with the phosphate groups of the nucleotide sugar donor, as well as with a DxD motif. The mechanisms of retaining GTs have not yet been elucidated. The retention of sugar configuration may be the result of a two-step double inversion process.

The enzymes involved in O-glycosylation all have at least one DxD motif (see Table 3) or its variant, which occurs in a short loop connecting a large and a small β-strand at the C-terminal end of the nucleotide sugar-binding domain [19]. By comparison, the inverting enzyme C2GnT1 does not require divalent cation for activity and does not utilize its conserved DxDxD sequence near the C-terminal end in the same fashion [22].

Generally, a flexible, disordered loop appears to be important for substrate binding in GT-A enzymes. The nucleotide sugar-binding domain is similar among several GTs examined. The binding of sugar donor substrate in this domain appears to induce a conformational change in the enzyme that forms a lid over the nucleotide sugar, and opens a pocket that facilitates acceptor substrate binding [23–25].

The structural basis for donor specificity has been elucidated for a number of GTs involved in glycoprotein biosynthesis. The crystal structures of the blood group A α3GalNAc-transferase and blood group B α3Gal-transferases that differ by only four critical amino acids has revealed the nucleotide binding sites. Not more than two amino acids appear to be in contact with the sugar moiety of the nucleotide sugar, but only one of these controls the donor specificity [26]. Analysis of the crystal structure of another α3-Gal-transferase that synthesizes the "linear B" xenotransplantation antigen (Table 1) confirmed that Glu317 is involved in acceptor substrate binding and catalysis [27], while His280 is crucial in determining the UDP-Gal binding specificity [28].

The crystal structures of GTs (Table 3) often do not reveal all of the interactions between acceptor substrates and amino acids in the active site, especially since large acceptors are bound in extended and accessory binding sites. In α3/4-Fuc-transferase III, it was shown that mainly one amino acid (Trp111) controls the acceptor specificity and enzyme activity [29]. The crystal structures of the ubiquitous β4GalT1 indicate the presence of a deep catalytic pocket that binds the donor substrate UDP-Gal as well as Mn^{2+} through Asp252 and 318, Gly292 and 315, and Glu317 residues [23]. The binding of UDP-Gal induces a conformational change in the enzyme by closing a loop which creates the acceptor binding pocket that could accommodate a large GlcNAc-terminating oligosaccharide. GlcNAc acceptor substrate is bound by Phe280 and 360, Tyr286, Arg259, and Ile363. The binding of GlcNAc acceptor substrate in α-anomeric configuration is blocked by Tyr286, but GlcNAcβ with large aglycone groups can be accommodated. This explains the acceptor specificity of the enzyme that favors GlcNAcβ-substrates [30].

3. POLYPEPTIDE GalNAc-TRANSFERASES

All O-glycan structures are initiated by members of a large family of polypeptide GalNAc-transferases (ppGalNAcT) that synthesize the tumor-associated Tn antigen (see Fig. 2) [31,32]. More than 20 ppGalNAcT genes have been identified. Orthologous genes are conserved from nematodes to mammals. The expression of the individual members of this family is cell type specific, and all mammalian cells appear to express at least one of the isoenzymes. The ppGalNAcT proteins are unusual in that they contain a ricin B chain-like lectin domain at the C-terminus, which may aid in binding glycopeptide substrates that have already been glycosylated.

The acceptor specificities of ppGalNAcT have been extensively studied [31–33]. No specific amino acid sequence motif is required for O-glycosylation. However, the amino acids surrounding the glycosylation site and the presence of oligosaccharides in the acceptor substrate have a significant influence on the rate of GalNAc transfer. Proline residues in the acceptor promote activity, possibly by exposing the Ser or Thr residues to the enzyme. Although GalNAc residues in acceptors may promote binding of the enzyme via the lectin domain, the presence of multiple GalNAc residues, as well as larger oligosaccharide structures in the glycopeptide substrate can inhibit the activity of ppGalNAcT1 by steric interference [33].

3.1 Polypeptide GalNAc-transferase structure

The ppGalNAcTs are retaining transferases of the CAZy GT27 family, with a ricin B chain lectin-like domain at the C-terminus. Mouse ppGalNAcT1 was expressed as a soluble protein in which the TM and cytoplasmic domains at the N-terminus were replaced with a maltose-binding protein. The crystal structure with Mn^{2+} complexed between the lectin and the catalytic domains showed a central eight-stranded β-sheet flanked by α-helices, and a typical nucleotide binding domain, spatially separated from the lectin domain [24]. Three repeat sequences (α,β,γ) of the lectin domain form a trefoil structure, expected to extend into the Golgi lumen. The deletion of the individual lectin domain repeat sequences, or of 12 amino acids from the C-terminus abolished ppGalNAcT1 activity towards deglycosylated bovine submaxillary mucin and also reduced the expression of the enzyme [34].

The lectin and catalytic domains are flexible but seem to have no contact except for the connecting peptide. ppGalNAcT4 lacking the lectin domain is similar in its action towards peptide but has a higher K_m towards glycopeptides substrates, indicating that the function of the lectin domain is to bind to acceptor substrate via GalNAc residues. The presence of high concentrations of free GalNAc reduced the ability to act on glycopeptides [35].

Instead of a DxD motif, mouse ppGalNAcT1 has a conserved DxH motif. The His211 and Asp209 residues of this DxH motif, as well as the conserved His344 complexes Mn^{2+} ion in a deep UDP-GalNAc-binding pocket. Mutations in His211, Asp209, or His344 to Ala abolish activity of the homologous bovine ppGalNAcT1 [36]. A mutation in Glu127 (to Gln) also drastically reduces activity; thus Glu127 may coordinate with phosphate oxygens through a water molecule.

Human ppGalNAcT2 containing its catalytic and lectin domains was crystallized with UDP and the acceptor peptide PTTDSTTPAPTTK from the VNTR of rat submandibular mucin [37]. The catalytic domain is very similar to that of mouse ppGalNAcT1. In the presence of bound UDP, peptides containing nine amino acid residues in extended conformation can be accommodated in the extended acceptor binding site [24,37].

The crystal structures with and without acceptor, but having bound UDP, the reaction product, in the active site, vary by the presence of a loop that forms a lid over UDP. In the absence of the acceptor the lid is folded back, opening the UDP-binding cleft and enabling UDP to leave (and presumably allowing UDP-GalNAc to bind). In analogy to mouse ppGalNAcT1, Asp224 and His226 within the DxH motif, as well as His359, bind the Mn^{2+} and phosphate groups of UDP in human ppGalNAcT2.

3.2 Polypeptide GalNAc-transferase function

The gene expression of ppGalNAcT can be up- or downregulated upon treatment of cells with cytokines and growth factors, and in disease. For example, the expression levels of ppGalNAcT1, T2, and T3 were found to be increased in colorectal carcinoma tissues. In particular, the expression levels of T3 are increased in several types of cancers [38]. The levels of ppGalNAcT3 enzyme protein are high in well-differentiated tumors, and are an indicator of a good prognosis in colorectal cancer patients. A mutation in ppGalNAcT3 causes a human condition called familial tumor calcinosis [39]. The mutations are in the linkage region between the catalytic domain and the lectin domain of the enzyme, and this may affect phosphate metabolism resulting in hyperphosphatemia.

The expression of the various members of the ppGalNAcT family is regulated in a developmental- and tissue-specific fashion in *Drosophila* [40]. Gene deletion studies [32] indicate that specific ppGalNAcT are involved in viability of the organism. A deletion of several individual ppGalNAc-transferases in mice, however, did not cause distinct phenotypes [31,32]. This suggests that there is functional redundancy of ppGalNAcT in mice, possibly because *O*-glycosylation is essential for survival.

4. CORE 1 ß3-GAL-TRANSFERASE

GalNAc-Ser/Thr (Tn antigen) is converted to the core 1 structure (T antigen) by core 1 β3Gal-transferase (C1GalT). Both the Tn and T antigens are cancer-associated and indicate a lack of enzyme activities involved in the extension of *O*-glycans [2,4].

C1GalT activity is ubiquitous and the gene is widely expressed [3,40] in human tissues, *Drosophila melanogaster*, *Caenorhabditis elegans*, and other species. An unusual association of C1GalT with a specific chaperone protein, Cosmc, has been discovered in humans and mice [41]. It appears that in vertebrates the enzyme is only active when Cosmc is coexpressed, which prevents degradation of the enzyme and ensures its Golgi localization.

C1GalT is an inverting transferase of the CAZy GT 31 family. The enzyme has a DxD motif and requires divalent metal ion such as Mn^{2+} for activity. C1GalT purified from rat liver, as well as the recombinant soluble epitope-tagged enzyme exist mainly as the disulfide-bonded homodimer. Purified C1GalT transfers Gal to the GalNAc-Ser/Thr residues of glycoproteins as well as to GalNAc derivatives with hydrophobic aglycone groups. GalNAcα-benzyl penetrates into cells in culture and serves as an alternative substrate for C1GalT. Therefore, it reduces core 1 synthesis on natural glycoproteins. In addition, sialylated complex O-glycans can be built up on GalNAc-benzyl, and this reduces overall mucin sialylation and glycosylation [42].

Although C1GalT can convert core 6, GlcNAcβ1–6GalNAc, to core 2 (see Table 1) *in vitro*, it does not act on sialylα2–6GalNAc-mucin. The sialyl-T antigen, sialylα2–6(Galβ1–3)GalNAcα-Ser/Thr, is therefore synthesized by sialyltransferase action on core 1 (see Table 1, Fig. 2). Using glycopeptides as substrates, we have shown that C1GalT activity is greatly influenced by the amino acids and glycosylation patterns juxtaposed to the GalNAc acceptor site [43]. In similar studies, we found that the enzymes synthesizing core 2 and 3 structures, as well as β4Gal-transferase extending GlcNAc-terminating O-glycans are also influenced by distinct glycosylation patterns of glycopeptide substrates. This suggests that the synthesis, as well as further processing of O-glycan core structures, is peptide site directed and controlled by existing glycosylation.

The human T-lymphoblastoid cell line Jurkat and the human colon cancer cell line LSC lack C1GalT activity and therefore express the Tn antigen [2,3]. Although Jurkat cells express C1GalT, they have a mutation in the gene encoding Cosmc that disables its role as a chaperone [41]. Several human conditions are associated with a lack of C1GalT activity due to the absence of functional Cosmc. In IgA nephropathy, a low expression of Cosmc is present in B lymphocytes, while somatic mutations in the Cosmc gene occur in patients with the Tn syndrome [44]. A deletion of C1GalT in mice causes a lack of sialylated core 1 structures in endothelial, hematopoietic, and epithelial cells. Embryos develop large hemorrhages in the brain and spinal cord, with chaotic microvascular network. This suggests that C1GalT is essential for the development of the vascular system [45].

5. CORE 3 ß3GlcNAc-TRANSFERASE

In an alternative pathway, the Tn antigen is converted to the core 3 structure by core 3 β3-GlcNAc-transferase (C3β3GnT6). Like C1GalT, C3β3GnT6 is a member of the CAZy GT 31 family, has a DxD motif and requires Mn^{2+} for activity. The gene is homologous to that of other β3-GlcNAc-transferases [46], and is expressed primarily in the foveolar epithelium of the stomach, and in the colon and small intestine. Both C1GalT and C3β3GnT6 utilize and compete for GalNAc-R as an acceptor, and neither enzyme requires peptide in the substrate. In contrast to C1GalT, C3β3GnT6 does not act on core 6 to form core 4 [46]; it also does not act on sialylα2–6GalNAc-mucin [47]. The sialylated core 3 structures found in colonic mucins are therefore synthesized by the addition of sialic acid to core 3.

The expression and activity levels of C3β3GnT6 are reduced in colon cancer tissues that contain a mixture of cell types [2]. In several cultured colon cancer cell lines, the activity was undetectable [48]. Upon transfection of the C3β3GnT6 gene into human fibrosarcoma cells, T antigen expression was markedly decreased. Cells showed significantly lower migration activity, and lung metastases were suppressed [49]. This suggests that, in spite of the ability to carry Lewis antigens at their termini, core 3 structures can reduce the extravasation of cancer cells and could be a predictor of positive outcome in cancer.

6. SYNTHESIS OF CORES 2 AND 4

O-Glycan core 2 (see Table 1) provides a branch and additional backbone structure for chain extensions and the attachment of antigenic determinants and cell adhesion ligands, and is synthesized from core 1 by core 2 β6GlcNAcT (C2GnT1-3). C2GnT enzymes are members of a subfamily of β6GlcNAc-transferases that synthesize GlcNAcβ1-6Gal branches. These β6-GlcNAc-transferases belong to the GT family 14 (see Table 2). They utilize UDP-GlcNAc in the absence of divalent metal ions. However, the enzymes differ in their tissue distribution, properties, and biological functions. C2GnT1 is expressed in many mucin-secreting and non-mucin-secreting epithelial, endothelial, and hematopoietic cell types, and the activity from leukocytes has been named the L-type enzyme [50].

C2GnT1 is absolutely specific for the unmodified core 1 substrate. Thus, core 2 is synthesized before sialylation, sulfation, fucosylation, or extension of core 1. The C2GnT2 activity from mucin-secreting cells has been named the M-type enzyme [50]. C2GnT2 can synthesize core 2 and 4, and other branched GlcNAcβ1-6(GlcNAcβ1-3)Gal structures found in extended chains, and therefore has broader substrate specificity than C2GnT1. Another L-type isoenzyme (C2GnT3) that synthesizes only core 2 is expressed mainly in the thymus [51].

6.1 C2GnT1 function

The branched core 2 structure, with its highly flexible GlcNAcβ1-6 arm, can display important adhesion ligands; it is bulky and can thus mask underlying mucin peptide epitopes. In mammary cells, complex core 2 structures prevent the exposure of peptide epitopes of MUC1 mucin. The lack of C2GnT1 expression in many breast cancer cells is thus associated with the presence of smaller sialylated O-glycans and allows binding of anti-MUC1 peptide antibodies [16].

Several mucin-like glycoproteins contain ligands for selectin binding. The cell surface glycoprotein PSGL-1 is an example of a mucin-like cell surface glycoprotein that carries high-affinity ligands for P- and E-selectins attached to core 2 [52]. Not only leukocytes but also cancer cells have the ability to bind to selectins via their O-glycans, and this binding may be a first step in the extravasation of cancer cells into a secondary tissue site [6]. The expression of C2GnT plays a critical role in the formation of high-affinity selectin-binding ligands. CHO cells expressing PSGL-1 do not display SLe^x and do not bind with high affinity to P-selectin unless transfected with C2GnT1 [3]. *In vivo* studies in mice having a deletion in the C2GnT1 gene demonstrated reduced leukocyte rolling along the endothelium, reduced leukocyte recruitment into the peritoneum, and reduced lymphocyte migration to peripheral lymph nodes [53]. A combined deficiency in mice of C2GnT1 and GlcNAc6ST sulfotransferase, two enzymes involved in the synthesis of high-affinity selectin ligands on core 2, further reduced lymphocyte homing to peripheral lymph nodes [7].

C2GnT1 activity is regulated during activation of lymphocytes, cell differentiation, apoptosis, and cytokine stimulation, and several disease states including cancer are associated with altered C2GnT1 activity or expression [3,4,17]. For example, in lung cancer tissues C2GnT1 expression is associated with a poor prognosis and malignant potential which may be related to the ability of cancer cells to attach to the endothelium [54].

6.2 C2GnT1 structure

The crystal structure of mouse C2GnT1 lacking the cytoplasmic tail and the transmembrane region has been determined in complexes with and without the acceptor substrate Galβ1–3GalNAc [22]. The crystallized protein was a dimer, linked through disulfide bonds formed by Cys235. However, the recombinant soluble active enzyme expressed in insect cells is monomeric [55], and when the enzyme is expressed in CHO cells, only a small amount is present intracellularly as the dimer [14]. Each monomer of C2GnT1 has a stem region containing α-helices and 2 conserved N-glycosylation sites. The catalytic domain has the GT-A fold with α/β/α sandwich and a central six-stranded mixed β-sheet with 3,2,1,4,6,5 topology.

The activity of C2GnT1 is independent of the presence of a metal ion, and the enzyme is active with EDTA. In the crystal structure, the metal ion seems to be replaced by basic amino acids Arg378 and Lys401 (both conserved in C2GnT sequences), which could stabilize the pyrophosphate group of UDP-GlcNAc donor substrate.

Two conserved N-glycosylation sites of human C2GnT1 have an important role in stabilizing the enzyme [18]. In addition, the N-glycan of Asn95 contributes to the cis/medial Golgi localization of the full-length enzyme [56]. The crystal structure of the mouse homolog suggests that these glycans are highly flexible [22].

Eight of the nine conserved Cys residues of human C2GnT1 are essential for the activity [55]. A deletion of the ninth residue, Cys217, reduces the activity but also prevents the inactivation of the enzyme by Cys-specific reagents. Because the presence of UDP-GlcNAc prevented this inactivation, it was concluded that Cys217 was in the sulfhydryl form near the active site. Similarly, mouse C2GnT1 was shown to form four disulfide bonds, with a free thiol group at Cys217 [57]. The crystal structure confirmed the biochemical and mutational studies and showed that the conserved Cys217 is in the free sulfhydryl state, while eight other conserved Cys form disulfide bonds [22].

The mouse C2GnT1 structures with and without acceptor substrate did not reveal a significant conformational difference. The Galβ1–3GalNAc disaccharide fits well into the acceptor binding site; therefore, peptide or other aglycone groups linked to the GalNAc residue would protrude into the solution. Human C2GnT1 has an absolute requirement for the 4- and 6-hydroxyls of both Gal and GalNAc residues and the 2-acetamido group of GalNAc [50]. The crystal structure of mouse C2GnT1 shows that Glu320 of the invariant SPDE sequence binds to both O-4 and O-6 of GalNAc and may activate the 6-hydroxyl to induce a nucleophilic attack on the C-1 of the GlcNAc moiety of UDP-GlcNAc. While

Glu320 may be the critical catalytic base, Arg254 forms a hydrogen bond with the O-4 of Gal and Asp243 binds to O-4 and O-6 of the Gal residue [22]. These contacts are expected to be different in C2GnT2 which can accommodate GlcNAc instead of Gal. The N-acetyl group of GalNAc as well as O-2 of Gal are in contact with Tyr358 and Trp356 in C2GnT1. In C2GnT3, having the same narrow specificity as C2GnT1, Tyr358 is also present; however, in C2GnT2, the smaller Gly368 is present instead. This could explain the broader acceptor specificity of C2GnT2 which could accommodate GlcNAc as well as Gal at that site.

6.3 C2GnT2

The M-type enzyme C2GnT2 that synthesizes both cores 2 and 4 is highly active in colonic tissues and many colon-derived mucin-secreting cell lines [47,48,50]. The mRNA levels are high in colon and kidney but lower in small intestine and pancreas, and are variable in cancer cell lines [58]. The M-type enzyme C2GnT2 can incorporate a GlcNAcβ1-6 branch into the core 1 substrate to form core 2, into the core 3 substrate to form core 4, and into GlcNAcβ1-3Gal structures to form the branch of the I antigen, GlcNAcβ1-6(GlcNAcβ1-3)Gal (Table 1). Thus, the M-enzyme has the distal IGnT activity, while IGnTs of the related subfamily of β6-GlcNAc-transferases have the central IGnT activity. Once the core 3 structure is extended by a Gal residue, C2GnT2 does not synthesize core 4.

The overall structure and catalytic mechanism of C2GnT2 is expected to be similar to that of C2GnT1. However, the acceptor binding site of C2GnT2 must also accommodate a GlcNAcβ residue instead of Galβ, and a Galβ residue instead of GalNAcα. The expression and activity of C2GnT2 exhibit cyclic variations during the estrous cycle in oviductal tissues of the golden hamster. The enzyme in oviductal tissue is thought to function in the synthesis of the oviductin mucin which binds to zona pellucida glycoproteins and is involved in fertilization and the protection of oocytes [11]. Retinoic acid, IL-4, and IL-13 upregulate C2GnT2 expression in H292 human airway lung cancer cells, while EGF downregulates C2GnT2 expression and activity [59].

Colonic adenoma cells that have progressed to tumorigenic cells exhibit a much lower activity of C2GnT2 [60]. A less dramatic reduction has been observed in colon cancer tissue [2,61]. This suggests that there is a switch from C2GnT2 to C2GnT1 in colon cancer tissues but this is not always seen in cultured colon cancer cells which can have high C2GnT2 activities [48]. An overexpression of the gene in colon cancer cells HCT116 resulted in suppression of cell growth, cell adhesion, motility, and invasive properties. Similar protective effects were observed in nude mice, where C2GnT2-expressing cells exhibited suppressed tumor growth [61].

7. EXTENSION AND TERMINATION OF *O*-GLYCANS

Most of the extension and termination reactions in the O-glycosylation pathways involve families of transferases that synthesize O-glycans, the antennae of complex N-glycans, and glycolipids. A small number of these enzymes show a distinct preference for O-glycan structures (see Table 2). Transferases that extend

O-glycan chains include the β3GlcNAc-transferase family members, β6-GlcNAc-transferases that synthesize branched structures (IGnT), β3-Gal-transferases and β4-Gal-transferases. A ubiquitous elongation β3-GlcNAc-transferase (β3GnT3), found in pig gastric mucosa, is selective for O-glycans and elongates core 1 and 2 structures (see Fig. 2) [62,63]. Transfection of the β3GnT3 gene into mice lacking C2GnT1 demonstrated that extended core 1 structures can also display selectin ligands, independent of core 2 structures. Terminally acting enzymes involved in the synthesis of blood group and tissue antigens, which are especially common in mucins, include the blood group A and B transferases, linear B α3-Gal-transferase, Cad blood group β4-GalNAc-transferase, α2- and α3-Fuc-transferases, sulfotransferases, and sialyltransferases. In addition, Gal and GlcNAc residues can be sulfated by families of sulfotransferases. Gal3ST4 is a sulfotransferases that uses the O-glycan core 1 structure as its preferred substrate. The expression or activity levels of many of these enzymes have been found to be altered in disease such as cancer, and upon cytokine stimulation and as a consequence, glycoproteins have different glycan structures and antigens, and have altered biological behavior [3,4,17].

7.1 Sialyltransferases

O-Glycans can be terminated with sialic acid in α2-3 (but not 2-6) linkage to Gal residues, and in α2-6-linkage to GalNAc. The sialic acid-containing structures mask underlying antigens but are also recognized directly as determinants for cell adhesion molecules. Sialylation often stops chain growth and further branching [3]. Sialyltransferases are inverting GTs, members of the GT29 family, that have common motifs (L, S, VS, and III) in their sequences named sialylmotifs [64]. A large (L) sialylmotif is found in the middle of the catalytic domain where it contributes to the binding of the nucleotide sugar substrate CMP-βsialic acid. A short (S) motif is located near the C-terminus and contributes to the binding of both donor and acceptor substrates. According to the linkage they synthesize, sialyltransferases have been classified in subfamilies with characteristic additional sequence motifs of 6-20 residues that could account for the linkage-specific action of sialyltransferases [65]. Three subfamilies are involved in O-glycosylation: (1) ST3Gal, (2) ST6GalNAc-I and II, and (3) ST6GalNAc-III and IV.

Members of the α3-sialyltransferase (ST3Gal) family (ST3Gal-I, II, III, IV, and VI) can add sialic acid to Gal residues of O-glycans. ST3Gal-I and ST3Gal-IV are responsible for synthesizing the sialyl-T antigen (Table 1), i.e. sialylating the Galβ1-3 residue attached to GalNAc-Thr (Fig. 2). The biological importance of α3-sialylation has been demonstrated in mice lacking the ST3Gal-IV gene, which develop bleeding disorders and thrombocytopenia [66]. Inflammatory conditions can increase the expression of sialyltransferases, including ST3Gal-I [4,17]. The activity and mRNA levels of ST3Gal-I are also increased in breast cancer cells and tissues, which results in small and sialylated O-glycans of the cell surface mucin MUC1. Normal mammary MUC1 has large complex O-glycans of core 2 structure that mask underlying peptide epitopes. The increase in α3-sialylation results in the formation of smaller oligosaccharides and exposure of MUC1 mucin peptide epitopes that are characteristic of cancer [9,16].

Four α6-sialyltransferases, ST6GalNAc-I, II, III, and IV, act on the GalNAc residue of O-glycans. ST6GalNAc-I synthesizes the STn antigen (see Fig. 2) and requires peptide in the acceptor substrate. The enzyme also acts on core 1 and sialylα2–3Galβ1–3GalNAcα-peptide, and possibly on other modified core 1 or core 3 structures. In human breast cancer cells MDA-MB-231, the expression of ST6GalNAc-I is associated with decreased cell adhesion, increased cell migration, and tumorigenicity [67].

ST6GalNAcT-II acts preferably on core 1, and possibly on core 3, and also requires peptide in the acceptor substrate. The expression of ST6GalNAc- is correlated with poor survival and invasive potential of colorectal cancer [68]. ST6GalNAc-III and IV do not require peptide in the substrate but are absolutely specific for the sialylα2–3Galβ1–3GalNAc structure. Treatment of human endothelial cells with tumor necrosis factor alpha (TNFα) significantly upregulates the expression of ST6GalNAc-III [17].

8. CONCLUSIONS

The growing knowledge of GT structure and function, and of the dimensions and properties of the enzyme active sites, allows the design of specific substrate analogs as glycosyltransferase inhibitors. Because of the important roles of O-glycans in many biological phenomena, these inhibitors would be useful for studying glycan functions and help in the development of therapeutic drugs against cancer, inflammatory disease and many other conditions. Future challenges lie in determining the structural basis of enzyme localization, the amino acids involved in defining GT substrate specificities, and the factors regulating glycodynamics (i.e. the environmental factors affecting GT expression and activities).

Acknowledgments

The writing of this manuscript was supported by the Canadian Cystic Fibrosis Foundation and Queen's University.

REFERENCES

1. Hollingsworth MA & Swanson BJ (2004) *Nat. Rev.* **4**, 45–60.
2. Brockhausen I (1999) *Biochim. Biophys. Acta* **1473**, 67–95.
3. Brockhausen I, Schutzbach J & Kuhns W (1998) *Acta Anat.* **161**, 36–78.
4. Brockhausen I (2006) *EMBO Rep.* **7**, 599–604.
5. Brockhausen I & Kuhns W (1997) *Glycoproteins and Human Disease*. Medical Intelligence Unit, CRC Press and Mosby Year Book, Chapman & Hall, New York.
6. Kojima N, Handa K, Newman W & Hakomori S (1992) *Biochem. Biophys. Res. Commun.* **182**, 1288–1295.
7. Hiraoka N, Kawashima H, Petryniak B et al. (2004) *J. Biol. Chem.* **279**, 3058–3067.
8. Blottiere HM, Burg C, Zennadi R et al. (1992) *Int J Cancer* **52**, 609–618.
9. Taylor-Papadimitriou J, Burchell JM, Plunkett T et al. (2002) *J. Mammary Gland Biol. Neoplasia* **7**, 209–221.

10. Talbot P, Shur BD & Myles DG (2003) *Biol. Reprod.* **68**, 1-9.
11. McBride DS, Brockhausen I & Kan FW (2005) *Biochim. Biophys. Acta* **1721**, 107-115.
12. Sorensen AL, Reis CA, Tarp MA *et al.* (2006) *Glycobiology* **16**, 96-107.
13. Brockhausen I (2003) *Biochem. Soc. Trans.* **31**, 318-325.
14. El-Battari A, Prorok M, Angata K *et al.* (2003) *Glycobiology* **13**, 941-953.
15. Zerfaoui M, Fukuda M, Langlet C *et al.* (2002) *Glycobiology* **12**, 15-24.
16. Dalziel M, Whitehouse C, McFarlane I *et al.* (2001) *J. Biol. Chem.* **276**, 11007-11015.
17. Garcia-Vallejo JJ, Van Dijk W, Van Het Hof B *et al.* (2006) *J. Cell. Physiol.* **206**, 203-210.
18. Toki D, Sarkar M, Yip B *et al.* (1997) *Biochem. J.* **325**, 63-69.
19. Breton C, Snajdrova L, Jeanneau C, Koca J & Imberty A (2006) *Glycobiology* **16**, 29R-37R.
20. Vrielink A, Rueger W, Driessen HPC & Freemont PS (1994) *EMBO J.* **13**, 3413-3422.
21. Chiu CP, Watts AG, Lairson LL *et al.* (2004) *Nat. Struct. Mol. Biol.* **11**, 163-170.
22. Pak JE, Arnoux P & Zhou S *et al.* (2006) *J. Biol. Chem.* **281**, 26693-26701.
23. Ramakrishnan B, Balaji PV & Qasba PK (2002) *J. Mol. Biol.* **318**, 491-502.
24. Fritz TA, Hurley JH, Trinh LB, Shiloach J & Tabak LA (2004) *Proc. Natl Acad. Sci. USA* **101**, 15307-15312.
25. Qasba PK, Ramakrishnan B & Boeggeman E (2005) *Trends Biochem. Sci.* **30**, 53-62.
26. Patenaude SI, Seto NO, Borisova SN *et al.* (2002) *Nat. Struct. Biol.* **9**, 685-690.
27. Gastinel LN, Bignon C, Misra AK, Hindsgaul O, Shaper JH & Joziasse DH (2001) *EMBO J.* **20**, 638-649.
28. Zhang Y, Swaminathan GJ, Deshpande A *et al.* (2003) *Biochemistry* **42**, 13512-13521.
29. Dupuy F, Petit JM, Moolicone R, Oriol R, Julien R & Maftah A (1999) *J. Biol. Chem.* **274**, 12257-12262.
30. Brockhausen I, Benn M, Bhat S *et al.* (2006) *Glycoconj. J.* **23**, 523-539.
31. Ten Hagen KG, Fritz TA & Tabak LA (2003) *Glycobiology* **13**, 1R-16R.
32. Schwientek T, Bennett EP, Flores C *et al.* (2002) *J. Biol. Chem.* **277**, 22623-22638.
33. Brockhausen I, Toki D, Brockhausen J *et al.* (1996) *Glycoconj. J.* **13**, 849-856.
34. Tenno M, Saeki A, Kezdy FJ, Elhammer AP & Kurosaka A (2002) *J. Biol. Chem.* **277**, 47088-47096.
35. Hassan H, Reis CA, Bennett EP *et al.* (2000) *J. Biol. Chem.* **275**, 38197-38205.
36. Wragg S, Hagen FK & Tabak LA (1997) *Biochem. J.* **328**, 193-197.
37. Fritz TA, Raman J & Tabak LA (2006) *J. Biol. Chem.* **281**, 8613-8619.
38. Landers KA, Burger MJ, Tebay MA *et al.* (2005) *Int. J. Cancer* **114**, 950-956.
39. Topaz O, Shurman DL, Bergman R *et al.* (2004) *Nat. Genet.* **36**, 579-581.
40. Ju T, Brewer K, D'Souza A, Cummings RD & Canfield WM (2002) *J. Biol. Chem.* **277**, 178-186.
41. Ju T & Cummings RD (2002) *Proc. Natl Acad. Sci. USA* **99**, 16613-16618.
42. Delannoy P, Kim I, Emery N *et al.* (1996) *Glycoconj. J.* **13**, 717-726.
43. Granovsky M, Bielfeldt T, Peters S *et al.* (1994) *Eur. J. Biochem.* **221**, 1039-1046.
44. Ju T & Cummings RD (2005) *Nature* **437**, 1252.
45. Xia L, Ju T, Westmuckett A *et al.* (2004) *J. Cell Biol.* **164**, 451-459.
46. Iwai T, Inaba N, Naundorf A *et al.* (2002) *J. Biol. Chem.* **277**, 12802-12809.
47. Brockhausen I, Matta KL, Orr J & Schachter H (1985) *Biochemistry* **24**, 1866-1874.
48. Vavasseur F, Yang J, Dole K, Paulsen H & Brockhausen I (1995) *Glycobiology* **5**, 351-357.
49. Iwai T, Kudo T, Kawamoto R *et al.* (2005) *Proc. Natl Acad. Sci. USA* **102**, 4572-4577.
50. Kuhns W, Rutz V, Paulsen H *et al.* (1993) *Glycoconj. J.* **10**, 381-394.
51. Schwientek T, Yeh JC, Levery SB *et al.* (2000) *J. Biol. Chem.* **275**, 11106-11113.
52. Bernimoulin MP, Zeng XL, Abbal C *et al.* (2003) *J. Biol. Chem.* **278**, 37-47.
53. Sperandio M, Thatte A, Foy D, Ellies LG, Marth JD & Ley K (2001) *Blood* **97**, 3812-3819.
54. Machida E, Nakayama J, Amano J & Fukuda M (2001) *Cancer Res.* **61**, 2226-2231.
55. Yang X, Qin W, Lehotay M *et al.* (2003) *Biochim. Biophys. Acta* **1648**, 62-74.
56. Prorok-Hamon M, Notel F, Mathieu S, Langlet C, Fukuda M & El-Battari A (2005) *Biochem. J.* **391**, 491-502.
57. Yen TY, Macher BA, Bryson S *et al.* (2003) *J. Biol. Chem.* **178**, 45864-45881.
58. Schwientek T, Nomoto M & Levery SB (1999) *J. Biol. Chem.* **274**, 4504-4512.
59. Beum PV, Bastola DR & Cheng PW (2003) *Am. J. Respir. Cell Mol. Biol.* **29**, 48-56.
60. Vavasseur F, Dole K, Yang J *et al.* (1994) *Eur. J. Biochem.* **222**, 415-424.

61. Huang MC, Chen HY, Huang HC *et al.* (2006) *Oncogene* **25**, 3267–3276.
62. Brockhausen I, Williams D, Matta KL, Orr J & Schachter H (1983) *Can. J. Biochem. Cell Biol.* **61**, 1322–1333.
63. Yeh JC, Hiraoka N, Petryniak B *et al.* (2001) *Cell* **105**, 957–969.
64. Harduin-Lepers A, Mollicone R, Delannoy P & Oriol R (2005) *Glycobiology* **15**, 805–817.
65. Patel RY & Balaji PV (2006) *Glycobiology* **16**, 108–116.
66. Ellies LG, Ditto D & Levy GG (2002) *Proc. Natl Acad. Sci. USA* **99**, 10042–10047.
67. Julien S, Adriaenssens E, Ottenberg K *et al.* (2006) *Glycobiology* **16**, 54–64.
68. Schneider F, Kemmer W, Haensch W *et al.* (2001) *Cancer Res.* **61**, 4605–4611.

CHAPTER 19
Glycosylation in development
Shaolin Shi and Pamela Stanley

1. INTRODUCTION

Glycosylation is a common posttranslational modification of proteins that adds various sugar moieties (glycans) to asparagine residues via N-linkage or to serine/threonine residues via O-linkage or to ceramide to form glycolipids. Glycosylation affects the function of glycoproteins and glycolipids due to effects on their physical properties or due to specific recognition of sugars by glycan-binding proteins. Glycosylation status depends on the complement of glycosyltransferases and other glycosylation activities expressed by a cell, and changes with cell differentiation and during embryonic development. Here we summarize glycosylation events and glycan structures important for development using mainly mouse models as examples.

2. GLYCOSYLATION AND OOGENESIS

Several genes that encode glycosyltransferases have been specifically inactivated in primordial oocytes with the use of Cre recombinase under the control of the ZP3 promoter. In a female carrying a ZP3Cre transgene, any portion of a gene that is flanked by LoxP sites will be deleted early in oogenesis, resulting in oocytes and eggs that lack the function of that gene. The first glycosyltransferase gene to be inactivated in oocytes was UDP-GlcNAc: polypeptide β-N-acetyl-glucosaminyltransferase encoded by the Ogt gene which resides on the X chromosome. Ogt transfers N-acetylglucosamine to threonine or serine residues of cytoplasmic and nuclear proteins including many transcription factors such as c-Myc, c-Fos, c-Jun, p53, and Sp1. O-GlcNAc addition sites often overlap with phosphorylation sites in proteins, and O-GlcNAc modification can compete with phosphorylation, thereby potentially modulating essential and conserved cell signaling pathways. Targeted inactivation of Ogt results in early embryonic lethality at about embryonic day 5.5 (E5.5) [1]. Conditional inactivation of this gene in oocytes via ZP3Cre gives rise to infertile females [2]. The lack of Ogt may impair the developmental competence of mutant eggs or the development of the placenta in which genes on the paternal X chromosome are silent [3].

Complex or hybrid *N*-glycans are also required for oogenesis [4]. The *Mgat1* gene encodes GlcNAcT-I, which initiates the synthesis of complex and hybrid *N*-glycans. Females homozygous for floxed *Mgat1* alleles and carrying a ZP3Cre transgene ovulate fewer eggs compared with control mice. Among ovulated eggs, approximately half are developmentally incompetent, giving rise to embryos arrested during blastogenesis. We hypothesize that complex *N*-glycans are required for an early stage of oogenesis. When the *Mgat1* gene is inactivated efficiently, impaired development results, but if it is eliminated later the oocyte is unaffected [4]. Oocyte inactivation of the *PigA* gene which initiates glycophosphatidylinositol (GPI) synthesis also results in defective oocytes [5]. In this case, oocytes are refractory to sperm fusion and females are infertile. Surprisingly, inactivation of the *Pofut1* gene in oocytes does not affect oogenesis, ovulation, fertilization, blastogenesis, or implantation [6]. *Pofut1* encodes protein *O*-fucosyltransferase 1, an essential component of the canonical Notch signaling pathway [7]. Notch signaling was proposed on the basis of gene expression studies, to be involved in mouse preimplantation development. Thus, it is important to examine the properties of embryos that lack both maternal and zygotic transcripts of glycosyltransferase genes.

3. GLYCOSYLATION AND SPERMATOGENESIS

Targeted mutation of the gene encoding ganglioside GM2/GD2 synthase results in mice with no complex gangliosides [8]. The major consequence is that males are sterile and aspermatogenic. Gangliosides specifically bind testosterone and the mutant mice have low testosterone levels in blood and seminiferous tubules. Gene targeting of UDP-galactose:ceramide galactosyltransferase (Cgt) in mice demonstrated that seminolipid is also essential for spermatogenesis [9]. Seminolipid derived from the sulfation of galactosylalkylacylglycerol is synthesized by Cgt, and is upregulated during spermatogenesis. Spermatogenesis in $Cgt^{-/-}$ males is affected at the pachytene stage. The importance of seminolipid for spermatogenesis is further demonstrated by targeted inactivation of another gene, cerebroside sulfotransferase (*Cst*), which sulfates Gal to form seminolipid. $Cst^{-/-}$ males are infertile due to arrest of spermatogenesis before the first meiotic division [10].

Another example of the importance of glycosylation in spermatogenesis was shown by targeted disruption of *Man2a2* which encodes α-mannosidase-IIx (MX), an enzyme involved in the synthesis of *N*-glycan intermediates. Male MX-null mice have small testes and are infertile [11]. Germ cells fail to adhere to Sertoli cells in the seminiferous tubules, resulting in the premature release of developing germ cells from the seminiferous epithelium to the epididymis. MX in germ cells is involved in the synthesis of a GlcNAc-terminated triantennary and fucosylated *N*-glycan structure, which may play a key role in the adhesion between germ cells and Sertoli cells.

4. GLYCOSYLATION AND FERTILIZATION

O-Linked oligosaccharides on ZP3, one of three major proteins of the zona pellucida, and the Lewis X determinant on glycans of ZP2 and ZP3 and even N-glycans have been proposed as sperm receptors in different species. An alternative model is that ZP3 and ZP2 form a three-dimensional structure that serves as a receptor that determines the taxon-specificity of sperm–egg binding [12]. However, a critical role for N- or O-glycans on ZP3 in sperm binding is suggested by the fact that mouse oocytes expressing only human ZP3 with mouse ZP1 and ZP2 bind mouse sperm but not human sperm [13], suggesting that human ZP3 protein is not sufficient for human sperm binding. Mass spectrometric analysis revealed that the O-glycans of the human ZP3 from mouse eggs are identical to O-glycans in native mouse ZP3 [14]. Thus it remains possible that sugars on O-glycans mediate sperm–egg binding. However, terminal sugars on N-glycans are now known to be dispensable for sperm–egg binding since eggs lacking complex and hybrid N-glycans are fertilized and bind sperm [4,15].

5. GLYCOSYLATION AND PREIMPLANTATION DEVELOPMENT

In vitro culture of mouse embryos at the two-cell stage with tunicamycin, an antibiotic that specifically inhibits N-glycosylation of proteins, prevents blastogenesis because the embryos fail to compact. When morula or blastocyst stage embryos are cultured with tunicamycin, trophoblast cells die while inner cell mass development is less affected. Thus targeted disruption of the *Gpt* gene, encoding UDP-GlcNAc:dolichol phosphate N-acetylglucosamine-1-phosphate transferase required for the production of N-glycans *in vivo* was expected to result in preimplantation arrest. However, *Gpt*-null embryos developed normally to implantation, presumably due to the presence of maternal *Gpt* transcripts in early embryos [16]. Oocyte-specific elimination of the *Gpt* gene would determine if maternal and zygotic null Gpt blastocysts would develop. Complex and hybrid N-glycans are synthesized in the Golgi and are required for development beyond mid-gestation [17,18]. They were expected to also be important for preimplantation development. However, oocyte-specific deletion of the Mgat1 gene gives eggs lacking complex N-glycans that are fertilized by Mgat1⁻ sperm and blastocysts devoid of complex and hybrid N-glycans develop normally [4].

In another surprising result, embryos with an inactive *Fut9* gene that synthesize very little stage-specific embryonic antigen 1 (SSEA-1), also develop normally [19]. The SSEA-1 glycan determinant was previously proposed to be important for compaction of the blastocyst. Similarly, Notch signaling was proposed to be important for preimplantion development based on microarray data showing the expression of Notch pathway genes but embryos lacking maternal and zygotic Pofut1, an essential fucosyltransferase required for canonical Notch signaling, develop normally, implant and undergo gastrulation [6].

6. GLYCOSYLATION AND IMPLANTATION

Many experiments with inhibitors have implicated a role for glycan recognition in the adhesion of trophblasts to the uterine epithelium. In humans the expression of L-selectin in trophoblasts is dramatically increased when the embryo hatches. At the same time, the expression of sulfated selectin ligands is strongly upregulated in uterine epithelium during the window of receptivity [20]. However, this does not occur in mice. For mouse blastocysts it is also clear that complex and hybrid N-glycans are not required for adhesion to uterine epithelium [4]. However, several heparan sulfate proteoglycans have been implicated in implantation in mice. For example, perlecan is expressed on the surface of trophoectoderm cells at the time when mouse embryos have acquired attachment competence [21]. Another proteoglycan, syndecan-4 is expressed on the surface of uterine cells. Perlecan and syndecan-4 and perhaps other proteoglycans may be important for the creation of a microenvironment that allows embryos to implant [22].

7. GLYCOSYLATION AND POSTIMPLANTATION DEVELOPMENT

Following implantation, embryos undergo dramatic developmental changes including placentation, germ layer formation, gastrulation, and organogenesis. In these processes, many cell and tissue types are formed and complicated cell–cell interactions occur. Interference of glycosylation status by targeted inactivation of different glycosyltransferase genes has provided direct and convincing evidence of roles for a variety of glycans in postimplantation development.

7.1 *N*-Glycans and mucin *O*-glycans

Inactivation of *Mgat1*, the gene encoding *N*-acetylglucosaminyltransferase I (GlcNAcT-I), results in embryonic lethality at ~ E10 with retarded growth, defective neural tube formation, impaired vasculogenesis, heart looping, and randomized left–right asymmetry in tail positioning [17,18]. In addition, $Mgat1^{-/-}$ mutant ES cells do not contribute to the organized epithelial layer of bronchi, suggesting a requirement for complex N-glycans in the formation of bronchial epithelium [23].

To explore roles for complex or hybrid N-glycans in brain, Syn-1Cre transgenic mice were used to delete *Mgat1* and *Mgat2* genes specifically in neurons. In the absence of *Mgat1*, neurons have no hybrid or complex N-glycans, while in the absence of *Mgat2*, neurons have truncated hybrid N-glycans but no complex N-glycans [24]. Mice lacking GlcNAcT-I in neurons exhibit severe locomotor defects, tremors, paralysis, and early postnatal death. However, the lack of GlcNAcT-II in neurons does not cause neuronal and locomotor phenotypes, demonstrating that hybrid N-glycans, but not complex N-glycans, are required for neuronal development and functions. However, $Mgat2^{-/-}$ mice have several defects, including dysmorphic facial features, severe locomotor deficits, reduced muscular development, malformation and dysfunction of the skeleton, hematologic abnormalities (including blood coagulopathy and anemia), testicular atrophy, and spermatogenic failure [25]. Interestingly $Mgat2^{-/-}$ mice express a novel structure with a Lewis^x antenna on the bisecting GlcNAc. This new structure might contribute in a gain-of-function manner to mutant phenotypes. Some of the phenotypes in *Mgat2*-null mice resemble defects exhibited by humans with the congenital disorder of glycosylation CDG-IIa.

Other *N*-glycan branching glycosyltransferases have been deleted with much milder effects. Deletion of the *Mgat3* gene that encodes GlcNAcT-III and transfers the bisecting GlcNAc to *N*-glycans causes no obvious phenotype, but the progression of chemically induced liver tumors is retarded in the absence of GlcNAcT-III [26]. Deletion of the *Mgat5* gene that encodes GlcNAcT-V and transfers a β1,6 branch to *N*-glycans results in T cell activation and reduced tumor development and metastasis in a mammary tumor model [27]. Very interestingly, the lack of GlcNAcT-V reduces the residency time of several growth factor receptors on the cell surface and results in reduced signal transduction [28]. A related defect in cell surface residency was observed for Glut2 in pancreatic β cells of mice lacking GlcNAcT-IVa [29] which consequently become diabetic.

Polysialic acid (PSA) on N-glycans is important in brain development where it exists on the *N*-glycans of the neural cell adhesion molecule NCAM. PSA is a linear homopolymer synthesized by ST8sialV and ST8siaII, which may contain up to 200 sialic acid residues in α2,8-linkage. The expression of PSA is significantly upregulated during brain development. It modulates the adhesion properties of NCAM and affects neuronal and glial cell migration, outgrowth or sprouting of axons, and axon branching and fasciculation [30]. Double knockout of ST8sialV and ST8siaII gives rise to a severe phenotype, including specific brain wiring defects, progressive hydrocephalus, postnatal growth retardation, and precocious death [31]. These symptoms were rescued by further removal of NCAM in a triple knockout, suggesting that the severe phenotype is due to gain-of-function of NCAM lacking PSA. Therefore, PSA on NCAM plays a role in controlling and coordinating the functions of NCAM to ensure proper development of the brain.

Core fucosylation of *N*-glycans is catalyzed by an α1,6-fucosyltransferase (Fut8). Targeted inactivation of *Fut8* in mouse results in severe growth retardation and early postnatal lethality [32]. Overexpression of matrix metalloproteinases might be responsible for the observed progressive emphysema-like changes in lungs of $Fut8^{-/-}$ pups. The phenotype resembles that of a transforming growth factor beta (TGFβ)-null mutation, and TGFβ1 receptor appears to require a core fucose in complex *N*-glycans for activity [32]. Core fucosylation is also important for epidermal growth factor (EGF) receptor-mediated signaling. In embryonic fibroblasts from *Fut8*-null embryos EGF-induced phosphorylation of the EGF receptor is blocked, and EGF receptor-mediated JNK and ERK activation are inhibited. The binding of EGF to its receptor is eliminated in *Fut8*-null cells [33].

Fucosylation in the sialyl Lewis^x determinant is required for selectin-dependent leukocyte trafficking and its loss leads to leukocyte adhesion deficiency type II (also known as CDG-IIc) in humans. Deletion of the genes encoding *Fut7*, *Fut4*, or core2 GlcNAcT inhibits selectin ligand synthesis and reduces inflammatory responses [34]. Inactivation of different GlcNAcTs and sulfotransferases affects ligands for L-selectin and the homing of leukocytes [35]. Targeted inactivation of the *FX* gene in the mouse inhibits the de novo pathway for GDP-fucose synthesis, and results in embryonic lethality at ~E12.5. The likely cause of death at this early stage is due to a lack of O-fucose on Notch receptors [7]. Some pups are born and survive, exhibiting severe neutrophilia, myeloproliferation, lack of leukocyte selectin ligand expression, and infertility and pups may be rescued with fucose supplementation [36]. Several glycosyltransferases that transfer sialic acid to both N- and O-glycans or GalNAc to mucin O-glycans have been deleted in the mouse with various, relatively mild consequences [37]. However, deletion of the gene encoding β4GalT-1 results in neonatal lethality [35], although in at least one genetic background, a lack of β4GalT-1 is tolerated [38]. Deletion of T synthase that adds Gal to GalNAc on mucin O-glycans has more severe consequences [39]. Embryos lacking T synthase die at ~E13.5 with defects in vascularization.

7.2 O-Fucose glycans

O-Fucose glycans are present on proteins containing EGF-like domains with the consensus $C^2X_{4-5}(S/T)C^3$ (where C^2 and C^3 are the second and third cysteine residue in an EGF-like domain). Proteins known to carry O-fucose glycans include clotting factors VII and IX, u-PA (urokinase-type plasminogen activator), t-PA (tissue-type plasminogen activator), Cripto, Notch receptors, and Notch ligands. The enzyme that adds O-fucose to EGF repeats is protein O-fucosyltransferase 1 encoded by the *Pofut1* gene. O-Fucose on Notch receptors can be elongated to form a disaccharide (GlcNAcβ3Fuc), a trisaccharide (Galβ4GlcNAcβ3Fuc), or a tetrasaccharide (SAα3Galβ4GlcNAcβ3Fuc). Targeted inactivation of the *Pofut1* gene in mouse eliminates canonical Notch signaling resulting in embryonic lethality at ~E9.5 with defects in somitogenesis, vasculogenesis, neurogenesis, and cardiogenesis [7]. When the lunatic Fringe gene (*Lfng*), which encodes a β3GlcNAcT that adds GlcNAc to O-fucose on EGF repeats, is mutated, mice have defective somitogenesis and some pups die at birth. However, others survive and females lacking Lfng have a block at meiosis II and are infertile [40]. Lfng enhances Notch signaling induced by Delta1, and inhibits Jagged1- and Jagged2-induced Notch signaling. Galactose and β4GalT-1 were shown in a coculture reporter assay to be required for Fringe to inhibit Jagged1-induced Notch signaling [41]. Although *B4galt1* mutant mice do not exhibit overt Notch signaling defects, the expression of Notch signaling target genes such as *Hes5* and *Mesp2* is significantly reduced in *B4galt1*-null embryos. In addition, *B4galt1*-null embryos have an extra lumbar vertebra, consistent with aberrant Notch signaling [42].

7.3 O-Mannose glycans

Dystroglycan is the best-characterized glycoprotein with O-mannose glycans. The O-mannose glycans on the α-dystroglycan subunit are essential for dystroglycan to function [63]. Thus, mutations in human *POMT1* which encodes a subunit of the *O*-mannosyltransferase that adds Man to Ser/Thr in the mucin domain of α-dystroglycan, lead to muscular dystrophy and defects in brain development. *Pomt1* gene ablation in the mouse results in embryonic lethality at E7.5–9.5 with growth retardation and disruption of Reichert's membrane formation. Mutations in *POMGnT1*, which encodes the *O*-mannose β1,2N-acetylglucosaminyl-transferase which adds GlcNAc to Man on α-dystroglycan, also cause muscular dystrophy, as well as defects in eye and brain. The *O*-mannose disaccharide is further elongated with a Gal and a sialic acid residue to form a tetrasaccharide. A hereditary muscle disease is associated with mutations in GNE that encodes a bifunctional enzyme required for sialic acid synthesis [43]. Besides these known glycosyltransferase genes, there are several other putative glycosyltransferases (Large, Fukutin, and Fukutin-related protein, Fkrp) that contribute to the functional glycosylation of α-dystroglycan and in which mutations lead to muscular dystrophy. The *Large* gene encodes a potentially bifunctional glycosyltransferase and mutations give rise to muscular dystrophy in mice and humans. However, the reaction(s) catalyzed by Large remain unknown. Overexpressed Large is able to modify complex *N*-glycans and mucin-type *O*-glycans on α-dystroglycan [44], which may explain how it rescues α-dystroglycan glycosylation in muscle cells lacking POMT1 [45]. The disruption of another gene, *Fukutin*, also results in embryonic lethality at ~E6.5–7.5. Chimeric mice derived from ES cells bearing two disrupted *Fukutin* alleles develop severe muscular dystrophy, deficiency of α-dystroglycan and laminin binding, hippocampal, and cerebellar dysgenesis and defects in the eye [46].

7.4 Glycosaminoglycans (GAG) and proteoglycans

Glycosaminoglycans consist of repeating disaccharide units, and are divided into several classes: heparan sulfate (HS), chondroitin sulfate (CS), hyaluronan (HA), heparin, dermatan sulfate (DS), and keratan sulfate (KS). Except for hyaluronan they are found attached to proteins to form proteoglycans. Proteoglycans are present on the cell surface and in the extracellular matrix, and contribute to development by interacting with multiple signaling pathways including Wnt, Hedgehog (Hh), FGF, and TGFβ pathways.

Hyaluronan

Hyaluronan is an unsulfated GAG made of up to 10 000 disaccharide units (GlcNAcβ1,3GlcAβ1,4) not covalently attached to protein. Some proteins bind to hyaluronan to function (e.g. versican and CD44 which modulates Smad1 activation in the BMP-7 signaling pathway) [47]. Targeted disruption of *has2*, which encodes hyaluronan synthase 2, the major hyaluronan synthase expressed during embryogenesis, causes growth retardation, cardiovascular defects, and death at ~E9.0. Hyaluronan is required for two pivotal events in heart development, matrix expansion and the initiation of cell migration. In addition, hyaluronan is involved in signaling via ErbB2-ErbB3, which is required for epithelial–mesenchymal transformation during cardiac morphogenesis [48].

Heparan sulfate proteoglycans (HSPGs)

The GAG chains on HSPGs consist of repeating GlcNAcα1,4GlcAβ1,4 units of various length with modifications including sulfation. This group of proteoglycans is divided into several families according to their core proteins (i.e. glypicans, syndecans, and perlecans). The involvement of HSPGs in development was revealed in *Drosophila* in which mutation of a gene, *sgl*, encoding UDP-glucose dehydrogenase causes cuticle defects that are similar to those of Wnt and Hh signaling deficiencies [49]. Additional *Drosophila* mutants in GAG biosynthesis including *sfl* (N-deacetylase/N-sulfotransferase), *ttv (EXT1)*, *sotv (EXT2)*, *botv (EXTL3)*, *frc* (UDP-sugar transporter), and *slalom* (adenosine 3'-phosphate 5'-phosphosulfate transporter) exhibit phenotypes associated with Wnt, Hh, FGF, and TGFβ signaling [50]. In vertebrates, similar results were obtained [51]. Inactivation of *Ext1* in mouse by gene targeting results in embryos which die at E8.5 with a loss of HS, mesoderm, and extraembryonic tissues. Hedgehog (Hh) binding to mutant embryos is reduced indicating that *Ext1* assists Hh association with the cell surface [52]. Mice with *Ext1* deleted specifically in brain using a nestin-Cre transgene exhibit the loss of olfactory bulbs, have a small cerebral cortex, and the major commissures at the forebrain are absent due to defective axon guidance which also affects retinal axons. These and other observations led to the proposal of a "heparin sulfate code" such that HS modifications occurring in a region-specific manner result in enhancing or inhibiting ligand–receptor interactions [53].

Ext2 has also been inactivated in mouse, and gives phenotypes similar to *Ext1*-null embryos [54]. NDSTs (N-deacetylase/N-sulfotransferases) are HS-modifying enzymes that replace the acetyl group of N-acetylglucosamine with a sulfate group. *Ndst1*-null mice are unable to produce lung surfactants, pups suffer from lung failure and die shortly after birth. *Ndst2*-null mice have connective tissue-type mast cell defects. The number of granules containing protease, histamine, and inflammatory mediators are much reduced, and granule contents are either lacking or reduced. *Ndst1* and *Ndst2* double mutants exhibit embryonic lethality at an early stage [55]. When 2-O-sulfotransferase, another HS-modifying enzyme, is inactivated renal agenesis and defects in eye and skeleton, as well as cleft palate and polydactyly are observed with neonatal lethality [56].

Null mutations of certain core proteins of HSPGs (e.g. *glypican-3*) also give rise to severe developmental phenotypes, including developmental overgrowth, perinatal death, cystic and dysplastic kidneys, and abnormal lung development [57]. Multiple proteoglycans are upregulated in nephrogenesis, including syndecan-1, -2, and -4, glypican-1, -2, and -3, versican, decorin, and biglycan. Further it is HS, but not CS, that is required for ureteric bud branching as shown by in vitro experiments in which HSPGs were reduced by sodium chlorate (inhibiting sulfation of GAGs) and heparitinase (degrading GAG chains) [58].

7.5 Glycosphingolipids (GSLs)

Roles for GSLs in development have also been identified by gene targeting. Glucosylceramide synthase is the enzyme that transfers glucose from UDP-glucose to ceramide to form glucosylceramide. Targeted inactivation of the *Ugcg* gene results in embryonic lethality at E7.5 with greatly enhanced apoptosis in the ectoderm, which appears to be a primary cause of embryonic death [59]. Further experiments showed that the absence of GSLs leads to poor cell differentiation. To avoid embryonic lethality *Ugcg* alleles were floxed by *Lox*P sites and excised in the brain by a nestin-Cre transgene specifically expressed in neurons. Mice died 3 weeks after birth, exhibiting defects in the cerebellum, peripheral nerves with abnormal structures, and reduced axon branching of Purkinje cells [60]. The deficiency of neuronal GSLs was also shown to lead to downregulation of some genes that are involved in brain development and homeostasis.

Gangliosides are sialylated forms of GSLs. To investigate the role of this group of GSLs in the central nervous system (CNS), both *Siat9* (GM3 synthase gene) and *Galgt1* (GM2 synthase gene), were disrupted by gene targeting, in order to block the synthesis of all gangliosides. Mutant mice are born but develop striking vacuolar pathology in the white matter of the CNS with axonal degeneration and perturbed axon–glia interactions, showing that gangliosides are essential for the development of the CNS [61].

8. MOLECULAR BASES OF ROLES FOR GLYCOSYLATION IN DEVELOPMENT

Glycosylation of protein and lipid is clearly required for the development of vertebrates. However, glycosylation affects a multitude of glycoproteins, glycolipids, and proteoglycans, making it difficult to identify the primary basis of a glycosylation-defective phenotype. It is also difficult to determine the precise structural changes caused by altering glycosylation in a cell or tissue. Several hundred genes encode glycosylation-related enzymes, adding further complexity to the analysis. Nevertheless, in some cases glycosylation has been shown to be directly responsible for the biological activity of a glycoprotein. Thus deletion of the *Pofut1* gene prevents *O*-fucosylation of Notch receptors and leads to a lethal Notch signaling phenotype [7]. Since Notch lacking *O*-fucose does not bind Notch ligands [62], it may be that the ligands recognize *O*-fucose in the context of EGF repeats. Notch ligand binding is also affected by the addition of the GlcNAc to *O*-fucose on Notch receptors by Fringe enzymes [62]. In this case, the binding of the ligand Delta is enhanced whereas the binding of the ligand Jagged is reduced. The absence of Lfng leads to skeletal defects and this is thought to be primarily because of reduced Notch signaling via Delta1.

Another example in which the functional glycoprotein substrate of several glycosyltransferases is known is the *O*-mannose glycans on α-dystroglycan [63]. Specific glycoproteins are also affected by glycosyltransferase mutations in the branching GlcNAc transferases of *N*-glycans. Thus loss of GlcNAcT-V in T cells of $Mgat5^{-/-}$ mice leads to activation of T cells due to the reduced ability of galectin-3 to bind to T cell receptors [64]. More recently it was shown that mammary tumor cells from $Mgat5^{-/-}$ mice have reduced growth factor and cytokine signaling because of their reduced ability to be maintained on the cell surface through binding to galectin-3 [28]. Another example in which the cell surface residency of a glycoprotein was found to depend on having fully branched *N*-glycans is Glut2 in pancreatic β cells of mice lacking GlcNAcT-IVa [29]. In this case, galectin-9 was implicated in being at least partially responsible for maintaining Glut2 on the cell surface. In addition, this requirement was cell type specific as it did not occur in liver cells that also express Glut2. Thus, molecular interactions responsible for phenotypes that arise from a specific glycosyltransferase gene knockout are being defined. It is now abundantly clear that the inactivation of glycosyltransferase genes in the mouse is a necessary first step to obtaining mechanistic insights. Such mutant mice provide a wealth of tissues, cells and cell lines that can be used in a variety of strategies to identify glycosylation specific functions.

REFERENCES

1. Shafi R, Iyer SP, Ellies LG et al. (2000) *Proc. Natl Acad. Sci. USA* **97**, 5735–5739.
2. O'Donnell N, Zachara NE, Hart GW & Marth JD (2004) *Mol. Cell. Biol.* **24**, 1680–1690.
3. Harper MI, Fosten M & Monk M (1982) *J. Embryol. Exp. Morphol.* **67**, 127–135.
4. Shi S, Williams SA, Seppo A et al. (2004) *Mol. Cell. Biol.* **24**, 9920–9929.
5. Alfieri JA, Martin AD, Takeda J, Kondoh G, Myles DG & Primakoff P (2003) *J. Cell Sci.* **116**, 2149–2155.
6. Shi S, Stahl M, Lu L & Stanley P (2005) *Mol. Cell. Biol.* **25**, 9503–9508.
7. Shi S & Stanley P (2003) *Proc. Natl Acad. Sci. USA* **100**, 5234–5239.
8. Takamiya K, Yamamoto A, Furukawa K et al. (1998) *Proc. Natl Acad. Sci. USA* **95**, 12147–12152.
9. Fujimoto H, Tadano-Aritomi K, Tokumasu A et al. (2000) *J. Biol. Chem.* **275**, 22623–22626.
10. Honke K, Hirahara Y, Dupree J et al. (2002) *Proc. Natl Acad. Sci. USA* **99**, 4227–4232.
11. Akama TO, Nakagawa H, Sugihara K et al. (2002) *Science* **295**, 124–127.
12. Hoodbhoy T & Dean J (2004) *Reproduction* **127**, 417–422.
13. Rankin TL, Coleman JS, Epifano O et al. (2003) *Dev. Cell* **5**, 33–43.
14. Dell A, Chalabi S, Easton RL et al. (2003) *Proc. Natl Acad. Sci. USA* **100**, 15631–15636.
15. Hoodbhoy T, Joshi S, Boja ES, Williams SA, Stanley P & Dean J (2005) *J. Biol. Chem.* **280**, 12721–12731.
16. Marek KW, Vijay IK & Marth JD (1999) *Glycobiology* **9**, 1263–1271.
17. Ioffe E & Stanley P (1994) *Proc. Natl Acad. Sci. USA* **91**, 728–732.
18. Metzler M, Gertz A, Sarkar M, Schachter H, Schrader JW & Marth JD (1994) *EMBO J.* **13**, 2056–2065.
19. Kudo T, Kaneko M, Iwasaki H et al. (2004) *Mol. Cell Biol.* **24**, 4221–4228.
20. Genbacev OD, Prakobphol A, Foulk RA et al. (2003) *Science* **299**, 405–408.
21. Smith SE, French MM, Julian J, Paria BC, Dey SK & Carson DD (1997) *Dev. Biol.* **184**, 38–47.
22. San Martin S, Soto-Suazo M & Zorn TM (2004) *Am. J. Reprod. Immunol.* **52**, 53–59.
23. Ioffe E, Liu Y & Stanley P (1996) *Proc. Natl Acad. Sci. USA* **93**, 11041–11046.
24. Ye Z & Marth JD (2004) *Glycobiology* **14**, 547–558.
25. Wang Y, Tan J, Sutton-Smith M et al. (2001) *Glycobiology* **11**, 1051–1070.
26. Stanley P (2002) *Biochim. Biophys. Acta* **1573**, 363–368.
27. Dennis JW, Pawling J, Cheung P, Partridge E & Demetriou M (2002) *Biochim. Biophys. Acta* **1573**, 414–422.
28. Partridge EA, Le Roy C, Di Guglielmo GM et al. (2004) *Science* **306**, 120–124.
29. Ohtsubo K, Takamatsu S, Minowa MT, Yoshida A, Takeuchi M & Marth JD (2005) *Cell* **123**, 1307–1321.
30. Kleene R & Schachner M (2004) *Nat. Rev. Neurosci.* **5**, 195–208.

31. Weinhold B, Seidenfaden R, Rockle I et al. (2005) J. Biol. Chem. **280**, 42971–42977.
32. Wang X, Inoue S, Gu J et al. (2005) Proc. Natl Acad. Sci. USA **102**, 15791–15796.
33. Wang X, Gu J, Ihara H, Miyoshi E, Honke K & Taniguchi N (2006) J. Biol. Chem. **281**, 2572–2577.
34. Becker DJ & Lowe JB (2003) Glycobiology **13**, 41R–53R.
35. Kawashima H, Petryniak B, Hiraoka N et al. (2005) Nat. Immunol. **6**, 1096–1104.
36. Smith PL, Myers JT, Rogers CE et al. (2002) J. Cell Biol. **158**, 801–815.
37. Lowe JB & Marth JD (2003) Annu. Rev. Biochem. **72**, 643–691.
38. Nishie T, Miyaishi O, Naruse C, Hashimoto N & Asano M (2004) Nephrology (Carlton) **9**(Suppl 2), A48.
39. Xia L, Ju T, Westmuckett A, An G et al. (2004) J. Cell Biol. **164**, 451–459.
40. Hahn KL, Johnson J, Beres BJ, Howard S & Wilson-Rawls J (2005) Development **132**, 817–828.
41. Chen J, Moloney DJ & Stanley P (2001) Proc. Natl Acad. Sci. USA **98**, 13716–13721.
42. Chen J, Lu L, Shi S & Stanley P (2006) Gene Expr. Patterns **6**, 376–382.
43. Huizing, M Rakocevic G, Sparks SE et al. (2004) Mol. Genet. Metab. **81**, 196–202.
44. Patnaik SK & Stanley P (2005) J. Biol. Chem. **280**, 20851–20859.
45. Barresi R, Michele DE, Kanagawa M et al. (2004) Nat. Med. **10**, 696–703.
46. Takeda S, Kondo M, Sasaki J et al. (2003) Hum. Mol. Genet. **12**, 1449–1459.
47. Peterson RS, Andhare RA, Rousche KT et al. (2004) J. Cell Biol. **166**, 1081–1091.
48. McDonald JA & Camenisch TD (2002) Glycoconj. J. **19**, 331–339.
49. Hacker U, Lin X & Perrimon N (1997) Development **124**, 3565–3573.
50. Selleck SB (2001) Semin. Cell Dev. Biol. **12**, 127–134.
51. Iozzo RV. (2005) Nat. Rev. Mol. Cell Biol. **6**, 646–656.
52. Lin X, Wei G Shi Z et al. (2000) Dev. Biol. **224**, 299–311.
53. Holt CE & Dickson BJ (2005) Neuron **46**, 169–172.
54. Stickens D, Zak BM, Rougier N, Esko JD & Werb Z (2005) Development **132**, 5055–5068.
55. Kjellen L (2003) Biochem. Soc. Trans. **31**, 340–342.
56. Wilson VA, Gallagher JT & Merry CL (2002) Glycoconj. J. **19**, 347–354.
57. Forsberg E & Kjellen L (2001) J. Clin. Invest. **108**, 175–180.
58. Steer DL, Shah MM, Bush KT et al. (2004) Dev. Biol. **272**, 310–327.
59. Yamashita T, Wada R, Sasaki T et al. (1999) Proc. Natl Acad. Sci. USA **96**, 9142–9147.
60. Jennemann R, Sandhoff R, Wang S et al. (2005) Proc. Natl Acad. Sci. USA **102**, 12459–12464.
61. Yamashita T, Wu YP, Sandhoff R et al. (2005) Proc. Natl Acad. Sci. USA **102**, 2725–2730.
62. Haines N & Irvine KD (2003) Nat. Rev. Mol. Cell Biol. **4**, 786–797.
63. Barresi R & Campbell KP (2006) J. Cell Sci. **119**, 199–207.
64. Demetriou M, Granovsky M, Quaggin S & Dennis JW (2001) Nature **409**, 733–739.

SECTION 7
Glycobiology and medicine

CHAPTER 20
Glycobiology of mucins in the human gastrointestinal tract

Anthony P. Corfield

1. MUCINS AND GLYCOBIOLOGY

The mucins, or mucus glycoproteins, are ubiquitous molecules found at mucosal surfaces in many organisms, underlining their biological significance in mucosal protective processes. This short review focuses on recent developments in current thinking in this area. They have been stimulated over many years by the continuing work of Nathan Sharon in what has now become glycobiology.

1.1 Mucin gene structure

The mucins comprise a family of molecules known as the MUC genes. At the present time there are 19 known members in this family and they can be divided into two major groups [1–3]. The secreted mucins form viscous and largely gelforming extracellular mucus bilayers, while the membrane-associated mucins make up a larger group of molecules that possess typical membrane-spanning domains and have characteristic mucin properties. The arrangement and genomic organization of mucin genes has been well defined [4]. This has given a major insight into the function of mucin molecules and to the relationship between their tissue and cellular expression and its regulation.

The major secreted mucin genes code for very large molecules, typically $2-20 \times 10^5$ Da, which can form oligomers and aggregate to yield gel networks [1–3]. Typically the secreted mucin genes contain one or more central domains, rich in serine, threonine, and proline, which exist as variable number tandem repeats (VNTR domains) and which carry the bulk of the O-linked glycan chains typical of these macromolecules. At both N- and C- termini protein domains with homology to von Willebrand factor D and C domains are found and mediate dimerization and oligomerization.

Other shared protein motifs include cysteine-rich domains distributed through the *MUC2*, *MUC5AC*, and *MUC5B* genes. These protein domains are also the target for selective proteolytic cleavages responsible for the normal *in vivo* processing of mucins and a GDPH (glycine; aspartic acid; proline; histidine) autocatalytic proteolytic site is located within the C-terminal von Willebrand D4 domain in MUC2 and MUC5AC [5].

Membrane-associated mucins also contain the serine/threonine-rich tandem repeat domains, and apart from MUC4 lack the von Willebrand motifs. Instead they have transmembrane and cytoplasmic tail domains. In keeping with their membrane-associated functions, suggestive of extensive cell surface interactions, sea-urchin sperm protein, enterokinase, and agrin (SEA) and epidermal growth factor (EGF) domains are commonly found at the C-terminus and enable a range of specific proteolytic cleavages and membrane signaling events, distinguishing them from the secreted mucins [6,7]. These events are characteristic for the membrane-associated mucins which undergo an initial proteolytic cleavage to generate a non-covalently linked heterodimer consisting of mucin-type and membrane-type subunits and which allow the multiple functions of these mucins at the apical membrane surface. Release of the mucin subunit as a soluble molecule without the membrane subunit occurs after a second proteolytic cleavage, but is still poorly understood [8]. In addition, a signal sequence is located at the N-terminus and targets the mucin apoprotein to the endoplasmic reticulum and ultimately to secretion vesicles or apical cell surface membranes [9].

Although many of these domains have significant sequence homology with other well-defined proteins, and clearly have function properties in line with these molecules, each mucin has its own sequence identity.

1.2 Mucin glycan structure

The mucins have long been identified by their high content of *O*-linked glycan chains, frequently contributing 60–80% of the dry weight of these macromolecules. The arrangement of the glycan chains in the VNTR domains generates a novel molecular pattern for each mucin, donating precise structural identity and rheological properties [10,11].

A library of *O*-linked glycan chain structures has been established over many years. The expression of these chains in mucins correlates with their tissue location so that the same mucin gene protein may carry different *O*-glycans depending on the tissue in which it is synthesized. Furthermore, many of the glycan structures found in mucins are shared with other non-mucin glycoproteins and glycolipids [3,10].

The biosynthetic pathways for the formation of the *O*-linked glycans are well known and constitute an important non-template mechanism for post-translational protein modification. Thus, although *O*-glycan sequence is generated without a template code, the fidelity of synthesis is high [10].

The range of *O*-glycan chains found on mucins is widely varied, but can be identified as products of the known *O*-glycosylation pathways, forming and extending the *O*-glycan core sequences. The majority of these structures have not been associated with any particular biological function, other than bulk carbohydrate content and charge contributing to rheological properties. Examination of the databases of known *O*-glycan sequences reveals examples with biological significance. Examples of biologically relevant *O*-glycans include cancer markers such as the Tn antigen (GalNAc-α-Ser/Thr) and its sialylated form

(Neu5Ac-α2-6GalNAc-α-Ser/Thr); the Thomsen Friedenreich antigen (Gal-β1-3GalNAc-α-Ser/Thr) and the sialylated glycans sialyl Lewis[a] (CA 19-9) and sialyl Lewis[x] [3,10,12].

Besides these examples the mucins are also known to carry the ABO and Lewis blood group structures [12,13]. The expression of these antigens serves to illustrate a sophisticated biological regulation closely related to the O-glycan sequence. The epithelial synthesis of blood group antigens is linked with the secretor system. This means that only those individuals who are positive for the secretor gene are able to express the ABO antigens on glycoconjugates on epithelial surfaces. The biological significance of these events has been demonstrated in the gastrointestinal tract, where the turnover and recycling of mucins is mediated by the mucinase activity produced by the resident enteric microflora [14–16]. The properties of the microflora with regard to mucin degradation correlates strongly with the secretor and blood group status of the individual. The terminal non-reducing monosaccharides of the blood group antigens in secretor positive mucin O-glycans are the first structures to be removed. The appropriate glycosidases carry out this cleavage and are induced in the enteric bacterial strains found in each individual. The presence of an inappropriate microflora (e.g. blood group A in a blood group H individual) would lead to extensive unregulated degradation of the mucus in the mucosal barrier and underlines the biological specificity of this system.

Further support for the sophistication and biological consequence of blood group antigen manifestation in mucins has been reported recently. The secreted salivary mucin MUC5B carries ABO antigen-containing O-glycans in secretor-positive individuals, while non-secretors had no blood group antigens but increased levels of sialylated glycans with sialyl Lewis[a] [13].

The determination and evaluation of the large number of O-glycan structures has presented a considerable technical demand, which has been met in recent years through improvements in isolation and detection with increasingly smaller amounts of material culminating in recent advances in mass spectrometric techniques where mixtures can be reliably investigated [17–20]. This has put the detailed analysis of samples of physiological relevance within grasp and has allowed an assessment of the potential structure–function relationships at a new level.

1.3 Mucin structure–function relationships

The ascribed physiological functions of mucins are due largely to the glycoprotein nature of these macromolecules. Many properties depend on the range of protein domain arrangements, while others relate to the glycan sequence.

The rheological properties of the mucins in forming viscoelastic gels at mucosal surfaces are vital for the creation of a regulated microenvironment [1,21]. These properties are generated through the abundance of O-glycans attached to the VNTR domains. In some cases the negative charge due to the sialic acid and glycosulfate contributes to viscoelasticity. The secreted mucus gel created has a variety of assets facilitating protection from and interaction with the external environment in the gastrointestinal lumen [21]. It creates a network which is able to bind water, exchange ions, act as a filtration medium, provide arrays of glycan sequences in localized areas [1,2], and provides a scaffold for the attachment and storage of growth factors. It enables interactions with the mucosal cell apical membrane glycocalyx [22] and affords an optimal medium for the enteric microflora to interact with the host at the mucosal surface and to repel pathogens [23]. Close examination of the nature of the mucus gel has led to the identification of a bilayer arrangement with a strong gel in close contact with the mucosal cell surface and a sloppy, weaker gel on top [21]. The wealth of potential functions is closely linked with the need to simultaneously carry out a variety of protective and dietary roles and to remain active in a continuous, synchronized, and dynamic manner.

The mucosal barrier includes the membrane-associated mucins situated in the apical surface membranes [1,2]. The protein domains found in these mucins correlate well with their cell surface functions. Anchoring through the transmembrane domain permits positioning of the mucin member of the heterodimer in an external luminal position and allows signal transduction through the cytoplasmic tail [24,25].

Strong evidence for the participation of mucin EGF domains in cell signaling has come from an assortment of experiments showing the cell surface interaction of members of the EGF receptor family with mucins. The best-characterized membrane signaling events have been reported for MUC4 [24]. One of the EGF domains carried by MUC4 has been shown to interact with ErbB2 and thus mediate signaling from the external environment. As many of the membrane-associated mucins also contain EGF domains in a similar arrangement to those seen in MUC4 it is likely that such signaling extends to these mucins as well and underlines the broad consequence of these multifunctional membrane-located mucins in monitoring the external environment [24].

Individual glycans have been found to play specific roles in the biosynthesis of secretory and membrane-associated mucins. The presence of *N*-glycans is required for the transport of some secretory mucins (MUC2 but not MUC5AC) from the endoplasmatic reticulum to the Golgi apparatus prior to *O*-glycosylation [9]. The mucin precursors interact with chaperonins, in particular calreticulin and this binding is mediated by *N*-glycans [26]. In addition, the dimerization of MUC2 depends on the cysteine knot domain at the C-terminus of the molecule. A series of cysteine residues form disulfide bridges leading to dimer formation. *N*-Glycans are also found in this domain and mediate cysteine knot disulfide bridge formation [27].

Subcellular targeting of proteins is mediated in part through the recognition of glycans. Among the membrane mucins *N*-glycosylation is required for surface membrane localization of MUC17 [28]. *C*-Mannosylation of the cysteine-rich domains of MUC5AC and MUC5B is necessary for the transport of mucin precursors out of the endoplasmic reticulum [29]. Relatively little is known about *C*-mannosylation and its biological consequences and these observations add to the growing number of glycan structures mediating mucin manipulation and their biological activity within cells.

2. MUCINS IN THE GASTROINTESTINAL TRACT

The gastrointestinal mucosal surface shows considerable variation throughout its length. Many studies on the secreted mucus obtained from defined regions throughout the tract have reflected this variation. Chemical investigation of total mucus fractions isolated from different regions of the tract formed the basis of the glycan sequence databases which have been built up since [30–32]. Attempts to match the regional glycosylation with individual MUC gene products has led to the identification of glycoforms associated with individual MUC gene products in the gastrointestinal tract [13,33,34]. However, these examples remain small in number and it is largely the well-expressed MUC genes that have been analyzed so far. In addition, the number of studies generating glycan sequence data is small and the majority of information regarding glycosylation patterns has come from combinations of histochemical and immunohistological work with lectins and antibodies. Although few in number, the glycan sequence investigations have supplied an important base for comparison.

Early studies on gastric mucus glycosylation from different blood group individuals gave a view of the size range and sequence found in these mucins [32]. Subsequent analyses with lectins and antibodies provided an indication of partial sequence information and demonstrated the specific cellular location of these structures [35]. A chemical examination of the range of O-glycan structures associated with regions of the intestinal tract has shown gradients of glycan structures and negative charge due to sialic acid and ester sulfate [17,36]. Some of these structures and their locations are shown in Table 1.

Table 1 Structure of mucin glycans from different regions of the intestinal tract

Glycan structure	Type	Ileum	Caecum	Transverse colon	Sigmoid colon	Rectum
Neutral glycans						
GlcNAcβ3GalNAcol	core 3	+	+	+	+	+
GalNAcα3GalNAcol	core 5	+	+	+	+	+
Galβ3[GlcNAcβ6]GalNAcol	core 2	+	+	+	+	+
GalβGlcNAcβ3GalNAcol	core 3	+	+	+	+	+
GlcNAcβ3Galβ3GalNAcol	core 1	+	+	+	+	+
Galβ3(Fucα4)GlcNAcβ3GalNAcol	core 3	+	+	+	+	+
Fucα2Gal3GalNAcol	core 1	–	–	–	+	–
Galβ4GlcNAcβ3[GlcNAcβ6]GalNAcol	core 4	+	+	–	+	+
Glycans with one sulfate						
(SO$_3^-$)3Galβ4GlcNAcβ3GalNAcol	core 3	–	+	–	+	+
Galβ4(SO3$^-$)GlcNAcβ3GalNAcol	core 3	+	+	–	+	+
(SO$_3^-$)3Galβ4(Fucα3)GlcNAcβ3GalNAcol	core 3	–	+	+	+	+
Galβ3[(SO$_3^-$)3Galβ4GlcNAcβ6]GalNAcol	core 2	–	–	+	+	+
Galβ3[Galβ4(SO$_3^-$)GlcNAcβ6]GalNAcol	core 2	–	–	+	–	+
(SO$_3^-$)3Galβ4GlcNAcβ3Galβ3GalNAcol	core 1	–	+	+	–	–
Glycans with one sialic acid						
NeuAcα6GalNAcol	sialyl-Tn	+	+	+	+	+
Galβ3[NeuAcα6]GalNAcol	core 1	+	+	+	–	+

Structure	Core type					
GlcNAcβ3[NeuAcα6]GalNAcol	core 3	+	+	+	+	+
GlcNAcβ3Galβ3[NeuAcα6]GalNAcol	core 1	+	+	+	+	+
Galβ4(Fucα3)GlcNAcβ3[NeuAcα6]GalNAcol	core 3	+	+	+	+	+
NeuAcα3Galβ3[GlcNAcβ6]GalNAcol	core 2	−	−	−	+	−
GalNAcβ4(NeuAcα3)Galβ4GlcNAcβ3GalNAcol	core 3 Sd[a]	−	−	+	+	+
GalNAcβ4(NeuAcα3)Galβ3GlcNAcβ3GalNAcol	core 3 Sd[a]	−	−	+	+	+
Glycans with two acidic residues						
(SO₃⁻)3Galβ4GlcNAcβ3[NeuAcα6]GalNAcol	core 3	−	−	+	+	+
NeuAcα3Galβ3[NeuAcα6]GalNAcol	core 1	+	+	+	+	−
(SO₃⁻)3Galβ4(Fucα3)GlcNAcβ3[NeuAcα6]GalNAcol	core 3	+	+	+	+	+
NeuAcα3Galβ3[(SO₃⁻)3Galβ4GlcNAcβ6]GalNAcol	core 1	−	−	−	+	+
NeuAcα3Galβ4GlcNAcβ3[NeuAcα6]GalNAcol	core 3	+	+	+	+	+
(SO3⁻)3Galβ4(Fucα3)GlcNAcβ3Galβ3[NeuAcα6]GalNAcol	core 1	−	−	+	−	−
NeuAcα3Galβ4(Fucα3)GlcNAcβ3[NeuAcα6]GalNAcol	core 3	−	+	+	+	+
(SO₃⁻)3Galβ4GlcNAcβ3Galβ4GlcNAcβ3[NeuAcα6]GalNAcol	core 3	−	+	−	+	−
GalNAcβ4(NeuAcα3)Galβ4GlcNAcβ3[NeuAcα6]GalNAcol	core 3 Sd[a]	−	+	+	+	+
GalNAcβ4(NeuAcα3)Galβ3GlcNAcβ3[NeuAcα6]GalNAcol	core 3 Sd[a]	−	+	+	+	+

Individual monosaccharides are abbreviated as: NeuAc, sialic acid; Gal, galactose; GlcNAc, N-acetylglucosamine; GalNAc, N-acetylgalactosamine; Fuc, fucose; SO₃⁻, sulfate.
The type of glycan chain refers to the structure linked to the peptide chain: core 1 Galβ3GalNAcol; core 2, Galβ3[GlcNAcβ6]GalNAcol; core 3 GlcNAcβ3GalNAcol core 4, GlcNAcβ3[GlcNAcβ6]GalNAcol core 5, GalNAcα3GalNAcol.
The upper arm of the glycan is shown in bold.
Data were adapted from Robbe et al. (2004) [36].

Demonstration of such selective distributions implies a biological role for the glycan sequences expressed. The obvious target for interaction in the gastrointestinal tract is the enteric bacterial microflora and evidence is now accumulating that this interaction is considerably more sophisticated that first imagined [23]. The plethora of glycan structures provides the variety to enable a dynamic and specific recognition system throughout the tract that is well adapted to the requirements of the host with bacterial populations.

In addition to the characteristic MUC gene expression, the regional glycosylation variations provide a further level of adaptation and specificity and contribute to the stability of the protective barrier, donate flexibility to the mucosal response to the external environment, and enable a dynamic interaction with microflora populations.

In keeping with the regional variation in glycosylation discussed above for the secreted component of the mucous barrier, the mucosal cell apical membrane glycocalyx also has a characteristic identity at each location along the gastrointestinal tract. This is largely apparent from the histological investigations performed on tissue sections [37,38]. Structural knowledge of individual components of the gastrointestinal glycocalyx is currently sporadic and must embrace a wider range of glycoconjugates including mucins, glycoproteins, and glycolipids to be informative for biological function. The introduction of proteomics has offered an improvement to assess these membrane glycoconjugates [20].

2.1 Mucins in gastrointestinal development

The presence of a mucous barrier is apparent during fetal development and continues to mature at birth and through lactation, weaning, and early adulthood [3,39].

Fetal expression of the MUC genes has been reported in humans and follows distinct patterns [3,39]. Most data are available for mRNA levels and shows differences between the secreted and membrane-associated mucin genes. The earliest appearance of MUC genes is found at 6.5 weeks of gestation (MUC3 and MUC4). A pattern of gestational age-specific appearance along the gastrointestinal axis has been reported. The major adult mucin in the intestine, MUC2, is first detected at high levels at 10 weeks in the duodenum and in the intestine from 9 weeks. A typical adult pattern is found and maintained from week 23 in the small intestine and week 27 in the colon. Weak expression is also seen in the antral glands of the stomach at 26 weeks, although this is uncommon in the normal adult stomach [3,39].

In the developing foregut MUC5AC is found together with MUC5B in the superficial cells at 8 weeks. MUC5AC is present in the duodenal crypts from 16 weeks of gestation and is maintained through to birth and adulthood. It also appears transiently in the large intestine at 17 weeks and is not seen in the normal colon. However, it does appear as a cancer marker early in the adenoma carcinoma sequence in adults. Disappearance of MUC5B at 26 weeks is followed by its reappearance in the distal intestine in keeping with its adult pattern. MUC6 shows weak gastric expression at 8 weeks of gestation and is located in the superficial epithelium. The location changes to deeper glands with progressing gestation and finally matches the adult pattern. In addition, MUC6 is found in Brunner's glands of the small intestine at 26 weeks [3,39].

The membrane-associated mucins MUC3 and MUC4 are found very early in fetal gut development and may play a role in the establishment of the primitive gut itself. MUC3 and MUC4 are detected at 6.5 weeks of gestation when stratification is limited and before differentiation has been initiated. MUC1 is detected as a weak signal in the stomach at 8 weeks and later, at 18 weeks of gestation, in the colon [3,39].

There are very few studies on the patterns of glycosylation in fetal gut. Sulfation and O-acetyl-sialic acid expression has been found in the colon at 12 weeks of gestation [40], but a systematic study remains to be done to chart the expression and relevance of fetal glycan structure and mucin synthesis.

Mucins are present during fetal gut development before the major morphogenesis and cytodifferentiation events occur generating the established regions identifiable in the neonatal and adult gastrointestinal tract. Little consideration has been made so far regarding the functional role of these molecules in the specific processes involved in fetal gut development. However, it appears that the establishment of an adult-type MUC gene expression pattern is a basic requirement for the functioning and adaptation of the mucosal protective barrier at birth. In spite of the many adaptations required in the neonatal period and during lactation and weaning, the same pattern of MUC gene expression is maintained.

Studies on glycosylation, either in general or specifically for individual mucins at locations throughout the gastrointestinal tract during fetal development, have not been carried out and must now represent a major target for future work. Important nutritional and microbiological changes occur at this time.

3. MUCINS AND MUCOSAL PROTECTION

3.1 Concept of the protective barrier

Mucosal surfaces throughout the mammalian body are adapted to interact with the external environment in order to achieve a variety of functions dictated by their location. At the same time there is a need to preserve a continuous and effective protection against the many aggressive agents encountered in the external environment at these surfaces. In general the mucosal protective barrier consists of a secreted mucus gel, which interacts with an apical membrane-anchored glycocalyx. The barrier has properties that enable both innate and adaptive immune systems to function effectively.

3.2 Mucus as a physical viscoelastic shield

The secreted mucus gel provides a network that allows gel filtration, ion exchange, and hydration at the mucosal surface and thus the interaction of the external milieu with the host epithelial cells. It also constitutes a physical, viscoelastic barrier to the exterior and shows rheological properties necessary to maintain continuous protection. In the gastrointestinal tract the demands are for lubrication of particles in the diet, mechanical stability during peristalsis, resistance to mechanical forces generated by movement of food through the tract, and chemical stability against gastric acid, proteolytic digestion, and the action of bile salts [1,21,41]. The molecular structure of the mucins within the gel network adds arrays of regular glycan structures carried on the VNTR domains, allowing specific recognition sites for many lectins, adhesion molecules, bacteria, and protective proteins functioning in mucosal defense. This molecular arrangement has been termed stoichiometric power [1], and enables multivalent binding of these proteins which are integrated in a selective manner into the mucin gel.

3.3 Mucus as a medium for protective proteins

The mucus bilayer provides a medium for many proteins that play roles in mucosal defense. These include secretory IgA, lactoferrin, lysozyme, defensins, trefoil factor family peptides, protease inhibitors, galectins, and a number of others [1,3]. These proteins afford antibacterial action, repair and restitution tasks in addition to a variety of binding properties that may have direct roles in the mechanical stability of the mucus gel. In addition there are a number of peptides and growth factors, which may be sequestered in the mucus gel by virtue of the stoichiometric power of the glycan arrays in the mucin VNTR domains. These include the trefoil peptides, a range of interleukins, epidermal growth factor, and transforming growth factor alpha (TGFα) [1,3]. A role for these factors in mucosal defense has been proposed and fits well with mediation of inflammatory responses at the mucosal surface. The mucins in the mucus gel act as a scaffold for binding and permit the creation of reservoirs of these important factors enabling rapid and site-specific responses throughout the gastrointestinal tract [23].

3.4 Glycocalyx mucins and signaling

The mucosal barrier creates and maintains a specific microenvironment throughout the length of the gastrointestinal tract. The barrier relies on cell surface-anchored molecules to interact with luminal changes and to relay information back to the host cells to allow adaptation. The membrane-associated mucins are one of the families of molecules that are equipped to carry out such transfer of information through established signaling pathways [1,24]. In this way mediation of epithelial cell–cell adhesion, differentiation, and proliferation in response to external conditions is possible. Such signaling roles have been demonstrated for MUC1 and MUC4 [24]. The glycan-rich mucin domains of these molecules are instrumental in monitoring the external environment and this remains an area of continuous development.

4. THE MUCOUS BARRIER AND THE INTESTINAL MICROFLORA

4.1 Symbiotic/commensal microflora

The interaction of the enteric mucosal flora with the host mucosal surface has long been appreciated as a vital factor in the normal function and protection in the gastrointestinal tract [23,42–45]. Much work has attempted to unravel the intricacies of the interactions necessary to allow dynamic bacterial populations to operate beneficially with the host. Recent work has afforded a new insight into the sophistication of these interactions and given a focus for mucin glycobiology [23,42–44].

The ideal system relies on a symbiotic or commensal relationship of the enteric bacterial flora with the host. The microflora in each individual is required to provide nutritional advantage through provision of non-existing factors such as vitamins, and the ability to turn over the mucous barrier in a regulated manner. The development of a human gut microflora which is adapted to each individual and serves to preserve mucosal protection while allowing all aspects of digestion and nutrition to progress efficiently has resulted from the coevolution in both host and bacteria [44,45].

The establishment of bacterial populations during neonatal life accompany changes in both the number and range of bacterial species and development of the host glycan library at the mucosal surface [46]. The developing gut needs to generate a medium, a biofilm, to enable the establishment of stable residency for symbionts. The properties of such a biofilm must include (1) a matrix comprising a high molecular weight polymer, (2) a mechanism to allow bacteria to attach and prevent loss from the biofilm, and (3) an accessible nutrient source.

Considerable evidence shows that the intestinal mucus gel layer in the intestine satisfies these criteria. The mucins contribute the polymeric element to the biofilm matrix and the arrays of glycans provide targets for manipulation through bacterial glycohydrolases and creation of binding sites to facilitate residence [23,42–44]. The glycan-rich nature of the mucins offers the source of nutrition and the bacterial glycomes examined to date contain many enzymes and proteins that are designed to remove and degrade the carbohydrates to produce energy (see also http://www.afmb.univ-mrs.fr, http://www.glycosciences.org.uk and http://functionalglycomics.org).

Studies on the organization and interaction of bacteria in the gastrointestinal tract have used the mouse and *Bacteroides thetaiotaomicron*, an obligate anaerobe found in the normal adult colon. *B. thetaiotaomicron* has evolved the means to bind to and harvest mucus-associated glycans and can outcompete other microbes without the potential to utilize glycoconjugates [23,42–44]. Examination of the genes responsible for glycan recognition in *B. thetaiotaomicron* reveals that enzymes to degrade the host mucus glycans are present as well as binding proteins for the bacterial capsule polysaccharide. Therefore it is well equipped to forage for glycans and utilize non-host polysaccharide and host mucus as sources of dietary carbohydrate while at the same time creating an intestinal niche for attachment and colonization [47]. The significance of these observations with regard to active biological function is underlined by the demonstration that the expression pattern of the capsule polysaccharide-binding proteins varies as a function of bacterial growth phase and nutrient availability.

Adaptation of the human enteric microflora at the individual level includes recognition of secretor status and the ABO blood group system. The microflora is representative for secretor-positive individuals depending on their blood group such that the array of glycohydrolases induced are able to release each terminal blood group determinant sugar in order to degrade the rest of the glycan [15]. The glycohydrolase activity found in blood group A, B, or AB microflora rapidly destroys the mucus found in O or non-secretor individuals and emphasizes the existence of a regulated mechanism adapted to the needs of the individual as part of a general program for bacteria–host interaction [14,15].

The sophistication of this system has been demonstrated in mice for the utilization of fucose and has been termed glycan legislation [43]. A chemical bacterial signal is secreted and induces fucosyltransfer to a cell surface glycan carried by a cell surface glycoconjugate, and may include the membrane and secreted mucins in the intestinal epithelium. Fucosidases released by the microbe cleave terminal fucose from these molecules and this sugar is taken up by the bacteria through a specific fucose transporter. Once the internal bacterial concentration of fucose rises sufficiently a series of genes are induced which catalyze the breakdown of fucose to generate energy. At the same time the gene coding for the chemical signal is repressed. This sensor system is ideally adapted to a host–microbe symbiotic interaction [43].

The abundance and range of glycan structures found at the intestinal mucosal surface may have been generated as a result of the selective pressure to accommodate gut microflora populations in host niches and to resist pathogens [23,48]. The identification of specific mucin glycan structure expression on a regional basis in the intestine affords a molecular basis for the direction of the microbiota to defined niches. The work with mice has demonstrated that the epithelial glycans synthesized during the perinatal period direct early colonizers to niches throughout the entire length of the developing intestine. Recent work in our labs has demonstrated glycan legislation in the adult human intestine. O-Acetylated sialic acid expression in the host intestinal mucosa was found to be regulated by the presence or absence of the fecal microbiota. A switch-off of certain glycotopes could be shown which confirms the existence of bacterial regulation of host glycan expression and demonstrates its selective nature.

5. FUTURE PERSPECTIVES OF INTESTINAL MUCIN GLYCOBIOLOGY

The combination of improved mapping of mucin glycotopes throughout the gastrointestinal tract demands powerful screening technology in order to investigate the molecular nature of biological interactions. Study of gastrointestinal glycomics will draw on the synthetic chemical arena to design appropriate glycoarrays to do this. The scope of this technology is broad, but lends itself well to the questions currently posed. Generation of gene array probe sets that can access the relevant genes of carbohydrate biosynthesis, modification, and degradation will back-up the glycoarrays. [49,50] (see also http://www.afmb.univ-mrs.fr, http://www.glycosciences.org.uk and http://functionalglycomics.org).

At the same time gene array technology may be used to identify the novel natural chemical entities that microbes have evolved to manipulate the host biology in ways that enforce health and prevent various diseases arising in or outside the gastrointestinal tract. Nathan Sharon continues to make a major contribution to the development of what we now call glycobiology. It is a pleasure to acknowledge his enthusiasm and encouragement over the years.

REFERENCES

1. Hollingsworth MA & Swanson BJ (2004) *Nat. Rev. Cancer* **4**, 45–60.
2. Dekker J, Rossen JW, Buller HA & Einerhand AW (2002) *Trends Biochem. Sci.* **27**, 126–131.
3. Corfield AP, Carroll D, Myerscough N & Probert CS (2001) *Front. Biosci.* **6**, D1321–57.
4. Porchet N, Buisine MP, Desseyn JL *et al.* (1999) *J. Soc. Biol.* **193**, 85–99.
5. Lidell ME, Johansson ME & Hansson GC (2003) *J. Biol. Chem.* **278**, 13944–13951.
6. Levitin F, Stern O, Weiss M *et al.* (2005) *J. Biol. Chem.* **280**, 33374–33386.
7. Perrais M, Pigny P, Copin MC, Aubert JP & Van Seuningen I (2002) *J. Biol. Chem.* **277**, 32258–32267.
8. Komatsu M, Arango ME & Carraway KL (2002) *Biochem. J.* **368**, 41–48.
9. Asker N, Axelsson MA, Olofsson SO & Hansson GC (1998) *J. Biol. Chem.* **273**, 18857–18863.
10. Brockhausen I (2003) *Adv. Exp. Med. Biol.* **535**, 163–188.
11. Taylor C, Allen A, Dettmar PW & Pearson JP (2003) *Biomacromolecules* **4**, 922–927.
12. Brockhausen I (1999) *Biochim. Biophys. Acta* **1473**, 67–95.
13. Thomsson KA, Schulz BL, Packer NH & Karlsson NG (2005) *Glycobiology* **15**, 791–804.
14. Corfield AP, Wagner SA, Clamp JR, Kriaris MS & Hoskins LC (1992) *Infect. Immun.* **60**, 3971–3978.
15. Hoskins LC, Agustines M, McKee WB, Boulding ET, Kriaris M & Niedermeyer G (1985) *J. Clin. Invest.* **75**, 944–953.
16. Hoskins LC (1993) *Eur. J. Gastroenterol. Hepatol.* **5**, 205–213.
17. Robbe C, Capon C, Maes E *et al.* (2003) *J. Biol. Chem.* **278**, 46337–46348.
18. Robbe C, Capon C, Coddeville B & Michalski JC (2004) *Rapid Commun. Mass Spectrom.* **18**, 412–420.
19. Dell A & Morris HR (2001) *Science* **291**, 2351–2356.
20. Goldberg D, Sutton-Smith M, Paulson J & Dell A (2005) *Proteomics* **5**, 865–875.
21. Atuma C, Strugala V, Allen A & Holm L (2001) *Am. J. Physiol. Gastrointest. Liver Physiol.* **280**, G922–G929.
22. Newton JL, Allen A & Westley BR (2000) *Gut* **46**, 312–320.
23. Sonnenburg JL, Angenent LT & Gordon JI (2004) *Nat. Immunol.* **5**, 569–573.
24. Carraway KL, Ramsauer VP, Haq B & Carothers Carraway CA (2003) *Bioessays* **25**, 66–71.
25. Hanisch F-G & Muller S (2000) *Glycobiology* **10**, 439–449.
26. McCool DJ, Okada Y, Forstner JF & Forstner GG (1999) *Biochem. J.* **341**, 593–600.
27. Bell SL, Xu G & Forstner JF (2001) *Biochem. J.* **357**, 203–209.
28. Ho JJ, Jaituni RS, Crawley SC, Yang SC, Gum JR & Kim YS (2003) *Int. J. Oncol.* **23**, 585–592.
29. Perez-Vilar J, Randell SH & Boucher RC (2004) *Glycobiology* **14**, 325–337.
30. Podolsky DK (1985) *J. Biol. Chem.* **260**, 8262–8271.
31. Podolsky DK (1985) *J. Biol. Chem.* **260**, 15510–15515.
32. Slomiany BL, Zdebska E & Slomiany A (1984) *J. Biol. Chem.* **259**, 2863–2869.
33. Stanley CM & Phillips TE (1999) *Am. J. Physiol.* **277**, G191–G200.
34. Linden S, Nordman H, Hedenbro J, Hurtig M, Boren T & Carlstedt I (2002) *Gastroenterology* **123**, 1923–1930.
35. Jiang C, McClure SF, Stoddart RW & McClure J (2004) *Glycoconj. J.* **20**, 367–374.
36. Robbe C, Capon C, Coddeville B & Michalski JC (2004) *Biochem. J.* **384**, 307–316.
37. McMahon RF, Panesar MJ & Stoddart RW (1994) *Histochem. J.* **26**, 504–518.
38. McMahon RF, Warren BF, Jones CJ *et al.* (1997) *Histochem. J.* **29**, 469–477.
39. Buisine MP, Devisme L, Savidge TC *et al.* (1998) *Gut* **43**, 519–524.
40. Reid PE, Owen DA, Magee F & Park CM (1990) *Histochem. J.* **22**, 81–86.
41. Allen A, Hutton DA, Pearson JP & Sellers LA (1990) In: *The Cell Biology of Inflammation in the Gastrointestinal Tract*, pp. 113–125. Edited by TJ Peters. Corners Publications, Hull.
42. Backhed F, Ley RE, Sonnenburg JL, Peterson DA & Gordon JI (2005) *Science* **307**, 1915–1920.
43. Hooper LV & Gordon JI (2001) *Glycobiology* **11**, 1R–10R.

44. Xu J & Gordon JI (2003) *Proc. Natl Acad. Sci. USA* **100**, 10452–10459.
45. Deplancke B & Gaskins HR (2001) *Am. J. Clin. Nutr.* **73**, 1131S–1141S.
46. Agarwal R, Sharma N, Chaudhry R *et al.* (2003) *J. Pediatr. Gastroenterol. Nutr.* **36**, 397–402.
47. Sonnenburg JL, Xu J, Leip DD *et al.* (2005) *Science* **307**, 1955–1959.
48. Gagneux P & Varki A (1999) *Glycobiology* **9**, 744–755.
49. Ratner DM, Adams EW, Disney MD & Seeberger PH (2004) *Chembiochemistry* **5**, 1375–1383.
50. Feizi T, Fazio F, Chai W & Wong CH (2003) *Curr. Opin. Struct. Biol.* **13**, 637–645.

CHAPTER 21
Mucosal glycoconjugates in inflammatory bowel disease and colon cancer

Barry J. Campbell, Lu-Gang Yu and Jonathan M. Rhodes

1. MUCOSAL GLYCOSYLATION CHANGES IN INFLAMMATORY BOWEL DISEASE AND COLON CANCER

Glycoconjugate abnormalities are commonly seen in cancer and affect intracellular, cell surface, and secreted glycoconjugates [1]. Using lectin histochemistry, similar glycosylation changes have been shown to occur in colon cancer, cancer precursors such as adenomatous and metaplastic colonic polyps, and in the inflammatory bowel diseases ulcerative colitis and Crohn's disease [2,3], both conditions that are themselves associated with a high cancer risk [4]. Changes include neo-expression of O-linked oncofetal carbohydrate antigens such as the Thomsen Friedenreich (TF) oncofetal carbohydrate antigen (galactoseβ1–3N-acetylgalactosamine, the core 1 structure in O-linked oligosaccharide chains α-linked to serine or threonine of the protein core of colonic mucins) [5] (Fig. 1). TF antigen is expressed in neonatal colon, but in adult colon the TF antigen is normally masked by further glycosylation, sialylation, or sulfation and is undetectable in normal colonic mucosa, unless sensitivity is enhanced by avidin–biotin amplification. Increased expression of sialyl Tn (sialylα2,6N-acetylgalactosamine α-linked to serine or threonine of the protein core) [6] has been shown to be a marker of high risk for cancer development in inflammatory bowel disease mucosa. However, glycosylation changes can often be found in the absence of dysplasia and often seem to predate cytological malignant change in inflammatory bowel disease.

Other glycan changes common to cancer and colitis include reduced mucosal sulfation [7,8], shortening of the O-linked oligosaccharide side-chains, increased expression of ABO blood group antigens [9], and increased sialylation of peripheral blood group structures resulting, for example, in disialyl and trisialyl Lewis antigen variants, particularly those with Galβ1,4GlcNAc linkage (Lewis[x] and

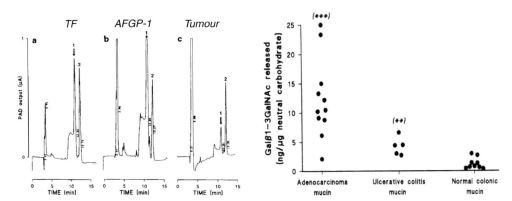

Figure 1. Increased expression of the oncofetal Thomsen Friedenreich (T or TF) carbohydrate antigen. From [5] with permission.

Lewis^y) [10]. Increased sialylation of mucins has also been documented by our own group in inflammatory bowel disease colonic mucosal explants [11].

2. MECHANISMS UNDERLYING ALTERED GLYCOSYLATION

The mechanisms that determine glycosylation abnormalities in colon cancer, colitis, and colitis-associated cancer are poorly understood. Possible explanations could include changes in the amino acid sequence affecting the glycosylation site of cell surface or secreted glycoproteins, altered substrate availability, alterations in glycosyltransferase or sulfotransferase mRNA expression, and/or catalytic activity or rearrangement of the Golgi apparatus. Another possibility is altered splicing of the protein component of cell surface glycoproteins. For example, cell surface expression of the TF oncofetal carbohydrate antigen on colon cancer cells tends to occur specifically on high molecular weight splice variants of the adhesion molecule CD44 [12]. This links colon cancer-associated glycosylation changes with the extensive literature on altered splicing of CD44 in colon cancer and inflammatory bowel disease [13]. Variant splicing of CD44 in turn may be a consequence of alterations in intron length [14] or the effect of proinflammatory cytokines [15,16]. The latter may represent an important mechanism for inflammation-induced alterations in cell surface glycosylation.

Although the association between CD44 splicing and glycosylation suggests that the nature of O-glycosylation may be determined, at least in part, by the amino acid sequence of the protein (CD44) undergoing glycosylation, it does not explain the simultaneous change in glycosylation of secreted mucins that is commonly seen in disease states. Here, alterations in mRNA expression and catalytic activities of the relevant glycosyltransferases and sulfotransferases would seem a likely explanation. For example, a colonic mucin sulfotransferase, which undergoes progressively reduced expression from adenoma to cancer, has as its preferred acceptor the TF antigen [17]. However, meticulous studies have shown that cancer-related changes in expression of the relevant glycosyl-, sialyl-, and sulfotransferases tend to correlate poorly with glycosylation changes [18].

Perhaps the most credible explanation for the constellation of glycosylation changes seen in cancer and inflammatory colonic disease is disruption of the Golgi. The relative positions in the Golgi of the different glycosyl-, sialyl- and sulfotransferases are critically important in determining O-glycan structure and these positions can be affected by intra-Golgi pH. Studies by our own research group have demonstrated that goblet cell-differentiated colon cell lines cultured in the presence of bafilomycin A_1, a macrolide which specifically disrupts the electrogenic vacuolar ATP-dependent Golgi proton pump (V-ATPase), thus raising intra-Golgi pH at subcytotoxic concentrations, causes glycosylation changes that mimic those seen in human colon cancer (i.e. reduced sulfation and increased expression of oncofetal TF antigen) [19] (Fig. 2). Hansson and colleagues have confirmed these findings and showed that they result from pH-induced changes in localization of the glycosyltransferases within the Golgi [20]. The effects of reduced Golgi acidification on N-glycosylation have been less studied, although N-glycans of β_1 integrin seem to be little affected [21]. A further report not only confirms that inhibition of Golgi acidification in cell lines causes glycosylation changes but shows that it leads to fragmentation of the Golgi stacks and that similar ultrastructural changes in the Golgi are seen in human colorectal cancer samples [22].

Similar O-glycosylation changes can also be induced in goblet-differentiated colon cell lines by proinflammatory cytokines [23]. We have therefore speculated

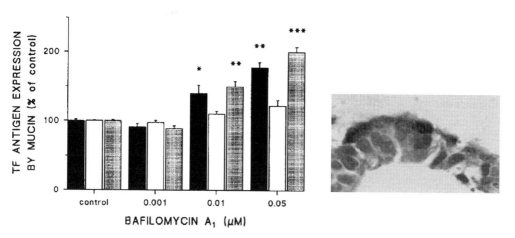

Figure 2. Demonstration of increased oncofetal carbohydrate (TF) expression in HT29-MTX polarized monolayer as a result of Golgi alkalinization. From [19] with permission.

that the alterations in *O*-glycosylation seen in colitis and in sporadic colon cancer might all result from reduced Golgi acidification, mediated by inflammatory cytokines although direct proof of this hypothesis is not yet available [24]. This is in keeping with the increasing recognition that the Golgi complex may represent a dynamic rather than fixed structure [25].

We need to keep an open mind regarding not only the nature of the external stimulus but also about the relevant ion transporter. Besides the Golgi proton pump, which is thought to be the main determinant of intra-Golgi pH in healthy cells, alterations in other ion transporters may be relevant in inflammation or cancer including the Golgi-associated chloride channel GOLAC-2, the Na^+/H^+ antiporter and the sodium-dependent and -independent chloride/bicarbonate exchangers such as *SLC26A3*/DRA (downregulated in adenoma) and a Golgi-associated Cl^-/HCO_3^- exchanger AE2 which has recently been shown to be necessary for the structural integrity of the Golgi apparatus [26]. There is a clear need to tackle some of these uncertainties by direct *in vivo* observations of intra-Golgi pH in the presence of cancer and inflammation.

3. INTERACTION OF MUCOSAL GLYCOCONJUGATES WITH DIETARY LECTINS

We have speculated that alterations in surface glycosylation of colon epithelial cells may allow changes in functionally important interactions with lectins of dietary or microbial origin [27]. Lectins are carbohydrate-binding proteins of non-immune origin and they are plentiful in fruit, vegetables, seeds, and nuts. Many of them are highly resistant to degradation by proteases in the gut and remain intact on passage through the gut [28,29]. Some dietary lectins are mitogenic. For example, the red kidney bean (*Phaseolus vulgaris*) lectin phytohemagglutinin (PHA), which recognizes terminal GalNAc, is a well-known mitogen [30] and its ingestion caused a marked increase of intestinal weight in mice [31,32]. It is also notorious for causing severe gastroenteritis when inadvertently eaten by humans, typically in a chilli con carne that has been slow cooked at too low a temperature. Soya bean (*Glycine max*) agglutinin (SBA), which also recognizes terminal GalNAc, similarly causes increased weight of the pancreas and colon when fed to rats [33].

All dietary lectins which recognize the TF antigen have so far been shown to affect cell proliferation. Peanut (*Arachis hypogaea*) agglutinin (PNA) [29] and amaranth (*Amaranthus caudatus*) lectin [34], as well as anti-TF antibodies [35] stimulate proliferation in human colon cancer cells *in vitro*. In animal feeding studies, ingested PNA stimulates intestinal growth in rats [36]. The trophic actions of such dietary lectins could have potential adverse implications for the development of gastrointestinal cancers whilst dietary oligosaccharides such as those terminated with galactose might inhibit these lectin-mediated effects and hence protect against cancer development [37]. As proof of concept, we have demonstrated that people who express TF antigen in their rectal mucosae have a 40% increase in rectal mitotic index after 7 days of daily ingestion of peanuts

[38]. A high intake of galactose-containing fibers, typically present in leafy green vegetables, has been shown in a case–control study to be associated with reduced risk of colon cancer [37]. The proliferative effects of some dietary lectins may also be indirectly mediated and ingestion of SNA, PHA, or PNA all result in increased serum concentration of cholecystokinin (CCK) [39].

Some dietary TF-binding lectins, such as those from the common edible mushroom *Agaricus bisporus* and jack fruit (jacalin) (see Chapter 13), are potent growth inhibitors for epithelial cells [34,40]. The growth inhibitory effect of mushroom lectin is dependent on its internalization and subsequent inhibition of the classical nuclear localization sequence-dependent nuclear protein import [41] (Fig. 3) by interaction with a truncated cytoplasmic form of the stress-related protein Orp150 [42]. Interaction of jacalin with epithelial cancer cells *in vitro* results in tyrosine phosphorylation of the tumor suppressor protein PHAPI/pp32, which activates protein phosphatase 2A with consequent dephosphorylation and inhibition of MAP kinase [43]. We have speculated, so far without direct proof, that these lectins may be binding to ligands of members of the galectin family of mammalian galactose-binding lectins.

The potential proinflammatory effects of dietary lectins deserve further investigation. Mannose-binding lectins from lentils (*Lens culinaris*) and broad

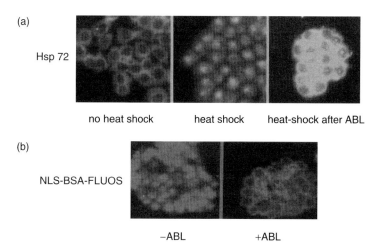

Figure 3. Common edible mushroom (*Agaricus bisporus*) lectin (ABL) inhibits nuclear import of (a) inducible Hsp72 and (b) NLS-peptide in digitonin-permalibilized cells. From [41] with permission.

beans and also PHA all induce the release of the proinflammatory cytokine interleukin 8 (IL-8) from human colon cancer Caco-2 cells *in vitro* [44]. Their role in inflammatory intestinal diseases has been little explored. There is also a potential for complex and sometimes harmful interactions between lectins and intestinal bacteria. PHA, for example, has been shown to be less toxic in germ-free animals [45].

4. INTERACTIONS BETWEEN INTESTINAL EPITHELIA AND MICROBIAL LECTINS

4.1 The mucus layer as a barrier to bacteria–epithelial interactions

Clinicians and pathologists are accustomed to seeing formaldehyde-fixed tissue sections in which the mucus barrier (see Chapter 20) is largely lost. As a consequence, too little attention has been given to the role of this barrier. Increasing evidence suggests that the normal mucus barrier is capable of keeping the luminal aspect of the surface epithelium almost sterile in the human colon. A recent study combining Alcian blue staining of fixed frozen mucosal biopsy sections with fluorescence *in situ* hybridization (FISH) for bacterial DNA has shown that in the normal colon there are usually no bacteria within or beneath the colonic mucus layer [46]. This is supported by studies showing that removal of the surface mucus by dithiothreitol followed by conventional microbial culture of the underlying mucosa generally yields no growth in the normal colon [47,48]. Moreover, the glycocalyx that immediately coats most intestinal epithelial cells has been shown to have an effective functional pore size of somewhere between 7.4 and 28.8 nm and prevents penetration by bacteria [49]. Thus, even potent pathogens such as *Salmonella* spp. and *Yersinia* spp. have been shown to require the presence of specialized M cells for invasion and pathogenicity [50]. M cells are specialized epithelial cells that populate the specialized dome epithelium that overlies Peyer's patches in the small intestine and the smaller, but similar, lymphoid follicles in the colon and these specialized epithelia lack goblet cells and have no significant overlying mucus layer or glycocalyx.

Thus, although several studies have looked at interactions between bacterial lectins and mucus glycoconjugates, it is likely that these interactions are a hindrance to bacterial encroachment onto the epithelium, rather than a help. Probably of greater relevance to pathogenicity is the ability of some bacteria to release mucus-degrading enzymes [51]. Secreted mucus in the human colon as well as in the small intestine is based largely on the MUC2 core. The only striking difference between normal human colonic and small intestinal mucus is the marked sulfation seen in the colon. This is likely to be functionally very important since the presence of ester sulfation renders the mucus oligosaccharides resistant to degradation by glycosidases [52]. Some bacteria are able to secrete sulfatases and glycosulfatases [53,54] that are able to initiate degradation of the oligosaccharides, and relapse of ulcerative colitis correlates with increased fecal concentration of such enzymes [55]. Rather disappointingly, no colonic mucosal disease has yet been shown to be convincingly due to an underlying alteration in mucus synthesis. Ulcerative colitis is associated with reduced mucin sulfation [7,8] and glycoconjugate changes have been reported in the mucosa of unaffected identical twins of individuals with ulcerative colitis [56] but these changes are confined to the surface epithelium and are probably the result of subtle acquired changes rather than underlying genetic abnormalities [57].

4.2 Interactions between bacteria, normal and diseased epithelial cells

As a consequence of the effectiveness of the normal colonic mucus layer as a barrier there is probably very little interaction *in vivo* between bacteria and the luminal aspect of normal colonic epithelial cells. However, colonic adenomas, which are premalignant, and carcinomas, are generally devoid of goblet cells and have no significant overlying mucus layer. They are thus much more likely to be in direct contact with bacteria [47]. Increased numbers of bacteria can also be found beneath the mucus layer in Crohn's disease [47,48,58] even though the mucus layer is, if anything, somewhat thicker than normal [59]. It is not clear whether these bacteria have penetrated the mucus layer directly although there is some evidence to support this [60]. Alternatively, they might have penetrated the mucosa via Peyer's patches or lymphoid follicles, which are arguably the point of initial lesion development in Crohn's disease and then spread through the mucosa as a consequence of defective macrophage or neutrophil clearance.

As already discussed, there are remarkably similar abnormalities in epithelial glycosylation in inflammatory conditions such as ulcerative colitis and Crohn's disease and in cancerous and precancerous polyps. These changes have the potential to affect bacteria–epithelial interaction. So far, however, there has been little evidence that the mucosa-associated microbiota are affected by this. We have identified only one example of an *E. coli*, designated HM44, that has selectivity for the TF antigen that is so commonly overexpressed in inflammatory and malignant colonic epithelia [47]. Interestingly, adherence of *Salmonella typhimurium* to Caco-2 colon cancer cells has also been shown to be mediated via TF [61] and is blockable by peanut lectin. This implies that eating peanuts, which has been shown to result in passage of intact lectin through to the colon [38], might protect against salmonellosis!

Regardless of the relative infrequency of disease-specific bacterial lectin–epithelial carbohydrate interactions, there is plentiful evidence of interaction between colonic bacterial isolates and colon cancer cells *in vitro*. Since the immediate vicinity of the mucosa is relatively well oxygenated, much of the focus has been on interaction between epithelial cells and microaerophilic organisms, particularly *E. coli*, rather than anerobes. *E. coli* isolated from the ileum of patients with Crohn's disease and from the colon of patients with either Crohn's disease or colon cancer generally lack markers of pathogenicity associated with known intestinal pathogenic forms of *E. coli* such as EPEC, EHEC, ETEC, EIEC, and EAEC but they have been shown to adhere to and invade colon cancer cells *in vitro*, a property that correlates with their ability to agglutinate human red blood cells [47,58]. There is also some limited evidence that colonic epithelial invasion occurs *in vivo*. The characterization of these interactions is lagging a few years behind the understanding of interactions between uro-epithelial *E. coli* and bladder epithelial cells but it is looking as though there are many similarities. This should not be surprising since uropathogenic *E. coli* do, of course, originate from the colon. Uropathogenic *E. coli* have four main types of adhesin: (i) FimH, expressed on type 1 pili [62], (ii) PapGi (I, II, III), expressed on P pili, (iii) SfaS, SfaA/FocH, expressed on S/F 1C pili, and (iv) Afa/Dr adhesins. About 80% of commensal *E. coli* fecal isolates encode FimH adhesins that bind only trimannose receptors, whereas about 70% of UPEC isolates are also able to bind monomannose residues. This has been shown to allow increased binding to glycoprotein receptors, such as uroplakin 1a carried on bladder epithelial cells. It is not yet known whether a similar interaction takes place in the colon. PapG adhesins bind to the glycolipid receptor globotriasylceramide, which consists of a Galα1–4Gal core linked by a β-glucose residue to a ceramide anchor. This has been shown to be involved in adherence of some *E. coli* strains to HT29 colon cancer cells [63]. Sfa adhesins interact with sialic acid residues on various receptors, including probably uroplakin. Afa-Dr adhesins bind to either DAF (decay accelerating factor), CEACAM (carcino-embryonic antigen cell adhesion molecule) family members expressed in the glycocalyx, the fuzzy coat that overlies the microvilli of normal intestinal cells underneath the mucus layer [64], or type IV collagen.

We have preliminary evidence that the hemagglutinating isolates from Crohn's disease and colon cancer often have characteristics of diffusely adherent *E. coli* (DAEC). They express some of the afa/Dr adhesins and produce diffusely adherent colonies on co-culture with HEP2 epithelial cells. Although the CEACAMs and other potential ligands are glycosylated, most of the known DAEC-epithelial adherence mechanisms such as Afa/Dr adhesin interactions with DAF and members of the CEACAM family have so far proved to be protein–protein interactions.

4.3 Interactions with M cells

The study of human intestinal M cells has been hampered by the lack of any totally convincing cell surface marker. In other mammals the fucose-binding *Ulex europaeus* lectin (UAE-1) can be used to identify and isolate M cells [65]. In humans there is a relative increase in surface expression of sialyl Lewis[a] antigen (identifiable using the tumor marker antibody CA19.9) [66]. M cells can be generated in the laboratory by co-culture of isolated intestinal epithelial cells and B-lymphocytes [67] but this is very laborious. More practical for most purposes is a model based on co-culture of Caco2 colon cancer cells and Raji B lymphoma cells [68]. The resulting epithelial cells bear most of the characteristics of human M cells including loss of surface microvilli, reduction in alkaline phosphatase activity, increased translocation of bacteria, and increased expression of sialyl Lewis[a] but may lack some of the specific features of M cells. Several groups are using this model in the hope that it will allow clarification of some of the lectin–carbohydrate as well as protein–protein and other interactions that are presumably involved in adhesion of and subsequent internalization of bacteria and viruses by M cells.

Rabbit and mouse M cells have been shown to interact with type 1 reovirus via a mechanism involving glycoconjugates that express alpha 2–3-linked sialic acid

[69]. It is intriguing that a fucose-binding *Aleuria aurantia* lectin has been recently identified as a selective ligand for human M cells, having been selected on the basis of its structural similarity to neuraminidases [70].

5. IMPLICATIONS FOR INFLAMMATORY BOWEL DISEASE AND COLON CANCER

There is increasing acceptance that the inflammatory bowel diseases ulcerative colitis and Crohn's disease represent an abnormal response to the normal gut microbiota. Despite a recent flurry of activity around the possibility of mycobacterial infection as a pathogenic mechanism for Crohn's disease, there has been no consistent demonstration of any pathogenic organism underlying either condition. In Crohn's disease several independent groups have shown an increase in mucosa-associated organisms, particularly, but not exclusively, *E. coli*. Direct immunohistochemistry, and also FISH, has also demonstrated the presence of bacteria, including *Listeria,* streptococci and, again, *E. coli*, within macrophages in Crohn's tissue. Crohn's disease patients, but not those with ulcerative colitis, are also much more likely than controls to have identifiable circulating antibodies to bacterial flagellin, of either anaerobic or enterobacterial origin. Efforts to identify any specific oligosaccharide receptor for the adhesive Crohn's *E. coli* isolates have so far failed but adhesion of the organisms to colon epithelial cells can be blocked by complex oligosaccharides such as mucins and soluble plantain fiber [47].

In ulcerative colitis there is much less evidence of involvement of any disease-specific mucosa-associated bacteria. Culture of mucosal biopsies after dithiothreitol treatment to remove surface mucus have variably shown either a modest increase or no increase in bacteria. There is also no suggestion that initial lesions originate over lymphoid follicles. It seems plausible that bacteria may play a more indirect role (e.g. by interaction between bacterial secreted products, including possibly flagellin, and epithelial receptors such as TLR5, resulting in IL-8 release and subsequent triggering of neutrophil recruitment). Chronicity would still have to be explained and here, in contrast to Crohn's disease, there is a reasonable body of evidence to support the plausibility of autoimmunity as a mechanism.

The possible role of bacteria in colon cancer is only just beginning to be investigated. NFκB activation has been shown to have a central role in determining inflammation-associated colon cancer in animal models [71] so bacteria–epithelial interactions leading to NFκB activation and thence to inhibition of apoptosis might well prove to have a critical role in cancer development, perhaps particularly in determining progression from adenoma to cancer [24]. If this hypothesis turns out to be correct this could represent the main target for dietary interventions to prevent cancer development.

6. INTERACTION OF MUCOSAL GLYCOCONJUGATES WITH ENDOGENOUS GALECTINS

Galectins (see Chapter 14) are a family of 15 (so far) naturally occurring carbohydrate-binding proteins which share consensus amino acid sequences and binding affinity for β-galactoside-terminated carbohydrates. They may be found inside cells, on the cell surface, and extracellularly [72]. Intracellular galectins-1 and -3 are involved in the regulation of mRNA splicing and apoptosis. Cell surface and extracellular galectins are involved in cell–cell and cell–matrix interactions. Increasing evidence suggests that galectins are important in the development and progression of colorectal cancer. Although several members of the galectin family (galectin-1, -2, -3,-4, -6, -9, and -11) may occur in gastrointestinal epithelial cells; much of our current knowledge of interactions between galectins and cancer is derived from studies of galectin-3.

Increased expression of galectin-3 in gastrointestinal cancer is associated with metastasis and poor prognosis [73]. Galectin-3 is overexpressed on the surface of adenoma cells compared with normal mucosa. Metastases express higher levels of galectin-3 compared with primary tumors from the same patients and, in animal cancer models, antisense reduction of galectin-3 expression is associated with decreased metastasis.

The mechanism of these effects of galectin-3 is not fully understood. Several basement membrane glycoconjugates, including laminin, fibronectin, and collagen are bound by galectin-3 [72]. Transfection of galectin-3 into breast cancer cells has been shown to enhance cell adhesion to laminin and other extracellular matrix proteins. Galectin-3 also stimulates tube formation by human umbilical endothelial cells grown in capillary tubes, an effect which is inhibited by competitive carbohydrates [75], and that suggests a role for galectin-3 in angiogenesis.

Galectin-3 also enhances tumor cell aggregation [76] and heterotypic adhesion to endothelial cells under static as well as flow conditions [77]. TF antigen-mimicking peptide or TF antigen-inhibitory peptides markedly inhibit these effects. We have recently found that cancer-associated TF antigen on MUC1, a highly glycosylated transmembrane mucin protein which carries increased copy numbers of shorter oligosaccharides such as TF disaccharide in epithelial cancer cells [78], is crucial in determining the increased epithelial cancer cell adhesion to endothelial cells that is induced by recombinant galectin-3 (Yu et al., submitted). The combination of the up to 5-fold increased concentrations of galectin-3 in the sera of patients with gastrointestinal cancer and the increased expression of TF by MUC1 on epithelial cancer cells may thus promote cancer metastasis. In support of this, co-injection of anti-galectin-3 antibody with human colorectal cancer RPMI1478 cells into the anterior mesenteric vein of nude mice resulted in 90% reduction in liver metastases [79].

In contrast to galectin-3, galectin-8 usually undergoes reduced expression in epithelial cancers, particularly in more invasive colorectal cancer cell lines. Again, functional explanations for these associations are not yet known.

Galectins have also been implicated in intestinal inflammation. Galectin-3 is distributed homogeneously in normal epithelia but is heterogeneously distributed in epithelial cells from Crohn's disease lesions [80]. Galectin-4 stimulates CD4$^+$ T-cells to enhance production of IL-6, which would potentially exacerbate chronic colitis and delay recovery from acute intestinal injury. In T-helper cell type 1-mediated experimental colitis induced by intrarectal administration of 2,4,6-trinitrobenzene sulfonic acid (TNBS) in mice, administration of human recombinant galectin-1 causes reduction of the number of hapten-activated spleen T-cells and production of inflammatory cytokines [81], suggesting a protective and immunomodulatory role for galectin-1 in TNBS-induced colitis. Similarly galectin-3 has been shown to induce synthesis of MUC2 [82]. Although the molecular mechanisms of the actions of these galectins are still unclear, it is likely that changes of cellular glycosylation associated with the inflammatory process may result in functionally important alterations in binding of the galectins.

Much work remains to be done to characterize the intriguing complexity of interaction between intestinal epithelial cell surface and mucin glycoconjugates and the dietary, microbial, and mammalian lectins which are in their environment [83]. It seems rare, at least *in vitro*, for any such interaction to fail to have impressive functional consequences for the epithelial cell.

Acknowledgments

Work in the authors' laboratory is supported by grants from the Medical Research Council, The Wellcome Trust, North West of England Cancer Research Fund, Cancer Research UK, the National Association for Colitis and Crohn's Disease, the NHS Research & Development Fund, the University of Liverpool, and the Mizutani Foundation for Glycoscience (Japan).

REFERENCES

1. Hakomori S (2001) *Adv. Exp. Med. Biol.* **491**, 369–402.
2. Rhodes JM, Black RR & Savage A (1986) *J. Clin. Pathol.* **39**, 1331–1334.
3. Rhodes JM, Black RR & Savage A (1988) *Dig. Dis. Sci.* **33**, 1359–1363.
4. Campbell BJ, Yu LG & Rhodes JM (2001) *Glycoconj. J.* **18**, 851–858.
5. Campbell BJ, Hounsell E, Finnie IA & Rhodes JM (1995) *J. Clin. Invest.* **95**, 571–576.
6. Karlen P, Young E, Brostrom O *et al.* (1998) *Gastroenterology* **115**, 1395–1404.
7. Raouf AH, Tsai HH, Parker N, Hoffman J, Walker RJ & Rhodes JM (1992) *Clin. Sci.* **83**, 623–626.
8. Corfield AP, Myerscough N, Bradfield N *et al.* (1996) *Glycoconj. J.* **13**, 809–822.
9. Kim YS, Gum Jr, J & Brockhausen I (1996) *Glycoconj. J.* **13**, 693–707.
10. Kim YS, Yuan M, Itzkowitz SH *et al.* (1986) *Cancer Res.* **46**, 598–609.
11. Parker N, Tsai HH, Ryder SD, Raouf AH & Rhodes JM (1995) *Digestion* **56**, 52–56.
12. Singh R, Campbell BJ, Yu L-G *et al.* (2001) *Glycobiology* **11**, 587–592.
13. Ponta H, Sherman L *et al.* (2003) *Nat. Rev. Mol. Cell Biol.* **4**, 33–45
14. Bell MV, Cowper AE *et al.* (1998) *Mol. Cell Biol.* **18**, 5930–5941.
15. Wimmenauer S, Steiert A, Wolff-Vorbeck G *et al.* (1999) *Tumour Biol.* **20**, 294–303.
16. Macdonald DC, Leir SH *et al.* (2003) *Eur. J. Gastroenterol. Hepatol.* **15**, 1101–1110.
17. Kuhns W, Jain RK, Matta KL *et al.* (1995) *Glycobiology* **5**, 689–697.

18. Yang JM, Byrd JC, Siddiki BB et al. (1994) Glycobiology 4, 873–884.
19. Campbell BJ, Rowe G, Leiper K & Rhodes JM (2001) Glycobiology 11, 385–393.
20. Axelsson MAB, Karlsson NG, Steel DM, Ouwendijk J, Nilsson T & Hansson GC (2001) Glycobiology 11, 633–644.
21. Skinner MA & Wildeman AG (2001) J. Biol. Chem. 276, 48451–48457.
22. Kellokumpu S, Sormunen R & Kellokumpu I (2002) FEBS Lett. 516, 217–224.
23. Flieger D, Hoff AS, Sauerbruch T & Schmidt-Wolf IGH (2001) Clin. Exp. Immunol. 123, 9–14.
24. Rhodes JM & Campbell BJ (2002) Trends Mol. Med. 8, 10–16.
25. Barr FA (2002) Trends Cell Biol. 12, 101–104.
26. Holappa K, Munoz MT, Egea G & Kellokumpu S (2004) FEBS Lett. 564, 97–103.
27. Rhodes JM (1996) Lancet 347, 40–44.
28. Brady PG, Vannier AM & Banwell JG (1978) Gastroenterology 75, 236–239.
29. Ryder SD, Smith JA & Rhodes JM (1992) J. Natl Cancer Inst. 84, 1410–1416.
30. Kordas K, Burghardt B, Kisfalvi K, Bardocz S, Pusztai A & Varga G (2000) J Physiol Paris 94, 31–36.
31. Sasaki M, Fitzgerald AJ, Grant G, Ghatei MA, Wright NA & Goodlad RA (2002) Aliment. Pharmacol. Ther. 16, 633–642.
32. Linderoth A, Biernat M, Prykhodko O et al. (2005) J. Pediatr. Gastroenterol. Nutr. 41, 195–203.
33. Grant G, Dorward PM, Buchan WC, Armour JC & Pusztai A (1995) Br. J. Nutr. 73, 17–29.
34. Yu L-G, Milton JD, Fernig DG & Rhodes JM (2001) J. Cell Physiol. 186, 282–287.
35. Yu LG, Jansson B, Fernig DG et al. (1997) Int. J. Cancer 73, 424–431.
36. Jordinson M, Goodlad RA, Brynes A et al. (1999) Am. J. Physiol. 276, G1235–1242.
37. Evans RC, Fear S, Ashby D et al. (2002) Gastroenterology 122, 1784–1792.
38. Ryder SD, Jacyna MR, Levi AJ, Rizzi PM & Rhodes JM (1998) Gastroenterology 114, 44–49.
39. Jordinson M, Playford RJ & Calam J (1997) Am. J. Physiol. 273, G946–950.
40. Yu L, Fernig DG, Smith JA, Milton JD & Rhodes JM (1993) Cancer Res. 53, 4627–4632.
41. Yu LG, Fernig DG, White MRH et al. (1999) J. Biol. Chem. 274, 4890–4899.
42. Yu LG, Andrews N, Weldon M et al. (2002) J. Biol. Chem. 277, 24538–24545.
43. Yu LG, Packman L, Weldon M, J Hamlett & Rhodes JM (2004) J. Biol. Chem. 279, 41377–41383.
44. Rodriguez-Juan C, Perez-Blas M, Suarez-Garcia E et al. (2000) Cytokine 12, 1284–1287.
45. Rattray EA, Palmer R & Pusztai A (1974) J. Sci. Food Agric. 25, 1035–1040.
46. Van der Waaij LA, Harmsen HJM, Madjipour M et al. (2005) Inflammatory Bowel Dis. 11, 865–867.
47. Martin HM, Campbell BJ, Hart CA et al. (2004) Gastroenterology 127, 80–93.
48. Swidsinski A, Ladhoff A, Pernthaler A et al. (2002) Gastroenterology 122, 44–54.
49. Frey A, Giannasca KT, Weltzin R et al. (1996) J. Exp. Med. 184, 1045–1059.
50. Cossart P & Sansonetti PJ (2004) Science 304, 242–248.
51. Hoskins LC, Agustines M, McKee WB, Boulding ET, Kriaris M & Niedermeyer G (1985) J. Clin. Invest. 75, 944–953.
52. Tsai HH, Gibson G, Hart CA & Rhodes JM (1991) Clin. Sci. 82, 447–454.
53. Rho J-J, Wright DP, Christie DL, Clinch K, Furneaux RH & Roberton AM (2005) J. Bacteriol. 187, 1543–1551.
54. Wright DP, Knight CG, Parkar SG, Christie DL & Roberton AM (2000) J. Bacteriol. 182, 3002–3007.
55. Tsai HH, Dwarakanath D, Hart CA & Rhodes JM (1995) Gut 36, 570–576.
56. Tysk C, Riedesel H, Lindberg E, Panzini B, Podolsky D & Jarnerot G (1991) Gastroenterology 100, 419–423.
57. Bodger K, Dodson AR, Tysk C et al. (2006) Gut 55, 973–977.
58. Darfeuille-Michaud A, Boudeau J, Bulois P et al. (2004) Gastroenterology 127, 412–421.
59. Pullan RD, Thomas GA, Rhodes M et al. (1994) Gut 35, 353–359.
60. Schultsz C, van den Berg FM, Ten Kate FN, Tytgat GNJ & Dankert J (1999) Gastroenterology 117, 1089–1097.
61. Giannasca KT, Giannasca PJ & Neutra MR (1996) Infect. Immun. 64, 135–145.

62. Sharon N (1987) *FEBS Lett.* **217**, 145–157.
63. Wold AE, Thorssen M, Hull S & Eden CS (1988) *Infect. Immun.* **56**, 2531–2537.
64. Frangsmyr L, Baranov V & Hammarstrom S (1998) *Tumor Biol.* **20**, 277–292.
65. Gebert A, Fassbender S, Werner K & Weissferdt A (1999) *Am. J. Pathol.* **154**, 1573–1582.
66. Giannasca PJ, Giannasca KT, Leichtner AM & Neutra MR (1999) *Infect. Immun.* **67**, 946–953.
67. Kerneis S, Bogdanova A, Kraehenbuhl J-P & Pringault E (1997) *Science* **277**, 949–952.
68. Neutra MR, Mantis NJ & Kraehenbuhl JP (2001) *Nat. Immunol.* **2**, 1004–1009.
69. Helander A, Silvey KJ, Mantis NJ *et al.* (2003) *J Virol.* **77**, 7964–7977.
70. Roth-Walter F, Bohle B, Scholl I *et al.* (2005) *Immunol. Lett.* **100**, 182–188.
71. Greten FR, Eckmann L, Greten TF *et al.* (2004) *Cell* **118**, 285–296.
72. Liu FT & Rabinovich GA (2005) *Nat. Rev. Cancer* **5**, 29–41.
73. Bresalier RS, Mazurek N, Sternberg LR *et al.* (1998) *Gastroenterology* **115**, 287–296.
74. Warfield PR, Makker PN, Raz A & Ochieng J (1997) *Invasion Metastasis* **17**, 101–112
75. Nangia-Makker P, Honjo Y, Sarvis R *et al.* (2000) *Am. J. Pathol.* **156**, 899–909.
76. Inohara H & Raz A (1995) *Cancer Res.* **55**, 3267–3271.
77. Glinsky VV, Glinsky GV, Rittenhouse-Olson K *et al.* (2001) *Cancer Res.* **61**, 4851–4857.
78. Baldus SE, Hanisch FG, Kotlarek GM *et al.* (1998) *Cancer* **82**, 1019–1027.
79. Inufusa H, Nakamura M, Adachi T *et al.* (2001) *Int. J. Oncol.* **19**, 913–919.
80. Jensen-Jarolim E, Gscheidlinger R, Oberhuber G *et al.* (2002) *Eur. J. Gastroenterol. Hepatol.* **14**, 145–152.
81. Santucci L, Fiorucci S, Rubinstein N *et al.* (2003) *Gastroenterology* **124**, 1381–1394.
82. Song S, Byrd JC, Mazurek N, Liu K, Koo JS & Bresalier RS (2005) *Gastroenterology* **129**, 1581–1591.
83. Sharon N & Lis H (2004) *Glycobiology* **14**, 53R–62R.

CHAPTER 22
Selectin-mediated metastasis of tumor cells: Alteration of carbohydrate-mediated cell–cell interactions in cancers induced by epigenetic silencing of glycogenes

Reiji Kannagi, Keiko Miyazaki, Naoko Kimura and Jun Yin

SUMMARY

Cell adhesion mediated by selectins and their carbohydrate ligands, sialyl Lewis^x and sialyl Lewis^a, plays an important role in cancer metastasis and tumor angiogenesis. Expression of these carbohydrate determinants is markedly enhanced in cancer cells compared with non-malignant epithelial cells. Our recent studies indicated the presence of further modified complex forms of sialyl Lewis$^{x/a}$ in non-malignant epithelial cells, which have additional sulfation or sialylation. Such complex determinants serve as specific ligands for another family of carbohydrate-recognizing molecules, siglecs, and maintain immunological homeostasis in normal mucous membranes. The impairment of these additional modifications in cancer cells, which we call "incomplete synthesis," leads to a considerable accumulation of sialyl Lewis$^{x/a}$ in cancer cells at early stages of malignant transformation. Epigenetic gene silencing through DNA methylation and/or histone deacetylation is proposed to confer the incomplete synthesis. In later stages of cancer progression, the increased transcription of several glycogenes responsible for the synthesis of sialyl Lewis$^{x/a}$ determinants by hypoxia-inducible factors further accelerates expression of these determinants on cancer cells. Hematogenous metastasis mediated by sialyl Lewis$^{x/a}$–selectin interactions determines the prognostic outcome of patients with advanced cancers. Cell adhesion molecules equipped with specific carbohydrate-recognition domains, previously called animal lectins, figure prominently in biological behaviors of cancer cells as well as in cell–cell interactions in normal mucosal membranes.

1. INTRODUCTION

Cell adhesion mediated by selectins on endothelial cells and their ligands, sialyl Lewis^x and sialyl Lewis^a on cancer cells, plays an important role in hematogenous metastasis of cancers [1–5]. Expression of these carbohydrate ligands is markedly enhanced in tumors, but the molecular mechanism that leads to cancer-associated expression of sialyl Lewis$^{x/a}$ has not been well elucidated. It has long been known that cell surface carbohydrate determinants undergo drastic changes during malignant transformation. Incomplete synthesis is a classical concept proposed in the early 1980s to characterize cancer-associated alteration of cell surface carbohydrates [6]. Recently, we proposed that the incomplete synthesis of normal carbohydrate determinants is involved in the accelerated expression of sialyl Lewis$^{x/a}$ in cancers [5,7]. Normal epithelial cells express various carbohydrate determinants, some of which have structures more complex than sialyl Lewis$^{x/a}$. Good examples are sialyl 6-sulfo Lewis^x and 2→3, 2→6 disialyl Lewis^a determinants, which are the further modified forms of sialyl Lewis$^{x/a}$ determinants (Fig. 1a,b). These determinants are strongly expressed on non-malignant colonic epithelial cells, and their expression is markedly reduced in colonic cancer tissues at early stages, suggesting that the impairment of GlcNAc 6-sulfation and α2→6 sialylation upon malignant transformation leads to the accumulation of sialyl Lewis$^{x/a}$ in colon cancer cells.

2. INDUCTION OF SIALYL LEWIS$^{A/X}$ EXPRESSION IN EARLY CANCERS THROUGH THE INCOMPLETE SYNTHESIS MECHANISM

2.1 Incomplete synthesis occurring in synthetic pathway for sialyl Lewis^a

Expression of sialyl Lewis^a is known to be increased in cancers of the digestive organs. The determinant serves as a ligand for E-selectin and mediates hematogenous metastasis of cancers. In contrast, disialyl Lewis^a, which has an extra sialic acid attached at the C-6 position of penultimate GlcNAc in sialyl Lewis^a, is preferentially expressed on non-malignant colonic epithelial cells (Fig. 1a,b) [8]. The significant decrease in disialyl Lewis^a expression in colonic cancer cells is associated with downregulated transcription of the gene for an α2→6 sialyltransferase responsible for disialyl Lewis^a synthesis. The downregulation of the gene expression was apparent at Dukes' A and B stages, indicating that it occurred in early stage cancers (Fig. 1c) [8]. Transfection of the gene encoding the α2→6 sialyltransferase to sialyl Lewis^a-positive cancer cells abrogated cell surface expression of sialyl Lewis^a, and led to the strong expression of disialyl Lewis^a (Fig. 1d) [8], suggesting that the disruption of α2→6 sialylation confers sialyl Lewis^a expression. Addition of a histone deacetylase inhibitor trichostatin A or DNA methylation inhibitor 5-aza-C to culture medium of sialyl Lewis^a-positive cancer cells induces significant transcription of the α2→6 sialyltransferase gene and surface expression of disialyl Lewis^a, suggesting that some epigenetic silencing mechanism for gene expression lies behind the decreased disialyl Lewis^a expression in cancer cells [8].

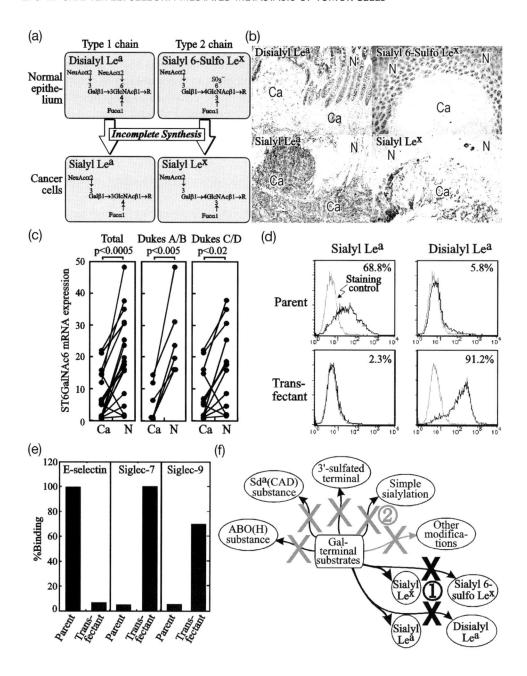

2.2 Functional consequences of incomplete disialyl Lewis[a] synthesis in cancer

Disialyl Lewis[a], which is preferentially expressed on non-malignant colonic epithelial cells, serves as a ligand for a lymphocyte inhibitory receptor, siglec-7/p75/AIRM1 and siglec-9 [8,9]. Both siglec-7 and -9 are known to be expressed on lymphocytes and monocytes, and carry ITIM domains, which suppress immune receptor-induced signal transduction in these leukocytes by recruiting protein phosphatases [10,11]. Ligand binding to these molecules inhibits the cytotoxic and proliferative activities of leukocytes. By confocal laser microscopy we recently showed a significant expression of siglec-7/-9 on numerous residential mononuclear leukocytes in the colonic mucosa, and *in situ* adhesion of these leukocytes to normal colonic epithelial cells bearing disialyl Lewis^a determinant [8,9]. Most of the siglec-7/-9-positive mononuclear leukocytes turned out to be of monocyte/macrophage origin. We propose that the physiological significance of disialyl Lewis^a expressed on normal gut epithelial cells is to suppress excessive activation of residential mononuclear leukocytes in the vicinity of normal epithelial cells, and thus to maintain immunological homeostasis in normal gut mucosal membranes. Normal gut mucous membranes are predisposed to various stimuli from intestinal bacterial flora and their toxic products, and several mechanisms to suppress excessive activation of residential leukocytes are proposed. We suggest that the disialyl Lewis^a–siglec-7/-9 interaction probably constitutes of one of such homeostatic mechanisms.

The disialyl Lewis^a determinant lacks a binding activity to selectins, while the sialyl Lewis^a determinant, which appears on cancer cells, lacks a binding activity to siglec-7/-9. Therefore the cancer-associated decrease in disialyl Lewis^a expression and concomitant appearance of sialyl Lewis^a determinant are accompanied by the loss of siglec-7/-9 binding activity of the cells, and a gain of E-selectin binding activity (Fig. 1e) [8,9]. Transition of carbohydrate determinants from the disialyl Lewis^a-dominant status to the sialyl Lewis^a-dominant status upon malignant transformation has dual functional consequences; the loss of normal cell–cell interactions between epithelial cells and residential mucosal mononuclear leukocytes on the one hand, and the acquisition of E-selectin binding ability on the other.

Figure 1. Expression of sialyl Lewis$^{x/a}$ is induced through the mechanism called incomplete synthesis in early cancers. (a) Schematic illustration of incomplete synthesis mechanism involving impairment of α2–6 sialylation in type 1 chain (left panel) and 6-sulfation in type 2 chain (right panel) for induction of sialyl Lewis$^{a/x}$ expression in cancers. (b) Typical distribution pattern of sialyl Lewis$^{a/x}$-related determinants in consecutive sections of colon cancer tissues. Disialyl Lewis^a determinant (left upper panel) is preferentially expressed in non-malignant epithelial cells, while the sialyl Lewis^a determinant (left lower panel) is specifically localized in cancer cells (adapted from [55]). Similarly, sialyl 6-sulfo Lewis^x determinant (right upper panel) is preferentially expressed in non-malignant epithelial cells, while the sialyl Lewis^x determinant (right lower panel) is specifically localized in cancer cells (adopted from [16]). (c) Real-time RT-PCR analyses of mRNA levels of ST6GalNAc6 in human colon cancer tissues (Ca) and non-malignant mucosa (N) stratified according to Dukes' stage classification [8]. (d) Flow-cytometric analyses of disialyl and sialyl Lewis^a determinants on human colon cancer SW1083 cells and the clone transfected with ST6GalNAc6 cDNA, indicating gain of disialyl Lewis^a and loss of sialyl Lewis^a by the transfection. (e) Binding of recombinant E-selectin, siglec-7 and -9 to SW1083 cells and the clone transfected with ST6GalNAc6 cDNA [8,9]. (f) Schematic illustration of incomplete synthesis mechanism involving impairment of 2–6 sialylation and 6-sulfation for induction of sialyl Lewis$^{a/x}$ expression at early cancers (1), followed by silencing of glycogenes for most of the other carbohydrate determinants, which leads to the further enhancement of sialyl Lewis$^{a/x}$ expression (2).

2.3 Incomplete synthesis occurring in synthetic pathway for sialyl Lewis^x

Similar to the case with sialyl Lewis^a, carbohydrate determinants with structures more complex than sialyl Lewis^x are present on normal epithelial cells. A typical example is sialyl 6-sulfo Lewis^x, where the GlcNAc residue in sialyl Lewis^x is further modified by 6-sulfation. The sialyl 6-sulfo Lewis^x determinant was first described as a specific L-selectin ligand expressed on high endothelial venules of peripheral lymph nodes, where it mediates homing of naïve T-lymphocytes bearing L-selectin [12–15]. Later, the determinant was found to be expressed also on normal gut epithelial cells [16]. The sialyl 6-sulfo Lewis^x determinant was preferentially expressed on non-malignant epithelial cells, and tended to disappear on cancer cells (Fig. 1a,b), suggesting an impairment of 6-sulfation occurring upon malignant transformation.

Initially downregulation of transcription of the genes for some 6-sulfotransferases was suspected to lie behind the impairment of 6-sulfation upon malignant transformation, since mRNA for I-GlcNAc6ST, a 6-sulfotransferase isoenzyme, was significantly decreased in cancer cells compared to non-malignant epithelial cells. Upon transfection to the appropriate cultured cell lines, however, I-GlcNAc6ST failed to confer sialyl 6-sulfo Lewis^x expression. Another 6-sulfotransferase isozyme in normal epithelial cells, GlcNAc6ST-1, was found to be capable of synthesizing sialyl 6-sulfo Lewis^x, but its mRNA level showed no significant change between cancer cells and non-malignant epithelial cells. To date the gene responsible for the decreased sialyl 6-sulfo Lewis^x expression in cancer cells has not been identified. Our preliminary results indicate that there are no significant changes between cancer cells and non-malignant epithelial cells in transcription of the genes for the other 6-sulfotransferases, 3'-phosphoadenosine-5'-phosphosulfate (PAPS) transporters, PAPS synthases, and relevant sulfatases so far tested, but only that of a sulfate transporter at the cell membrane is significantly decreased [17]. Whatever the responsible gene is, addition of histone deacetylase inhibitor to the culture medium recovers sialyl 6-sulfo Lewis^x expression on the cells, suggesting some epigenetic silencing mechanism lies behind the suppression of sialyl 6-sulfo Lewis^x expression upon malignant transformation.

Sialyl 6-sulfo Lewis^x may participate in the interaction of normal epithelial cells with mucosal resident cells. As it is an L-selectin ligand, it can mediate interaction with L-selectin-positive leukocytes. Recent results suggest the determinant serves as a ligand also for some siglecs [18,19].

2.4 Epigenetic gene silencing as the main mechanism for incomplete synthesis of carbohydrate determinants in cancers

Perhaps a most typical example of incomplete synthesis of carbohydrate determinants in cancers is loss of ABO(H) substances [20]. The synthesis of ABO(H) substances is the major metabolic pathway for galactose-terminal substrates at least in the right hemicolon, and the loss of ABO(H) substances will release a significant amount of excess substrates into synthetic pathways for other carbohydrate determinants. In contrast, the major carbohydrate determinant in the left hemicolon is known to be Sd$^\alpha$ antigen [21]. Incomplete synthesis of the Sd$^\alpha$ determinant has recently been reported to occur in gastric and colonic cancers, and a significant competition between the synthetic pathways for Sd$^\alpha$ and sialyl Lewis$^{x/a}$ determinants was ascertained in experimental studies employing cultured cancer cells [22]. The impaired synthesis of these major carbohydrate determinants would result in excess accumulation of precursor substrates, most of which will be utilized for synthesis of the second major carbohydrate determinant in the cells, which uses the same substrates as ABO(H) and Sd$^\alpha$ determinants. Simultaneous impairment in synthesis of the major and the second major carbohydrate determinants would, in turn, enhance expression of the third major carbohydrate determinant in the given cells and tissues. Competition between the synthetic pathways for 3'-sulfated determinants and sialyl Lewis$^{x/a}$ was also reported to occur in experimental studies using cultured cancer cells [23].

This substrate competition mechanism is another example of cancer-associated incomplete synthesis, and may well partly explain the accumulation of sialyl Lewis$^{x/a}$ determinants in cancers. However, the problem here is that the synthetic pathway for sialyl Lewis$^{x/a}$ determinants is not the second or third major, but only a very minor pathway in most normal tissues and organs, consisting of a small percentage of carbohydrate determinants synthesized from the common galactose-terminal substrates. Synthetic pathways for most of the other carbohydrate determinants need to be impaired until this substrate competition mechanism leads to enhanced expression of sialyl Lewis$^{x/a}$ in cancers. This is a quite different situation from that in cultured cancer cells commonly used in experimental studies, where the synthetic pathway for sialyl Lewis$^{x/a}$ determinant is already a major metabolic pathway and ready to compete with any synthetic pathway for other carbohydrate determinants.

On the other hand, with disialyl Lewis^a and sialyl 6-sulfo Lewis^x, incomplete synthesis of these determinants will directly result in the induction of sialyl Lewis$^{x/a}$ expression, because the impairment occurs within the same synthetic pathway as that for sialyl Lewis$^{x/a}$. Without impairments in α2→6 sialylation and 6-sulfation, what accumulates in the cells would be disialyl Lewis^a instead of sialyl Lewis^a, or sialyl 6-sulfo Lewis^x, instead of sialyl Lewis^x, even if the synthetic pathways for most of the other carbohydrate determinants were inactivated.

Figure 1f depicts a scheme for possible multistep progression of incomplete synthesis of carbohydrate determinants leading to induction and accumulation of sialyl Lewis$^{x/a}$ on cancers. First, epigenetic silencing occurs in the genes for modifications of sialyl Lewis$^{x/a}$, such as α2→6 sialylation and 6-sulfation, at the earliest stage of cancer, inducing significant expression of sialyl Lewis$^{x/a}$ (1). Second, along with cancer progression, synthetic pathways for normal carbohydrate determinants other than sialyl Lewis$^{x/a}$, such as ABO(H) substances, Sd$^\alpha$ and 3'-sulfated determinants become gradually impaired (2), which will facilitate further accumulation of sialyl Lewis$^{x/a}$.

2.5 Multistep progression of incomplete synthesis of carbohydrate determinants in cancers

Recent expansion in knowledge on the relationship between chromatin organization and gene transcription has highlighted the importance of epigenetic mechanisms, such as DNA methylation and histone deacetylation, in the initiation and progression of human cancers [24,25]. Epigenetic silencing is now recognized as a predominant mechanism for tumor suppressor gene inactivation in cancers and can affect gene function without genetic changes like somatic mutation, deletion or loss of heterozygosity. In fact, epigenetic silencing is observed to occur not only in tumor suppressor genes but in many other genes transcribed in normal cells. Recently increasing evidence has revealed that it is epigenetic gene silencing which is behind the incomplete synthesis of normal carbohydrate determinants in cancers. For instance, DNA methylation is proposed to be a main mechanism for the decrease of ABO(H) [26,27] or Sd^{α} determinants [22], and histone deacetylation for loss of disialyl Lewis^a [8] in cancers.

3. TUMOR HYPOXIA AND ENHANCEMENT OF SIALYL LEWIS$^{A/X}$ IN LOCALLY ADVANCED CANCERS

3.1 Malignant progression of cancers in locally advanced cancer nests

Progression of cancer is a long process, which sometimes spans a few years. Long after the initiation of sialyl Lewis$^{x/a}$ expression through incomplete synthesis in the early stages, cancer cells in the locally advanced stages accumulate genetic abnormalities, and more malignant cancer cells with higher infiltrative and metastatic activities evolve according to the principle of survival of the fittest. This process is called cancer progression, and expression of sialyl Lewis$^{x/a}$ determinants is further accelerated during the course of cancer progression.

One of the selection mechanisms for more malignant cells during cancer progression is their resistance to hypoxic conditions [28,29]. Because of the uncontrolled proliferation of cancer cells, delivery of oxygen is significantly reduced in solid tumors, and some of the cancer cells are always subject to a hypoxic environment (Fig. 2a). When normal cells are exposed to such an environment, α-chain of a transcription factor, HIF (hypoxia inducible factor), which is usually rapidly hydrolyzed at proteasomes in the cytoplasm under normoxic conditions, escapes degradation and undergoes a nuclear translocation. After forming the heterodimer with the β-chain constitutively present in the nucleus, it triggers transcription of several important cellular genes, which help the cells to adapt and/or cope with the hypoxic environment. This is a reversible reaction in normal cells. In cancers, however, this process leads to the evolution of cancer cell clones that acquire constitutive and irreversible expression of HIF, through the accumulation of genetic changes in oncogenes and/or anti-oncogenes.

One of the consequences of sustained HIF expression is a particular deviation in the intracellular carbohydrate metabolism, a metabolic shift from oxidative to elevated anaerobic glycolysis (the Warburg effect), which is correlated with the increased gene expression of some glycolytic enzymes and glucose transporters including *GLUT1* [7]. Another consequence is the facilitated production of vascular endothelial growth factor (VEGF), which supports tumor angiogenesis. Both enable cancer cells to survive and proliferate in hypoxic environments in locally advanced tumor nests. Natural selection of hypoxia-resistant cancer cells

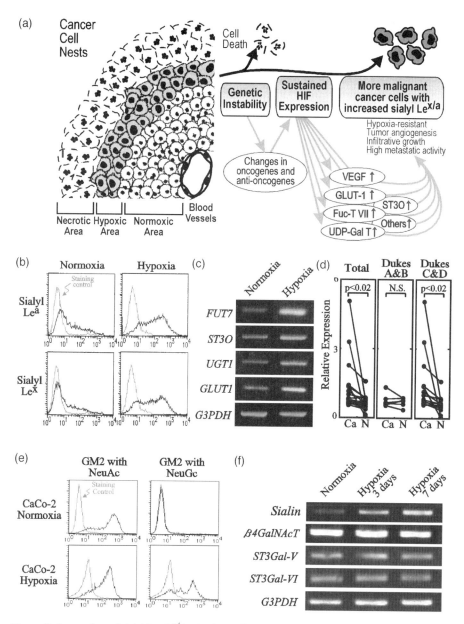

Figure 2. Expression of sialyl Lewis[x/a] is further enhanced during hypoxia-induced cancer progression in locally advanced cancers. (a) Schematic illustration of hypoxia-induced cancer progression and sialyl Lewis[x/a] accumulation in locally advanced cancer nests [56]. (b) Induction of sialyl Lewis[x/a] expression in SW480 colon cancer cells cultured under hypoxic conditions [30]. (c) Glycogenes induced in the same cells by hypoxic culture [30]. (d) Real-time RT-PCR analysis of mRNA levels of one of such glycogenes (*UGT1*) in human colon cancer tissues (Ca) and non-malignant mucosa (N) stratified according to Dukes' stage classification, indicating the preferential elevation in advanced cancers [31]. (e) Induction of GM2 carrying NeuGc in CaCo-2 colon cancer cells cultured under hypoxic conditions [40]. (f) Glycogenes induced in the same cells by hypoxic culture [40], indicating an increase in the expression of the gene for a sialic acid transporter (*sialin*), but not the genes for glycosyltransferases involved in synthesis of the GM2 core structure.

by low-oxygen environment in locally advanced tumor nests results in the clonal expansion of cancer cells with higher invasive and metastatic activities.

3.2 Transcriptional induction of glycogenes by hypoxia inducible factor-1α

We recently found a significant induction of sialyl Lewis$^{x/a}$ expression in cells cultured under hypoxic conditions (Fig. 2b), with a concomitant increase of E-selectin binding activity [30]. Participation of HIF in the process was evident from a complete abrogation of sialyl Lewis$^{x/a}$ induction by introducing a dominant negative HIF to the cells [30]. The genes induced by hypoxic culture included *GLUT1*, UDP-galactose transporter (*UGT1*), a fucosyltransferase (*FUT7*), a sialyltransferase (*ST3O*) and some other genes closely related to the carbohydrate metabolism as ascertained by DNA-microarray technique and reverse transcription polymerase chain reaction (RT-PCR) (Fig. 2c) [30].

These results indicate that augmentation of sialyl Lewis$^{x/a}$ expression on cancer cells is intimately involved in the process of cancer progression, and more malignant cancer cells tend to have an enhanced expression of these carbohydrate determinants. This mechanism contributes to the further enhancement of sialyl Lewis$^{x/a}$ expression in advanced cancer cells, which had already been predisposed to express these determinants by epigenetic gene silencing in the early stages of carcinogenesis. Expression of genes for *GLUT1*, *UGT1*, *FUT7*, and *ST3O* is significantly elevated in cancer cells prepared from surgical specimens of colon cancers, compared with non-malignant colonic epithelial cells taken from the same patients, especially those in the advanced stages such as Dukes' C and D stages (Fig. 2d) [30,31]. This suggests that the HIF-induced transcription of these genes indeed takes place in the actual cancer tissues of patients in advanced stages.

3.3 Hypoxia-resistant cancer cells in cancer therapy

The thus-selected hypoxia-resistant cancer cells (see Chapter 14) become predominant clones in locally advanced stage cancers, and are known to be refractory to radio- and chemotherapy [29,32]. Therefore, it is of particular importance to find a cell surface antigen selectively expressed on the hypoxia-resistant cancer cells, which can be used as a target for therapy of such cancer cells. It has been well documented that hypoxia-resistant cancer cells have a characteristic metabolic pattern and show enhanced secretion of some humoral factors, but the antigenic changes associated with hypoxia resistance have been little studied.

Our DNA-microarray analyses disclosed that transcription of a considerable number of carbohydrate-related genes is affected by hypoxia. The products of these genes are not limited to synthesis of sialyl Lewis$^{x/a}$, but involve synthesis of many other cell surface carbohydrate determinants. Hypoxia has a profound effect on the cell surface expression of carbohydrate determinants, some of which are good candidates for targeting therapy directed to hypoxia-resistant cancer cells.

Since the late 1970s it has been known that human cancers often express sialoconjugates containing *N*-glycolyl neuraminic acid (NeuGc) [33,34]. They were regarded as an example of illegitimate carbohydrate determinants in cancers, because humans are known to lack an enzyme required for NeuGc synthesis. The gene for the enzyme CMP-NeuAc hydroxylase was shown to be completely non-functional in humans because of a partial deletion in the sequence [35–37]; hence human cells and tissues lack NeuGc. Introduction of sialoconjugates of NeuGc-positive animals is known to elicit a strong immune reaction in humans. The responsible antigen for this immune reaction was termed Hanganutziu–Deicher (H-D) antigen, which is in essence NeuGc. Many human cancers are known to frequently express carbohydrate determinants containing NeuGc [33], and such determinants are regarded as good candidates for vaccination therapy of cancers [38,39], but how human cancers acquire NeuGc has remained enigmatic.

Recently, we have shown that tumor hypoxia induces gene expression of a sialic acid transporter, sialin, and this facilitates intracellular transportation of sialic acids, such as NeuGc, from the external milieu of the cells [40]. For instance, hypoxic culture of human colon cancer cells induces expression of GM2 ganglioside having NeuGc terminus, which can be targeted with a monoclonal antibody specific to NeuGc-GM2 (Fig. 2e,f) [40]. This indicates that acquisition of NeuGc-containing sialoconjugates on cancer cells is closely related to cancer progression driven by tumor hypoxia.

Expression of many other carbohydrate determinants is also induced by tumor hypoxia. What needs to be taken into consideration here is that hypoxia-induced change is not limited to malignant cells, but normal cells also respond to hypoxia as well. If cell surface determinants induced simply by hypoxia are chosen as a target of therapeutic attack, such determinants can be induced on the surface of hypoxic normal cells, which will also be open to attack. In this context, it is interesting to note that GM2 gangliosides, even those with normal sialic acid, are preferentially expressed on cancer cells, and regarded as good targets for immune therapy of cancers [41–43]. Genes for enzymes involved in the GM2 backbone synthesis, such as a GalNAcβ transferase, are not induced by hypoxia (Fig. 2f). The mechanism that leads to cancer-associated expression of GM2 ganglioside backbone *per se* is obviously independent from tumor hypoxia, and some oncogenes are proposed to induce transcription of the gene for the GalNAc transferase [44].

On the other hand, tumor hypoxia leads to secondary modification of the GM2 backbone with the abnormal neuraminic acid, NeuGc, through facilitating its intracellular transportation. In this aspect, GM2 ganglioside having NeuGc is unique in that it is synthesized only when the above two mechanisms are at work simultaneously, and therefore has a higher cancer specificity than the determinants inducible simply by hypoxia. Its immunogenicity in humans would make it an ideal candidate for a better antigen for cancer immunotherapy. Hypoxia-adapted cancer cells are known to show a poor response to conventional radio- and chemotherapeutic interventions. NeuGc-containing GM2 is therefore expected to be a better target for therapy of hypoxia-resistant highly aggressive cancers.

4. SIALYL LEWIS$^{A/X}$ IN HEMATOGENOUS METASTASIS OF CANCERS

4.1 Growth advantage of cancer cells that acquired enhanced sialyl Lewis$^{a/x}$ in hypoxic tumor environment in locally advanced cancer nests

Interaction of cancer cells with endothelial cells mediated by selectins and sialyl Lewis$^{x/a}$ determinants may have pathophysiological involvement not only in hematogenous metastasis of cancer as formulated earlier (Fig. 3a), but also in blood vessel formation of tumors. Some time ago we assessed the possible significance of this cell adhesion system in tumor vascularization by employing an *in vivo* model using a cultured endothelial cell line, which expressed selectins and could adhere to human cancer cells expressing sialyl Lewis$^{x/a}$ [45]. When such human cancer cells and the endothelial cell line cells at the ratio of 10:1 were subcutaneously cotransplanted into the back of nude rats, the tumors formed were extensively vascularized throughout by the blood vessel-like meshwork structures woven from the infused endothelial cells, the lumen of which contained the host blood cells (Fig. 3b, left panel) [45]. The administration of anti-Lewis$^{x/a}$ antibodies resulted in a marked reduction in the size of tumors, which were not vascularized and were accompanied by independent tiny remnant clumps composed of endothelial cells (Fig. 3b, right lower panel) [45]. These results served to corroborate that cell adhesion mediated by selectins and sialyl Lewis$^{x/a}$ determinants is significantly involved in tumor vascularization.

Cancer cells try to survive hypoxic conditions by acquiring sustained expression of HIF, thus altering their glucose metabolism from aerobic to anaerobic, and also by inducing endothelial cell growth with VEGF. The HIF-induced enhancement of sialyl Lewis$^{x/a}$ expression on cancer also seems to be a link in the chains of these unfolding events, given that sialyl Lewis$^{x/a}$ determinants promote tumor vascularization [45] and confer on the cells a growth advantage in hypoxic local tumor environments.

4.2 Clinical aspects of sialyl Lewis$^{x/a}$-mediated distant metastasis of cancers

Selective survival and expansion of the hypoxia-resistant aggressive cancer cell clones occur in a hypoxic milieu in locally advanced cancer nests. What happens next is vigorous infiltration and dissemination of the predominant aggressive cancer cell clones. Some of these cells undergo hematogenous metastasis to distant organs, which is a common feature of the terminal stage of malignant diseases. The sialyl Lewis$^{x/a}$ determinants, expression of which is more exacerbated during the natural selection of aggressive cancer cell clones in advanced tumors, mediate adhesion of cancer cells in the bloodstream to the vessel wall, thus facilitating hematogenous metastasis.

The numerous clinical statistics made available to date show that the degree of sialyl Lewis$^{x/a}$ expression on cancer cells significantly correlates with the prognostic outcome of the patients [4,34,46]. The frequency of hematogenous metastasis is the most important factor in determining the prognostic fate of patients with advanced cancers. The patient prognosis may well be predicted by

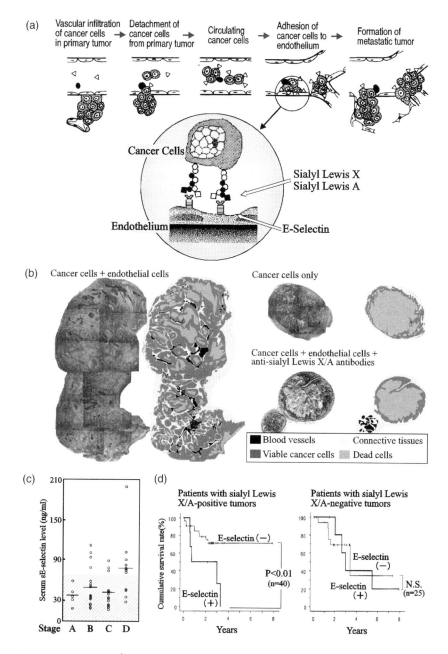

Figure 3. Sialyl Lewis$^{x/a}$ accumulated in advanced cancers facilitate tumor vascularization and hematogenous metastasis. (a) Schematic illustration for the role of sialyl Lewis$^{x/a}$ in adhesion of cancer cells to vascular endothelial cells during hematogenous metastasis of cancers [5]. (b) Experimental results indicating that sialyl Lewis$^{x/a}$ facilitate tumor vascularization. Gross appearances of *in vivo* tumors formed in nude rats by transplantation of human cancer cells with (left panel) or without (right upper panel) endothelial cell supplement, and those after treatment with anti-sialyl Lewis$^{x/a}$ antibodies (right lower panel) are shown [45]. (c) Serum E-selectin levels in patients with colorectal cancer stratified according to Dukes' stage classification. (d) Survival curves for the patients with non-small cell lung cancers having sialyl Lewis$^{x/a}$-positive (left) or -negative (right) tumors stratified by serum E-selectin levels [52].

evaluating the degree of sialyl Lewis$^{x/a}$ determinants on cancer tissues obtained at surgical resection.

The second important factor in this context in predicting patient prognosis is the level of E-selectin expression on the vascular bed of the patients. This can be evaluated by measuring soluble E-selectin in the sera of patients, because, once expressed on endothelial cells, E-selectin is proteolytically detached from the surface of endothelial cells and enters the general circulation. Levels of soluble E-selectin in the sera reflect the extent of E-selectin expression on the vessel walls in patients. E-selectin is a highly inducible protein; its expression is induced by a variety of stimuli, and the extent of its expression is highly variable in each patient. As cancer cells produce humoral factors that facilitate endothelial E-selectin expression, patients with advanced stage cancers generally have elevated serum E-selectin levels (Fig. 3c) [47–51]. Some patients have elevated levels even at early stages; this subset of patients is at greater risk of developing hematogenous metastasis. Interestingly, clinical statistics on patients with non-small cell lung cancer indicate a good correlation of E-selectin expression and prognostic outcome in patients with cancer cells expressing sialyl Lewis$^{x/a}$ determinants, while no significant correlation is observed in patients having sialyl Lewis$^{x/a}$-negative tumor cells (Fig. 3d) [52]. This finding indicates that the level of E-selectin expression can be taken into consideration in predicting patient prognosis only when the sialyl Lewis$^{x/a}$ determinants are significantly expressed on cancer cells.

Numerous other factors have been suggested to significantly influence the frequency of hematogenous metastasis by affecting the process of detachment of cancer cells from primary tumor into the bloodstream, such as cadherins, integrins, and metalloproteinases [53,54]. The overall effect of such factors would be reflected in the number of circulating cancer cells eventually, which can be evaluated either by flow cytometry using specific antibodies, or by RT-PCR using primers for epithelial cell-specific genes. At present, the following three factors should be taken into consideration in predicting the risk for hematogenous metastasis in patients with cancers: (i) degree of sialyl Lewis$^{x/a}$ expression on cancer cells; (ii) degree of E-selectin expression on vessel wall; and (iii) number of cancer cells circulating in the bloodstream.

REFERENCES

1. Phillips ML, Nudelman E, Gaeta FCA et al. (1990) Science 250, 1130–1132.
2. Takada A, Ohmori K, Takahashi N et al. (1991) Biochem. Biophys. Res. Commun. 179, 713–719.
3. Takada A, Ohmori K, Yoneda T et al. (1993) Cancer Res. 53, 354–361.
4. Kannagi R (1997) Glycoconj. J. 14, 577–584.
5. Kannagi R, Izawa M, Koike T, Miyazaki K & Kimura N (2004) Cancer Sci. 95, 377–384.
6. Hakomori S & Kannagi R (1983) J. Natl Cancer Inst. 71, 231–251.
7. Kannagi R (2004) Glycoconj. J. 20, 353–364.
8. Miyazaki K, Ohmori K, Izawa M et al. (2004) Cancer Res. 64, 4498–4505.
9. Miyazaki K, Ohmori K, Izawa M et al. Unpublished results.
10. Varki A & Angata T (2006) Glycobiology 16, 1R–27R.
11. Crocker PR (2005) Curr. Opin. Pharmacol. 5, 431–437.
12. Mitsuoka C, Sawada-Kasugai M, Ando-Furui K et al. (1998) J. Biol. Chem. 273, 11225–11233.

13. Kimura N, Mitsuoka C, Kanamori A et al. (1999) Proc. Natl Acad. Sci. USA **96**, 4530-4535.
14. Uchimura K, Gauguet JM, Singer MS et al. (2005) Nat. Immun. **6**, 1105-1113.
15. Kannagi R (2002) Curr. Opin. Struct. Biol. **12**, 599-608.
16. Izawa M, Kumamoto K, Mitsuoka C et al. (2000) Cancer Res. **60**, 1410-1416.
17. Kimura N, Miyazaki K, Ohmori K, Izawa M et al. Unpublished results.
18. Blixt O, Head S, Mondala T et al. (2004) Proc. Natl Acad. Sci. USA **101**, 17033-17038.
19. American Consortium_for Functional Glycomics at http://www.functionalglycomics.org/static/consortium/main.shtml
20. Hakomori S (1999) Biochim. Biophys. Acta Gen. Subj. **1473**, 247-266.
21. Capon C, Maes E, Michalski JC, Leffler H & Kim YS (2001) Biochem. J. **358**, 657-664.
22. Kawamura YI, Kawashima R, Fukunaga R et al. (2005) Cancer Res. **65**, 6220-6227.
23. Ikeda N, Eguchi H, Nishihara S et al. (2001) J. Biol. Chem. **276**, 38588-38594.
24. Baylin SB & Ohm JE (2006) Nat. Rev. Cancer **6**, 107-116.
25. Toyota M & Issa JP (2005) Semin. Oncol. **32**, 521-530.
26. Iwamoto S, Withers DA, Handa K & Hakomori S (1999) Glycoconj. J. **16**, 659-666.
27. Chihara Y, Sugano K, Kobayashi A et al. (2005) Lab. Invest. **85**, 895-907.
28. Semenza GL (2000) Crit. Rev. Biochem. Mol. Biol. **35**, 71-103.
29. Vaupel P (2004) Oncologist **9** (Suppl 5), 10-17.
30. Koike T, Kimura N, Miyazaki K et al. (2004) Proc. Natl Acad. Sci. USA **101**, 8132-8137.
31. Kumamoto K, Goto Y, Sekikawa K et al. (2001) Cancer Res. **61**, 4620-4627.
32. Semenza GL (2003) Nat. Rev. Cancer **3**, 721-732.
33. Malykh YN, Schauer R & Shaw L (2001) Biochimie **83**, 623-634.
34. Kannagi R (2003) In: *Immunobiology of Carbohydrates*, pp. 1-33. Edited by SYC Wong and G Arsequell. Kluwer Academic/Plenum Publishers, New York.
35. Irie A & Suzuki A (1998) Biochem. Biophys. Res. Commun. **248**, 330-333.
36. Irie A, Koyama S, Kozutsumi Y, Kawasaki T& Suzuki A (1998) J. Biol. Chem. **273**, 15866-15871.
37. Chou HH, Hayakawa T, Diaz S et al. (2002) Proc. Natl Acad. Sci. USA **99**, 11736-11741.
38. Carr A, Rodriguez E, Arango Mdel C et al. (2003) J. Clin. Oncol. **21**, 1015-1021.
39. Nakarai H, Chandler PJ, Kano K, Morton DL & Irie RF (1990) Int. Arch. Allergy Immunol. **91**, 323-328.
40. Yin J, Hashimoto A, Izawa M et al. (2006) Cancer Res. **66**, 2937-2945.
41. Livingston PO, Hood C, Krug LM et al. (2005) Cancer Immunol. Immunother. **54**, 1018-1025.
42. Livingston P (1998) Semin. Oncol. **25**, 636-645.
43. Hanai N, Nakamura K & Shitara K (2000) Cancer Chemother. Pharmacol. 46 Suppl, S13-S17.
44. Furukawa K, Akagi T, Nagata Y et al. (1993) Proc. Natl Acad. Sci. USA **90**, 1972-1976.
45. Tei K, Kawakami-Kimura N, Taguchi O et al. (2002) Cancer Res. **62**, 6289-6296.
46. Kannagi R (2003) In: *Carbohydrate-based Drug Discovery*, pp. 803-829. Edited by CH Wong. Wiley-VCH Verlag, Weinheim, Germany.
47. Ye C, Kiriyama K, Mitsuoka C et al. (1995) Int. J. Cancer **61**, 455-460.
48. Narita T, Kawakami-Kimura N, Kasai Y et al. (1996) J. Gastroenterol. **31**, 299-301.
49. Matsuura N, Narita T, Mitsuoka C et al. (1997) Anticancer Res. **17**, 1367-1372.
50. Matsuura N, Narita T, Mitsuoka C et al. (1997) Jpn. J. Clin. Oncol. **27**, 135-139.
51. Ito K, Ye CL, Hibi K et al. (2001) J. Gastroenterol. **36**, 823-829.
52. Tsumatori G, Ozeki Y, Takagi K, Ogata T & Tanaka S (1999) Jpn. J. Cancer Res. **90**, 301-307.
53. Hazan RB, Qiao R, Keren R, Badano I & Suyama K (2004) Ann. N. Y. Acad. Sci. **1014**, 155-163.
54. Hirohashi S & Kanai Y (2003) Cancer Sci. **94**, 575-581.
55. Itai S, Nishikata J, Yoneda T et al. (1991) Cancer **67**, 1576-1587.
56. Japanese Glycoforum at http://www.glycoforum.gr.jp/science/word/glycopathology/GD-A07E.html

CHAPTER 23
Glycosphingolipids in health and disease

Swetlana A. Boldin-Adamsky and Anthony H. Futerman

1. INTRODUCTION

It was with some trepidation that the authors of this chapter set out to write a review on the function of glycosphingolipids (GSLs) in a book to honor Professor Nathan Sharon's brilliant career. After all, GSLs are not lectins, and if they are not lectins, are they actually important? Fortunately for us, Professor Sharon does realize the importance of other molecules in biology, particularly those that contain sugars, such as GSLs. However had Professor Sharon chosen to work on GSLs when he began his scientific career in the mid-1950s, we have the feeling there would be very little left to learn about GSLs 50 years later!

We will now try to convince the reader that not only are GSLs important, but that there is much still left to learn about their roles in health (physiology) and disease (pathophysiology). GSLs are composed of a sphingoid long-chain base to which a fatty acid is attached via an amide bond at the C-2 position, and to which an oligosaccharide chain is attached at C-1. The oligosaccharide can be relatively simple (i.e. a mono- or disaccharide), or complex, containing a large number of both neutral and acidic sugars with a variety of branching patterns; gangliosides, perhaps the best-studied GSLs, are distinguished by the presence of one or more sialic acid residues in the oligosaccharide.

GSLs are synthesized in the endoplasmic reticulum (ER) and Golgi apparatus, reside mainly in the outer leaflet of the plasma membrane (PM), are internalized by endocytosis, and are degraded in lysosomes [1]. Significant variation exists in the types and levels of GSLs between different cells, and even within the same cell at different stages of development, implying that the regulation of GSL synthesis (and degradation) is cell type-specific and is developmentally regulated.

2. BIOLOGICAL FUNCTIONS OF GLYCOSPHINGOLIPIDS

GSLs have been implicated in many fundamental cell processes such as growth, differentiation, cancer pathogenesis [2], inflammation [3], cell recognition [4], and adhesion [5]. The carbohydrate portion of GSLs itself is involved in a number of recognition processes, like cell–cell, cell–substratum, and cell–pathogen interactions [6]. Based on their ability to interact with proteins or sugars, GSLs serve as binding sites for many pathogens, such as viruses [7], bacteria [8], and toxins [9]. It has also been suggested that GSLs play a role in human immunodeficiency virus (HIV) signaling, entry, and pathogenesis [10]. In addition, the importance of GSLs has been emphasized lately by their involvement in two processes – membrane rafts and apoptosis [11,12]. Numerous reports have suggested that some signaling machineries may reside within membrane rafts enriched in GSLs (mainly gangliosides), sphingomyelin (SM), and cholesterol [13], and ganglioside GD3 appears to contribute to mitochondrial damage, a crucial event during apoptosis [12]. Thus, overexpression of GD3 synthase increases in pathological conditions such as cancer and neurodegenerative disorders [14]. These examples indicate the importance of GSLs in cell physiology and pathophysiology.

A seminal study in the late 1990s demonstrated that deletion of glucosylceramide (GlcCer) synthase, producing a mouse totally lacking glucose-based GSLs, resulted in embryonic lethality [15]. Recently, the first disease in the pathway of GSL biosynthesis was identified [16]; it is likely that many other diseases exist in this pathway, although the essential nature of GSLs would suggest that defective synthesis, at least in early steps of the pathway, would not be compatible with life. In contrast, a number of inherited human metabolic disorders result from defects in the lysosomal degradation of GSLs. These GSL storage disorders (GSDs) (see Chapters 2 and 25) belong to a family of more than 40 known lysosomal storage disorders (LSDs), monogenic disorders leading to accumulation of undegraded substrates in lysosomes [17]. Interestingly, GSLs also accumulate in some LSDs secondarily to accumulation of the primary storage material [18]. For instance, brain storage of gangliosides GM2 and GM3 has been documented in Niemann–Pick type A disease, in which the primary storage material is SM, in Niemann–Pick type C disease, where the primary storage material is cholesterol, and also in mucopolysaccharidoses type I and III, where the primary storage materials are dermatan sulfate and heparan sulfate [19].

Below, we will discuss recent work from our laboratory attempting to delineate how GSL accumulation in various GSDs leads to cell pathology. Surprisingly, little is known about this, and it is our contention that understanding the mechanistic relationship between GSL accumulation and disease development will help not only in determining the underlying causes of the diseases, but also the roles of GSLs in normal cell physiology [19, 20].

3. THE PATHOPHYSIOLOGY OF GLYCOSPHINGOLIPID STORAGE DISORDERS

In our recent work, we have focused mainly on two diseases which display neurological symptoms, namely types 2 and 3 Gaucher disease [21], and Sandhoff disease (a type of Tay–Sachs disease) [22]. Gaucher disease is characterized by elevated levels of the simplest GSL, glucosylceramide (GlcCer), and Tay–Sachs and Sandhoff diseases are characterized by accumulation of ganglioside GM2, a complex GSL (Fig. 1).

Gaucher disease is the most common LSD, and is caused by a deficiency in glucosylceramidase (glucocerebrosidase, GlcCerase) activity. It is, however, somewhat atypical compared with other GSDs, as massive accumulation of storage material does not occur, although levels of GlcCer and its metabolite,

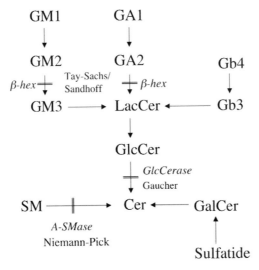

Figure 1. Glycosphingolipid (GSL) storage disorders. The defective enzymes are shown in italics. Sphingomyelin (SM) is a sphingolipid, not a GSL, but it is shown in the current scheme since GSLs also accumulate in Niemann–Pick A and B diseases. Cer, ceramide; GlcCer, glucosylceramide; LacCer, lactosylceramide; GalCer, galactosylceramide..

glucosylsphingosine (GlcSph; glucosylpsychosine) are significantly higher than normal [23]. GlcCer accumulation in symptomatic Gaucher disease patients is mainly restricted to macrophages, the so-called 'Gaucher cells'. Classification of Gaucher disease is based on age of onset or on signs of nervous system involvement. Type 1 (chronic non-neuronopathic) is the most common form (>90% of patients), in which the severity of symptoms varies widely and the age of diagnosis ranges from childhood until the eighth decade [24]. Type 2 (acute neuronopathic, infantile) and type 3 (chronic neuronopathic, juvenile) are classified according to the time of onset and rate of progression of the neurological symptoms [25]. Little is known about the molecular mechanisms by which GlcCer accumulation leads to neuropathology [26].

The GM2 gangliosidoses are caused by mutations in any of three genes, the *HEXA* gene, resulting in Tay–Sachs disease, the *HEXB* gene, resulting in Sandhoff disease, and the *GM2A* gene, resulting in GM2 activator deficiency [22]. The *HEXA* and *HEXB* genes code for β-hexosaminidase α- and β-subunits respectively, which dimerize to produce two forms of the enzyme, A ($\alpha\beta$) and B ($\beta\beta$), and a minor form, S ($\alpha\alpha$). In both Tay–Sachs and Sandhoff disease (α- and β-subunit deficiency, respectively), there is a deficit of hexosaminidase A ($\alpha\beta$), and as a consequence, massive accumulation of ganglioside GM2 in the brain. As in all LSDs, significant clinical heterogeneity is observed, varying from infantile-onset, rapidly progressive neurological disease culminating in death before 4 years of age, to late-onset, subacute or chronic progressive neurological disease [27].

A few years ago we observed that neurons that had accumulated GlcCer released more Ca^{2+} from intracellular stores in response to caffeine treatment than their untreated counterparts [28], which led to enhanced sensitivity to agents that induced neuronal cell death [28, 29]. Since Ca^{2+} release was evoked by caffeine, a known agonist of the ryanodine receptor (RyaR), a Ca^{2+}-release channel, we proposed that GlcCer accumulation in neurons somehow enhanced agonist-induced Ca^{2+} release via the ryanodine receptor, which is located in the ER. Moreover, subsequent analysis of Ca^{2+} release *in vitro* from isolated microsomes revealed that GlcCer was the only GSL tested that enhanced agonist-induced Ca^{2+} release via the ryanodine receptor [30]. Further studies demonstrated that GlcSph also stimulated Ca^{2+}-release, but via a different mechanism to GlcCer [31]. Finally, we demonstrated a correlation between levels of GlcCer accumulation in human brain tissue obtained post mortem from neuronopathic Gaucher disease patients (type 2) and levels of Ca^{2+} release [32]. Together, these findings suggested that defective Ca^{2+} homeostasis might be one of the mechanisms responsible for at least some of the neuropathophysiology in acute neuronopathic (type 2) Gaucher disease (Fig. 2).

Altered brain Ca^{2+} homeostasis was also observed in a mouse model (the $Hexb^{-/-}$ mouse) of Sandhoff disease [33], although the mechanism was completely different to that observed in Gaucher disease. Brain microsomes prepared from $Hexb^{-/-}$ mice exhibited a dramatic reduction in the rate of Ca^{2+} uptake, but no difference in the rate of Ca^{2+} release via the RyaR or the inositol 1,4,5-triphosphate receptor (IP_3R). Thapsigargin, a specific SERCA (sarco(endo)plasmic reticulum Ca^{2+}-ATPase) inhibitor, completely inhibited Ca^{2+} uptake into microsomes from the $Hexb^{-/-}$ mice at a concentration that caused only partial inhibition of Ca^{2+} uptake into control microsomes, indicating that SERCA activity is compromised. The change in the rate of Ca^{2+} uptake was not due to a reduction in levels of SERCA expression since mRNA and protein levels were unaltered, but rather a reduction in the V_{max} of SERCA. Moreover, exogenously added gangliosides GM2, GM1, and, to a lesser extent, GM3, inhibited Ca^{2+} uptake in normal brain microsomes, confirming that these gangliosides modify SERCA activity by a posttranscriptional mechanism. Previous studies had demonstrated that exogenously added gangliosides modulate SERCA activity in rabbit sarcoplasmic reticulum [34], but ours was the first to show a direct physiological link between endogenous ganglioside accumulation and neuronal cell death mediated via SERCA.

Recently, we have extended our studies to Niemann–Pick A disease. Niemann–Pick A (NPD-A) and B diseases are caused by a deficiency in lysosomal acid sphingomyelinase (A-SMase) [35]. In NPD-A, severe phenotypes are

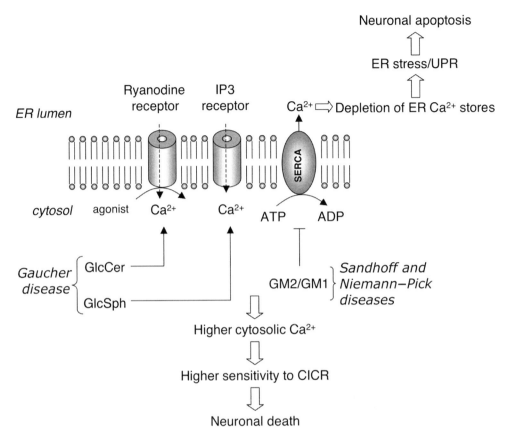

Figure 2. Neuronal Ca^{2+} homeostasis in Gaucher and Sandhoff diseases. In Gaucher disease, glucosylceramide (GlcCer) stimulates agonist-induced Ca^{2+} release via the ryanodine receptor (RyaR), and glucosylsphingosine (GlcSph) acts as an agonist of the inositol 1,4,5-triphosphate receptor (IP$_3$R). In Sandhoff and Niemann–Pick diseases, GM2 inhibits Ca^{2+} uptake via SERCA. As a consequence, cytosolic Ca^{2+} is elevated in all three diseases, which presumably leads to neuronal cell dysfunction and death, perhaps mediated via the unfolded protein response (UPR).

characterized by a progressive neurodegenerative course that leads to death in early infancy. We have shown that Ca^{2+} homeostasis is also altered in a mouse model of NPD-A disease, the A-SMase-deficient mouse (ASM$^{-/-}$) [36], with reduced rates of Ca^{2+} uptake via SERCA in the cerebellum of 6- to 7-month-old mice [37]. However, the mechanism responsible for defective Ca^{2+} homeostasis is completely different from that observed in the other two disease models. Levels of SERCA expression are significantly reduced in the ASM$^{-/-}$ cerebellum by 6–7 months of age, immediately prior to death of the mice, as are levels of the IP$_3$R, the major Ca^{2+}-release channel in the cerebellum. Systematic analyses of the time-course of loss of SERCA and IP$_3$R expression revealed that loss of the IP$_3$R preceded that of SERCA, with essentially no IP$_3$R remaining by 4 months of age, whereas SERCA was still present even after 6 months. These results suggest that dysregulated Ca^{2+} homeostasis plays a role in the neurodegeneration observed in a specific Purkinje cell population in NPD-A. Altered Ca^{2+} homeostasis and elevated cytosolic Ca^{2+} levels can lead to cell death by multiple pathways [38], and further investigation will be required to delineate which of these are involved in Purkinje cell death in ASM$^{-/-}$ mice, and presumably also in NPD-A.

The studies summarized above show that neuronal Ca^{2+} homeostasis is perturbed in Gaucher, Sandhoff, and Niemann-Pick A diseases, resulting in elevated cytosolic Ca^{2+} levels, which itself results in enhanced sensitivity of the neurons to stress (Fig. 2). The molecular mechanisms by which GM2 directly modulates SERCA, and by which GlcCer modulates the RyaR, are currently being investigated.

A number of biochemical pathways are known to be activated upon either depletion of ER Ca^{2+} stores or upon elevation of cytosolic Ca^{2+} levels, and we speculate that activation of one or other of these pathways may be downstream responses to GSL accumulation. For instance, upon depletion of ER Ca^{2+} stores, cells can enter a form of ER stress, the "unfolded protein response" (UPR), which causes suppression of global protein synthesis, activation of stress gene expression, and induction of apoptosis [39,40]. UPR induction was reported to play a role in many neurodegenerative disorders, such as Alzheimer's and Parkinson's diseases. Recently, ganglioside GM1 was shown to induce the UPR upon its accumulation in a mouse model of the GM1 gangliosidosis [41], and it has been suggested that ganglioside accumulation in GSDs may affect the integrity of mitochondrial membranes [42]. Together, these studies are beginning to provide the first mechanistic framework to delineate the biochemical pathways that are affected in GSDs.

4. TREATMENT OF GLYCOSPHINGOLIPID STORAGE DISORDERS

Therapeutic strategies for treatment of GSDs are discussed in a series of recent reviews [43–45] and may be broadly divided into enzyme replacement therapy (ERT), substrate reduction therapy (SRT), gene therapy, and chaperone-mediated therapy (CMT). The goal of both ERT and SRT is to reduce substrate storage, thus diminishing the deleterious effects caused by their accumulation. ERT achieves this by supplementing defective enzyme with active enzyme, whereas SRT works by lowering the rates of substrate synthesis, thus reducing GSL accumulation. CMT acts by stabilizing the structure of misfolded enzymes in the ER, and thus stimulating their delivery to lysosomes [44,46].

Two treatments for Gaucher disease are currently commercially available. ERT, using recombinant GlcCerase (Cerezyme®), alleviates many disease symptoms and is used by >3000 patients worldwide, and SRT, using the GSL synthesis inhibitor *N*-butyldeoxynojirimycin (NB-DNJ (Zavesca®)) (see Chapter 25), has been approved for patients for whom ERT is unsuitable, and is also being used in a number of clinical trials for other GSDs. Despite the impressive success of Cerezyme, and the recent availability of SRT, several issues remain unresolved concerning both treatments, consequently limiting their effectiveness and imposing burdens on patients [26,44]. Thus, we believe that significant improvements should be made in both ERT and SRT, and that new therapeutic approaches should be actively sought. With this in mind, we will briefly summarize below some of our recent work on three different LSDs that has provided some advances in our understanding of each of these treatments, and also proof of concept for the use of ERT, SRT, and gene therapy.

4.1 Enzyme replacement therapy

The elucidation of the three-dimensional structure of Cerezyme in our laboratory [47] provides the technological platform for the development of alternative or improved GlcCerase for ERT [26]. First, the Cerezyme structure should permit the design of structure-based drugs aimed at restoring the activity of defective GlcCerase. Second, it should provide the basis for engineering of more stable or more active GlcCerase, which could reduce the number of infusions and potentially also reduce cost. Site-directed mutagenesis and directed evolution are the main tools in this approach, and the availability of the three-dimentional structure will help in determining which parts of the protein to alter. Finally, the structure should permit a rational approach to the production of small molecules for use as pharmacological chaperones [26,46].

Furthemore, when we recently solved the structure of Cerezyme conjugated with an irreversible inhibitor, conduritol-B-epoxide (CBE) [48], we detected a significant structural change in two surface loops at the entrance to the active site which act as a lid controlling access to the active site. The role of one of the two surface loops is supported by the reduced activity of GlcCerase in various mutations known to cause Gaucher disease, such as V394L, R395P, and N396T [48] that are located in this loop. *In silico* mutational analysis is entirely consistent with the lid being in a closed conformation in the mutated proteins. The discovery of this lid regulating access to the GlcCerase active site provided the first mechanistic insight into how GlcCerase mutations result in reduced catalytic activity and as a consequence cause Gaucher disease.

4.2 Substrate reduction therapy

In an attempt to gain insight into the mechanism of action of SRT, we demonstrated that NB-DNJ reduces GM2 storage in hippocampal neurons from a mouse model of Sandhoff disease [33]. Of more importance, we also showed restored levels of Ca^{2+} uptake via SERCA after NB-DNJ treatment, demonstrating a direct correlation between the biochemical defect observed in Sandhoff neurons and levels of GM2 accumulation.

4.3 Gene therapy

Gene therapy could serve as a potential therapeutic approach for all LSDs, but in particular those in which the CNS is affected. A bicistronic lentiviral vector encoding both the hexosaminidase α- and β-chains (SIV.ASB) was shown to correct the β-hexosaminidase deficiency in Sandhoff fibroblasts and to reduce GM2 storage [49]. We have now demonstrated the potential of this SIV.ASB vector to reverse some of the pathophysiological events in cultured Sandhoff neurons [33,50]. When hippocampal neurons from embryonic Sandhoff mice were transduced with the lentivector, normal axonal growth rates were restored, as was the rate of Ca^{2+} uptake via SERCA and the sensitivity of neurons to thapsigargin-induced cell death, concomitant with a decrease in GM2 and GA2 levels [51]. These data reinforce the potential of this SIV.ASB vector as a possible treatment in Sandhoff disease.

5. PERSPECTIVES

In this chapter, we have summarized some of our recent work on defining the physiological and pathophysiological roles of GSLs, and outlined some of our studies on GSD therapy. One of the main challenges in the years ahead is to define the function of specific GSLs in defined biological events. By way of example, our studies have shown very specific effects of different GSLs on Ca^{2+} homeostasis, at least under pathophysiological conditions. Do GSLs affect Ca^{2+} homeostasis in normal physiology, and if so, how is the specificity determined? What roles do individual GSLs play in other events? In other words, why are there so many species of GSLs, and why do their levels change so dramatically during development and between cell types? It is to be hoped that these questions will have been answered by the time the next edition of this book is published to celebrate Nathan Sharon's 90th birthday.

Acknowledgments

Work in the authors' laboratory is supported by the Israel Science Foundation, the German-Israel Science Foundation, the Minerva Foundation, the National Gaucher Foundation, and the estate of Louis Uger, Canada. A.H. Futerman is The Joseph Meyerhoff Professor of Biochemistry at the Weizmann Institute of Science.

REFERENCES

1. Sandhoff K & Kolter T (2003) *Philos. Trans. R. Soc. Lond. B Biol. Sci.* **358**, 847–861.
2. Bieberich E (2004) *Glycoconj. J.* **21**, 315–327.
3. Pettus BJ, Chalfant CE & Hannun YA (2004) *Curr. Mol. Med.* **4**, 405–418.
4. Bucior I & Burger MM (2004) *Glycoconj. J.* **21**, 111–123.
5. Kannagi R, Izawa M, Koike T, Miyazaki K & Kimura N (2004) *Cancer Sci.* **95**, 377–384.
6. Fantini J, Maresca M, Hammache D, Yahi N & Delezay O (2000) *Glycoconj. J.* **17**, 173–179.
7. Arias CF, Isa P, Guerrero CA et al. (2002) *Arch. Med. Res.* **33**, 356–361.
8. Yuki N (2005) *Curr. Opin. Immunol.* **17**, 577–582.
9. Smith DC, Lord JM, Roberts LM & Johannes L (2004) *Semin. Cell Dev. Biol.* **15**, 397–408.
10. Viard M, Parolini I, Rawat SS et al. (2004) *Glycoconj. J.* **20**, 213–222.
11. Bektas M & Spiegel S (2004) *Glycoconj. J.* **20**, 39–47.
12. Morales A, Colell A, Mari M, Garcia-Ruiz C & Fernandez-Checa JC (2004) *Glycoconj. J.* **20**, 579–588.
13. Hakomori S (2004) *Glycoconj. J.* **21**, 125–137.
14. Malisan F & Testi R (2002) *Biochim. Biophys. Acta* **1585**, 179–187.
15. Yamashita T, Wada R, Sasaki T et al. (1999) *Proc. Natl Acad. Sci. USA* **96**, 9142–9147.

16. Simpson MA, Cross H, Proukakis C *et al.* (2004) *Nat. Genet.* **36**, 1225–1229.
17. Futerman AH & van Meer G (2004) *Nat. Rev. Mol. Cell. Biol.* **5**, 554–565.
18. Walkley SU (2004) *Semin. Cell Dev. Biol.* **15**, 433–444.
19. Raas-Rothschild A, Pankova-Kholmyansky I, Kacher Y & Futerman AH (2004) *Glycoconj. J.* **21**, 295–304.
20. Ginzburg L, Kacher Y & Futerman AH (2004) *Semin. Cell Dev. Biol.* **15**, 417–431.
21. Beutler E & Grabowski GA (2001) In: *The Metabolic and Molecular Bases of Inherited Disease*, pp. 3635–3668. Edited by CR Scriver, WS Sly, B Childs *et al.* McGraw-Hill, New York.
22. Gravel RA, Kaback MM, Proia R, Sandhoff K, Suzuki K & Suzuki K (2001) In: *The Metabolic and Molecular Bases of Inherited Disease*, pp. 3827–3876. Edited by CR Scriver, WS Sly, B Childs *et al.* McGraw-Hill, New York.
23. Nilsson O & Svennerholm L (1982) *J. Neurochem.* **39**, 709–718.
24. Cox TM & Schofield JP (1997) *Baillieres Clin Haematol* **10**, 657–689.
25. Vellodi A, Bembi B, de Villemeur TB *et al.* (2001) *J. Inherit. Metab. Dis.* **24**, 319–327.
26. Futerman AH, Sussman JL, Horowitz M, Silman I & Zimran A (2004) *Trends Pharmacol. Sci.* **25**, 147–151.
27. Mahuran DJ (1999) *Biochim. Biophys. Acta* **1455**, 105–138.
28. Korkotian E, Schwarz A, Pelled D, Schwarzmann G, Segal M & Futerman AH (1999) *J. Biol. Chem.* **274**, 21673–21678.
29. Pelled D, Shogomori H & Futerman AH (2000) *J. Inherit. Metab. Dis.* **23**, 175–184.
30. Lloyd-Evans E, Pelled D, Riebeling C *et al.* (2003) *J. Biol. Chem.* **278**, 23594–23599.
31. Lloyd-Evans E, Pelled D, Riebeling C & Futerman AH (2003) *Biochem. J.* **375**, 561–565.
32. Pelled D, Trajkovic-Bodennec S, Lloyd-Evans E, Sidransky E, Schiffmann R & Futerman AH (2005) *Neurobiol. Dis.* **18**, 83–88.
33. Pelled D, Lloyd-Evans E, Riebeling C, Jeyakumar M, Platt FM & Futerman AH (2003) *J. Biol. Chem.* **278**, 29496–29501.
34. Wang Y, Tsui Z & Yang F (1999) *FEBS Letts.* **457**, 144–148.
35. Schuchman EH & Desnick RJ (2001) In: *The Metabolic and Molecular Bases of Inherited Disease*, pp. 3589–3605. Edited by CR Scriver, WS Sly, B Childs *et al.* McGraw-Hill, New York.
36. Horinouchi K, Erlich S, Perl DP *et al.* (1995) *Nat. Genet.* **10**, 288–293.
37. Ginzburg L & Futerman AH (2005) *J. Neurochem.* **95**, 1619–1628.
38. Mattson MP & Chan SL (2003) *Nat. Cell Biol.* **5**, 1041–1043.
39. Kaneko M & Nomura Y (2003) *Life Sci.* **74**, 199–205.
40. Rutkowski DT & Kaufman RJ (2004) *Trends Cell Biol.* **14**, 20–28.
41. Tessitore A, del P Martin M, Sano R *et al.* (2004) *Mol. Cell* **15**, 753–766.
42. d'Azzo A, Tessitore A & Sano R (2006) *Cell Death Differ.* **13**, 404–414.
43. Jeyakumar M, Dwek RA, Butters TD & Platt FM (2005) *Nat. Rev. Neurosci.* **6**, 713–725.
44. Jmoudiak M & Futerman AH (2005) *Br. J. Haematol.* **129**, 178–188.
45. Platt FM, Jeyakumar M, Andersson U, Dwek RA & Butters TD (2005) *Adv. Exp. Med. Biol.* **564**, 117–126.
46. Fan JQ (2003) *Trends Pharmacol. Sci.* **24**, 355–360.
47. Dvir H, Harel M, McCarthy AA *et al.* (2003) *EMBO Rep.* **4**, 704–709.
48. Premkumar L, Sawkar AR, Boldin-Adamsky S *et al.* (2005) *J. Biol. Chem.* **280**, 23815–23819.
49. Arfi A, Bourgoin C, Basso L *et al.* (2005) *Neurobiol. Dis.* **20**, 583–593.
50. Pelled D, Riebeling C, van Echten-Deckert G, Sandhoff K & Futerman AH (2003) *Neuropathol. Appl. Neurobiol.* **29**, 341–349.
51. Arfi A, Zisling R, Richard E *et al.* (2006) *J. Neurochem.* **96**, 1572–1579.

CHAPTER 24
Pathogen–host interactions in *Entamoeba histolytica*

David Mirelman and William A. Petri Jr.

1. INTRODUCTION

Entamoeba histolytica is an intestinal protozoan parasite of humans that causes amebic colitis and amebic liver abscesses, which are diseases associated with significant levels of morbidity and mortality worldwide [1].

The organism has a simple life cycle existing as the microaerophilic motile trophozoite or as the detergent resistant cyst form. Infection begins by ingestion of cysts present in polluted water or vegetables. Cysts containing a chitin-based rigid cell wall are formed in the descending colon and this protects the metabolically dormant cells from the adverse conditions of the environment. Following ingestion, the cysts undergo an excystation process in the small intestine and the emerging motile trophozoites migrate and reside within the anaerobic confines of the human colon. The trophozoites reproduce by binary fission, they phagocytize large numbers of bacteria and food remnants. They lack mitochondria and derive energy from fermentation.

Epidemiological studies indicate that approximately 50 million people become infected with *E. histolytica* every year. Of these it is estimated that only 10% of them show symptoms of invasive disease (dysentery and extraintestinal lesions). These epidemiological data pose a number of important questions, for example why do only a fraction of the infected individuals develop symptomatic disease? One explanation for such a disparity in numbers is that there are considerable differences in the conditions found in the host and that these may be responsible in preventing the full expression of the trophozoite virulence factors needed to cause symptoms. Another reason could be differences in the pathogenicity of different strains of *E. histolytica*. The low virulence of some *E. histolytica* isolates resembles that of the closely related non-pathogenic species *E. dispar*.

During the past two decades, the tools of molecular biology have greatly contributed to the understanding of *E. histolytica* pathogenesis. The recent publication of the *E. histolytica* genome [2] provides remarkable new insights into the biology of *E. histolytica* and is becoming a powerful source for the understanding of the requirements for intestinal parasitism as well as for the identification and characterization of the molecular weapons and mechanisms which the parasite uses to damage the host tissues and kill mammalian cells [3]. From both *in vitro* and *in vivo* studies it was established that pathogenesis can be divided into three major stages: (1) adhesion of the trophozoites to distinct surface receptors on mammalian cells and induction of host cell responses related to the adhesion process; (2) proteolytic and hydrolytic destruction of extracellular matrix and mucus; and (3) contact-mediated killing and phagocytosis of target cells (Fig. 1).

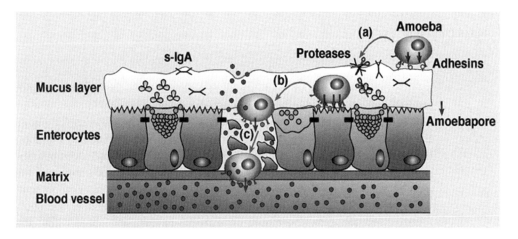

Figure 1. The invasive steps of *E. histolytica* trophozoites in the colon. (a) Adhesion and digestion of mucus and sIgA. (b) Penetration through the mucus layer, adherence to mucosal cells followed by insertion of amoebapore and killing of mucosal cells. (c) Passage through intestinal wall by formation of ulcerations and extraintestinal migration through the blood circulation.

The molecular basis for these events remains one of the vital areas of research in amebiasis. The rather simplistic model arising from these studies, however, is certainly incomplete, especially with regard to the well-established involvement of host responses such as inflammation as well as humoral and cellular immune components which the parasite then encounters. The aim of this review is to summarize the advances in understanding of the molecular factors and regulators that affect the virulence of *E. histolytica*.

2. VIRULENCE FACTORS

2.1 Adhesins – the Gal/GalNAc lectin

E. histolytica trophozoites adhere to colonic mucus and epithelial cells through interaction of a galactose and *N*-acetyl-D-galactosamine (Gal/GalNAc)-inhibitable lectin with host Gal/GalNAc-containing glycoconjugates [4–6]. Blockade of lectin activity with millimolar concentrations of Gal or GalNAc prevents the contact-dependent cytotoxicity for which the organism is named. Additionally, Chinese hamster ovary (CHO) cell glycosylation-deficient mutants lacking terminal Gal/GalNAc residues on N- and O-linked sugars are nearly totally resistant to amebic adherence and cytolytic activity. Importantly, the Gal/GalNAc lectin also mediates adherence to human neutrophils, colonic mucins, and epithelial cells, the *in vivo* targets of *E. histolytica*. The Gal/GalNAc lectin is composed of a 260 kDa heterodimer of disulfide-linked heavy (170 kDa) and light (35/31 kDa) subunits which is non-covalently associated with an intermediate subunit of 150 kDa (Fig. 2) [5]. The 170 kDa subunit contains a C-terminal cytoplasmic and transmembrane domain adjacent to a cysteine-rich extracellular domain. Five distinct genes (termed *hgl1* to *hgl5*) encoding the lectin's heavy subunit have been identified and sequenced. At least 89% sequence homology exists within this gene family and the number and location of every cysteine residue is conserved within the regions sequenced to date. The sequence of the *hgl* genes is nearly completely conserved in isolates of *E. histolytica* from different continents [7]. The carbohydrate recognition domain (CRD) is located within the cysteine-rich domain of the heavy subunit [8]. Interestingly the intermediate subunit is part of a large family of transmembrane kinases [7].

Disruption of the lectin (via inducible expression in the parasite of a dominant-negative mutant of the lectin) inhibited amebic adherence, cytotoxicity, and abscess formation in an animal model [9–11]. Inhibition of expression of the 35 kDa light subunit by antisense RNA did not significantly affect adhesion activity to mammalian or bacterial cells but strongly inhibited cytopathic activity, cytotoxic activity, and the ability to induce the formation of liver lesions in hamsters [12]. Amebae transfected with a truncated lgl subunit (in which the 55 N-terminal amino acids of the lgl were removed) showed a significant decrease in their ability to adhere to and kill mammalian cells as well as in their capacity to form rosettes with and to phagocytose erythrocytes. In addition, immunofluorescence confocal microscopy of this transfectant with anti-Gal-lectin antibodies showed an impaired ability to cap [13]. These results indicate that the light subunit has a role in enabling the clustering of Gal-lectin complexes and that its N-truncation affects this function, which is required for virulence. In contrast, disruption of heavy subunit function appears to have a greater effect on adherence [5].

2.2 Cysteine proteinases

Cysteine proteinases (CPs) are important virulence factors of various infective organisms and the main proteolytic enzymes in many protozoan parasites. As the name *E. histolytica* implies, amebae are well known for their great capacity to destroy host tissue and to degrade extracellular matrix proteins. Compelling evidence is now available which clearly shows that CPs are one of the main culprits for the pathology of the pathogen. Some of the CPs have been shown to be involved in pathogenicity such as destruction of host tissue or triggering an inflammatory response in the infected individual [14]. Other CPs were shown to be

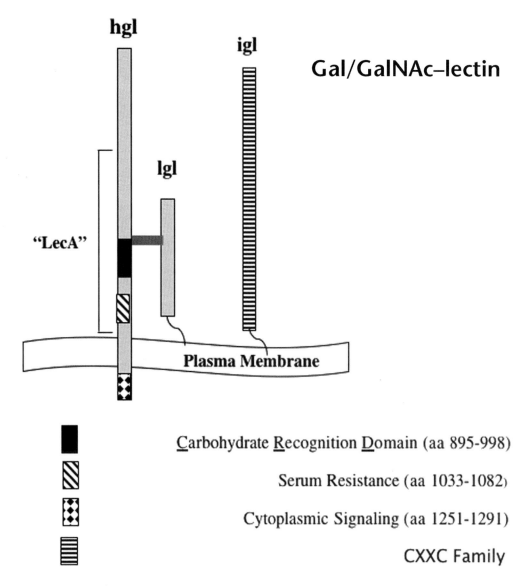

Figure 2. The Gal/GalNAc lectin of *E. histolytica.* The Gal/GalNAc lectin is a heterotrimeric molecule consisting of a transmembrane heavy chain (hgl) and GPI-anchored light (lgl) and intermediate (igl) chains. The "LecA" vaccine antigen comprises hgl amino acids 578–1154. Hgl and lgl are linked in a 1:1 molar ratio by disulfide bonds, whereas igl is non-covalently attached and is present at less than a 1:1 molar ratio to hgl–lgl. A carbohydrate recognition domain (CRD) is located in hgl at amino acids 895–998. The CRD region expressed in *E. coli* has Gal/GalNAc-binding activity. Antibodies which block serum resistance of the parasite bind to hgl amino acids 1251–1291. The cytoplasmic tail is implicated in the inside-out and outside-in signaling required for adherence and cytotoxicity. igl is a member of a large gene family of proteins containing CXXC motifs of unknown function.

essential for intracellular protein turnover and degradation as well as for life cycle processes such as en- and excystation [15]. The mucous layer lining the colonic epithelium is the first line of host defense against invasive pathogens such as *E. histolytica*. Inhibition of cysteine proteinases by specific inhibitors such as E-64 or laminin, prevents the disruption of the MUC2 polymer and the subsequent damage to the mucosal cells [16]. Cleavage of the C3 fragment of complement as well as digestion of IgA and IgG were also shown by the CPs of trophozoites and this may limit the effectiveness of the host humoral response and help the parasite evade the host defense systems [17]. Treatment of *E. histolytica* trophozoites with sublethal doses of CP inhibitors greatly reduced their ability to induce liver lesions in laboratory animals [18,19] and also blocked the process of excystation in *E. invadens* cysts [15]. Furthermore, trophozoites in which the expression of CPs was inhibited (about 90%) by the transcription of antisense mRNA, showed that they were incapable of forming liver abscesses in animal models [12,20]. CPs appear to be a major contributor to gut inflammation and tissue damage in amebiasis. Studies in a human intestinal xenograft model of disease [21] revealed that the trophozoites which were inhibited in the expression of CPs were defective in their ability to invade through the intestinal mucosa and caused significantly less damage to the intestinal permeability barrier. The CP-deficient trophozoites also failed to induce the intestinal epithelial cell production of the inflammatory cytokines interleukin 1 beta (IL-1β) and IL-8 and caused less intestinal inflammation and tissue damage [21].

Twenty full-length CP genes have been identified to date in the *E. histolytica* genome but only about half of them have been found to be expressed in trophozoites cultured under axenic conditions [22]. The relatively large number of CP genes appears not to be the result of gene amplifications since the various enzymes have only 10–86% sequence identity and some of them contain short introns. Four of the CPs (EhCP1, EhCP2, EhCP3, and EHCP5) have been purified and found to account for approximately 90% of the total CP activity present in lysates of axenically cultured trophozoites [22–25]. EhCP-5, which has significant homology to cathepsin L, appears to be of special importance for pathogenicity because it is the only CP that was found to be present on the surface of the trophozoite [24]. Purified recombinant EhCP-5 was found to cleave proIL-18 and mature IL-18 to biologically inactive fragments. suggesting that it may be involved in the control of the host response [26]. Interestingly in contrast to all the other cysteine proteinases, it was found that functional genes corresponding to EhCP-1 and EhCP-5 are absent in *E. dispar*. Although the genomic loci containing the gene for CP-5 was conserved in the two species, in *E. dispar* it was found to contain numerous deletions and stop codons in the reading frame. EhCP-1 (termed amoebiasin 1) is a chagasin-like CP which belongs to the CLAN A group of CPs. It appears to reside in larger vesicular structures and seems to be regulated by an endogenous 13 kDa inhibitor (ICP1) with homology to inhibitors of the chagasin-like family of other pathogens [27]. EhCP-2 (termed histolysain) appears to play only a minor or no role in the causation of tissue damage in experimental acute liver amebiasis [28]. Some of the proteinases appear to be secreted by the trophozoites such as the 56 kDa neutral proteinase. Another interesting secreted

CP proteinase is the EhCP112 that has 446 amino acids and which appears to be released as a complex with an adhesin-like protein EhCPADH giving it a molecular weight of 124 kDa [29]. Sera from human patients with amebiasis recognized recombinant EhCP112. Immunization of hamsters with a mixture EhADH112 and Ehcp112 cDNA inhibited liver abscess formation by virulent trophozoites.

Most of the current knowledge about the role and function of CPs in *Entamoeba* has been based on findings in axenically cultured trophozoites. Numerous reports have cited the increase in virulence and CP activity following passage through a hamster liver or co-cultivation with certain bacteria [30] but the effect on the expression of specific CPs was not determined. It is quite possible that many of the enzymes that are poorly or not expressed under such conditions or are in the inactive mixed disulfide form could be functional and important for the parasite during infection and invasion of the human host tissues or for feeding in the human host colon and for completion of the life cycle. The elucidation of the precise role of each of the various CP enzymes is a major challenge and in view of their important contribution to pathogenesis, their characterization will help us understand some of their unique properties which may be useful for the design of new therapies.

2.3 Amoebapores

Invasion occurs when trophozoites of pathogenic *E. histolytica* strains adhere to and subsequently kill intestinal epithelial cells, leading to the release of cytokines and the attraction of neutrophils and macrophages [6].

This contact-mediated cytolethal effect was found to be caused by the insertion of several small molecular weight proteins termed amoebapores which are spontaneously incorporated into the mammalian cell membranes to form depolarizing ion channels [31]. The pore-foming activity was originally detected in lysates of *E. histolytica* trophozoites over 23 years ago using artificial lipid bilayers [32,33]. It took a lot of effort, however, before the amoebapores were finally purified and their structure and mode of action was characterized [34,35]. Three isoforms of amoebapore (A, B, and C with an abundance ratio of 21:9:1) have been isolated and biochemically characterized. The primary structure of the 77 amino acid peptides was resolved by protein sequencing as well as by molecular cloning of the genes. Amoebapores exist as mature and potentially active peptides inside cytoplasmic granules and following the establishment of an intimate adhesion between the ameba and the target cell, the granules are discharged into the contact zone. After their insertion, the amoebapore molecules tend to oligomerize due to peptide–peptide interactions and form, by a barrel–stave mechanism, ion channels which eventually permeabilize the cell. The contact dependence and rapidity of the cytolytic reaction resembles that of cytotoxic lymphocytes [6]. An interesting structural and functional similarity was found between amoebapores and other toxic polypeptides such as NK-lysin and granulysin that were characterized in natural killer cells (NK) and cytotoxic T-cells [36]. The high-resolution, three-dimensional structure of the amoebapore was recently solved by NMR spectroscopy [31]. Molecular modeling indicates that structural features are conserved among the three isoforms and their hydrophobic character and electronegative charge for monomer–monomer interaction is provided by the surface orientation of the two helices and the location of a crucial histidine residue. The amoebapores analogs of the non-pathogenic species *E. dispar* have significantly less pore-forming activity [37] and this is probably due to the replacement of glutamic acid at position 2 by proline in *E. dispar* which abolishes one of the intermolecular electrostatic interactions [31]. Interestingly the amoebapore peptides are incapable of binding to the surface membranes of amebic trophozoites and this is what protects the parasite from its own toxic molecules [38]. Antibodies prepared against amoebapores, however, fail to protect mammalian cells from their cytotoxic effect.

The first convincing evidence for the role of amoebapores in the pathogenesis of *E. histolytica* infection *in vivo* was provided by experimenting with transfected trophozoites expressing decreased levels of amoebapore (<40%) due to the transcription of antisense mRNA. The amoebapore-deficient trophozoites clearly demonstrated a significant inhibitory effect on pathogenicity; they were unable to kill mammalian or induce liver lesions in hamsters [39]. A total and irreversible inhibition of amoebapore expression was recently achieved by an epigenetic gene silencing procedure [40,41]. Gene silencing in these trophozoites, termed G3, is at the transcription level and remained stable even after removal of the plasmid, resulting in virulence-attenuated amebae that are incapable of killing mammalian cells but still caused some colitis and gut inflammation in the SCID-HU-mice xenograft model [42]. Lysis of ingested bacterial cells by the G3 trophozoites was much slower but they were still capable of digesting them. Preliminary vaccination experiments of hamsters with live, virulence-attenuated trophozoites suggests that they may have a potential for immunoprotection. [43]. The recent demonstration of epigenetic gene silencing opens exciting new opportunities to prepare additional virulence-deficient epi-mutants which could be developed into attenuated vaccines to help control the disease in endemic areas.

2.4 The cell surface glycoconjugates

Most enteric pathogens have on their cell surfaces a layer of complex glycoconjugates or proteophosphoglycans which are highly antigenic and toxic to mammalian cells. Pathogenic *E. histolytica* trophozoites are covered with a layer of complex proteophosphoglycans which can be divided into two families, lipophosphoglycans (LPG) and lipophosphopeptidoglycans (LPPG). Their general structure is characterized by a glycosylphosphatidylinositol (GPI)-anchor, a serine-rich polypeptide backbone with branched glucosyl side-chains that are linked to the polypeptide by phophodiester bonds [44]. The LPPG and LPG molecules are very abundant on the trophozoite surface (8×10^7 copies/cell), they are recognized by the sera of human patients with amebiasis and their molecular mass and side-chain structures seem to vary in different strains, most likely due to the activity of different bacterial hydrolases [44]. Their role in pathogenicity is not clear but trophozoites of the non-pathogenic species *E. dispar* appear to have significantly lower amounts and their structure is different [45]. Furthermore, the non-virulent *E. histolytica* strain Rahman appears to be a naturally occurring mutant which synthesizes only one class, the LPPG, and this is only elaborated with short disaccharide chains of Glcβ1–6Gal [44]. Other virulence factors in strain Rahman, such as the Gal-specific lectin, the CPs, and AP are expressed at normal levels yet the trophozoites are incapable of killing mammalian cells, destroying tissue-cultured monolayers, or inducing liver lesions in hamsters [44].

A role for LPG in *E. histolytica* virulence was also proposed after antibodies against LPG were found to prevent trophozoite adherence and cytolysis of tissue cultured monolayers of mammalian cells. A monoclonal antibody EH5 prepared against LPG was also effective in protection against liver lesion formation in SCID mice [46] and in reducing the human intestinal inflammation and tissue damage that are normally seen in colonic xenografts in the SCID-HU-mice model [47]. The EH5 monoclonal antibody did not recognize the LPPG of non-virulent *E. histolytica* strain Rahman or that of *E. dispar*. Following their adherence to enterocytes, trophozoites of virulent *E. histolytica* were shown to transfer some of their LPG molecules to the apical side of the mammalian target cells and this may explain why sera from human patients of invasive amebiasis usually have high titers against LPG [48,49].

Very little is known about the biosynthetic pathways of these proteophosphoglycans. A promising step has been the development of a cell-free *in vitro* system which was capable of producing LPG-like chains of various sizes that were immunoprecipitated by an LPG-specific monoclonal antibody [50]. The elucidation of the biosynthetic pathway may help in the identification of novel potentially specific drug targets.

2.5 Additional virulence factors

The four classes of virulence factors described above have been widely recognized as the major ones and they are the best characterized until now. Their activity at the parasite cell level depends on the coordinated action of the acto-myosin cytoskeleton which allows for amebic motility, phagocytosis, and spatial reorganization of cellular components and virulence factors [51]. Trophozoites which invade into the submucosa come into contact with extracellular matrix proteins and this interaction induces alterations in actin polymerization, motility, and secretion via G protein-coupled receptors and phosphokinase A-dependent signaling [52].

Based on bioinformatic comparisons of the *E. histolytica* genome as well as comparisons of expression profiles with those of other non-virulent *Entamoeba*, it is obvious nowadays that there are many more factors that may play a role in virulence and/or in the interaction with the host but most of them are not, as yet, well characterized [53,54]. Included in such recently identified potential virulence factors are a group of serine and metalloproteases [55], a family of saposin-like proteins [56], carbohydrate hydrolases that are secreted and degrade colonic mucin oligosaccharides [57], membrane-bound acid phosphatases that help disrupt the cytoskeleton of host cells [58], and a family of transmembrane kinases that function in response to host conditions [7]. Furthermore, pathogenic trophozoites have a surface serine-rich protein [14] as well as a group of recently identified lysine/glutamic-rich proteins which bind to host cells [59] and a surface leucine-rich BspA protein that binds to fibronectin [54]. In addition, trophozoites were found to have an unusual surface peroxiredoxin molecule which protects the parasite from oxidant attacks generated by the host [60].

In conclusion, our knowledge about the regulatory mechanisms that control the expression of the different virulence factors as well as the host factors and conditions which affect these processes is still [55] very limited. Elucidation of such mechanisms will contribute greatly towards our understanding of the molecular regulation of pathogenesis of this parasite.

REFERENCES

1. Stanley SL Jr. (2001) *Trends Parasitol.* **17**, 280–285.
2. Loftus B, Anderson I, Davies R *et al.* (2005) *Nature* **433**, 865–868.
3. Petri WA, Jr. (2002) *Curr. Opin. Microbiol.* **5**, 443–447.
4. Petri WA, Jr, Smith RD, Schlesinger PH *et al.* (1987) *J. Clin. Invest.* **80**, 1238–1244.
5. Petri WA, Jr, Mann BJ & Haque R. (2002) *Annu. Rev. Microbiol.* **56**, 39–64.
6. Ravdin JI. and Guerrant RL (1982) *Rev. Infect. Dis.* **4**, 1185–1207.
7. Beck DL, Boettner DR, Dragulev B *et al.* (2005) *Eukaryot. Cell* **4**, 722–732.
8. Pillai DR, Wan PS. K, Yau YCW *et al.* (1999) *Infect. Immun.* **67**, 3836–3841.
9. Vines RR, Ramakrishnan G, Rogers J *et al.* (1998) *Mol. Biol. Cell* **9**, 2069–2079.
10. Tavares P, Rigothier mC, Khun H *et al.* (2005) *Infect. Immun.* **73**, 1711–1778.
11. Coudrier E, Amblard F, Zimmer C *et al.* (2005) *Cell. Microbiol.* **7**, 19–27.
12. Ankri S, Stolarsky T, Bracha R *et al.* (1999) *Infect. Immun.* **67**, 421–422.
13. Katz U, Ankri S, Stolarsky T *et al.* (2002) *Mol. Biol. Cell* **13**, 4256–4265.
14. Stanley SL Jr., Tian K, Joester JP *et al.* (1995) *J. Biol. Chem.* **270**, 4121–4126.
15. Makioka A, Kumagai M, Kobavashi S *et al.* (2005) *Exp. Parasitol.* **109**, 27–32.
16. Moncada D, Keller K & Chadee K (2003) *Infect. Immun.* **71**, 838–844.
17. Reed SL, Ember JA, Herdman DS *et al.* (1995) *J. Immunol.* **155**, 266–274.
18. Li E, Yang WG, Zhang T *et al.* (1995) *Infect. Immun.* **63**, 4150–4153.
19. Stanley SL Jr., Zhang T, Rubin D *et al.* (1995) *Infect. Immun.* **63**, 1587–1590.
20. Ankri S, Stolarsky T & Mirelman D (1998) *Mol. Microbiol.* **28**, 777–785.
21. Zhang Z, Wang L, Seydel KB *et al.* (2000) *Mol. Microbiol.* **37**, 542–548.
22. Bruchhaus I, Loftus BJ, Hall N *et al.* (2003) *Eukaryotic Cell* **2**, 501–509.
23. Bruchhaus I, Jacobs T, Leippe M *et al.* (1996) *Mol. Microbiol.* **22**, 255–263.
24. Jacobs T, Bruchhaus I, Dandekar T *et al.* (1998) *Mol. Microbiol.* **27**, 269–276.
25. Scholze H & Tannich E (1994) *Methods Enzymol.* **244**, 512–523.
26. Que X, Kim SH, Sajid M. *et al.* (2003) *Infect. Immun.* **1**, 1274–1280.
27. Riekenberg S, Witjes B, Saric M *et al.* (2005) *FEBS Lett.* **579**, 1573–1578.
28. Olivos-Garcia A, Gonzalez-Canto A, Lopez-Vancell R *et al.* (2003) *Parasitol. Res.* **90**, 212–220.
29. Ocadiz R, Orozco E, Carrillo E *et al.* (2005) *Cell Microbiol.* **1**, 221–232.
30. Padilla-Vaca F, Ankri S, Bracha R *et al.* (1999) *Infect. Immun.* **67**, 2096–2102.
31. Leippe M & Bruhn H (2005) *Trends Parasitol.* **21**, 5–7.
32. Young JD, Young TM, Lu LP *et al.* (1982) *J. Exp. Med.* **156**, 1677–1690.
33. Lynch EC, Rosenberg IM & Gitler C (1982) *EMBO J.* **1**, 801–804.
34. Leippe M, Andra J, Nickel R *et al.* (1994) *Mol. Microbiol.* **14**, 895–904.
35. Leippe M (1997) *Parasitol. Today* **13**, 178–183.
36. Andersson M, Gunne H, Agerberth B *et al.* (1995) *EMBO J.* **14**, 1615–1625.
37. Nickel R, Ott C, Dandekar T *et al.* (1999) *Eur. J. Biochem.* **265**, 1002–1007.
38. Andra J, Berninghausen O & Leippe M (2004) *FEBS Lett.* **564**, 109–115.
39. Bracha R, Nuchamowitz Y, Leippe M *et al.* (1999) *Mol. Microbiol.* **34**, 463–472.
40. Bracha R, Nuchamowitz Y & Mirelman D (2003) *Eukaryotic Cell* **2**, 295–302.
41. Anbar M, Bracha R, Nuchamowitz Y *et al.* (2005) *Eukaryotic Cell* **4**, 1775–1784.
42. Zhang X, Zhang Z, Alexander D *et al.* (2004) *Infect. Immun.* **72**, 678–683.

43. Bujanover S, Katz U, Bracha R et al. (2003) Int. J. Parasitol. 33, 1655–1663.
44. Moody-Haupt S, Patterson JH, Mirelman D et al. (2000) J. Mol. Biol. 297, 409–420.
45. Bhattacharya A, Arya R, Clark CG et al. (2000) Parasitology 120, 31–35.
46. Marinets A, Zhang T, Guillen N et al. (1997) J. Exp. Med. 186, 1557–1565.
47. Zhang Z, Duchene M & Stanley SL, Jr. (2002) Infect. Immun. 70, 5873–5876.
48. Lauwaet T, Oliveira MJ, Callewaert B et al. (2004) Int. J. Parasitol. 34, 549–556.
49. Prasad R, Tola M, Bhattacharya S et al. (1992) Mol. Biochem. Parasitol. 56, 279–287.
50. Arya R, Mehra A, Bhattacharya A et al. (2003) Mol. Biochem. Parasitol. 126, 1–8.
51. Tavares P, Rigothier M-C, Khun H et al. (2005) Infect. Immun. 73, 1771–1778.
52. Meza I (2000) Parasitol. Today 16, 23–28.
53. MacFarlane R, Bhattacharya D & Singh U (2005) Exp. Parasitol. 110, 196–202.
54. Davis PH, Zhang Z, Chen M et al. (2006) Mol. Biochem. Parasitol. 145, 111–116.
55. Barrios-Ceballos MP, Martinez-Gallardo NA, Anaya-Velazquez F et al. (2005) Ex. Parasitol. 110, 270–276.
56. Bruhn H & Leippe M (2001) Biochim. Biophys. Acta 1514, 14–20.
57. Moncada D, Keller K & Chadee K (2005) Infect. Immun. 73, 3790–3793.
58. Aguirre-Garcia MM, Anaya-Ruiz M & Talamas-Rohana P (2003) Parasitology 126, 195–202.
59. Seigneur M, Mounier J, Prevost M-C et al. (2005) Cell. Microbiol. 7, 569–579.
60. Choi MH, Sajed D, Poole L et al. (2005) Mol. Biochem. Parasitol. 143, 80–89.

SECTION 8
Industrial glycobiology

CHAPTER 25
The development of iminosugars as drugs

Bryan Winchester

1. INTRODUCTION

Iminosugars are polyhydroxylated alkaloids that resemble monosaccharides, in which the ring oxygen has been replaced by nitrogen (Figs 1-3). This substitution renders them metabolically inert but does not prevent their interaction with proteins that recognize carbohydrates. They can act as inhibitors or allosteric effectors of enzymes by mimicking the pyranosyl and furanosyl structures of their natural substrates. They can also behave as molecular chaperones to protect these enzymes during folding and may act as ligands for other proteins with carbohydrate-recognizing domains. These properties can be exploited to modulate the synthesis, breakdown distribution, or recognition of glycoconjugates, such as glycosphingolipids, glycogen, glycoproteins, and DNA/RNA. Therefore, iminosugars are very valuable reagents for studying glycobiology and as potential therapeutic agents [1-3]. Promising antiviral, antitumor, antifungal, antidiabetic, and immunosuppressive drugs, male contraceptives and drugs to alleviate inborn errors of metabolism have been developed based on these biological activities.

Nature also seems to have learnt this piece of chemistry, as many iminosugars occur naturally, particularly in plants and microorganisms, but also in insects and amphibians (chapter 2 in [4] and [5-7]).

In recognition of their pharmacological potential almost every possible iminosugar has been synthesized chemically together with derivatives designed to increase the cellular permeability or resemblance to the transition state of specific enzymes (chapters 4 and 5 in [4] and [8-11]). Iminosugars can be synthesized using carbohydrates such as monosaccharides, aminosugars, glyconolactones, and alditols or non-carbohydrate molecules as the starting material. The advantage of using a natural sugar as the starting material is that the stereochemistry of the hydroxyl groups is preselected. When using non-carbohydrate molecules as the starting material the stereochemistry of the polyhydroxylated carbon skeleton is built up chemically from smaller, readily available chiral molecules. Sometimes the exquisite specificity and catalytic efficiency of enzymes, such as aldolase, are used to join two small molecules to establish stereocenters early in the synthetic route [12]. An impressive list of iminosugars, their inhibitory properties, and key references has been compiled [4]. For reference to the early work on the discovery, isolation, and synthesis of iminosugars please look at cited reviews [4–11].

2. CLASSES OF IMINOSUGARS

2.1 Piperidines (Fig. 1)

The archetypal iminosugar and the first to be described was nojirimycin (5-amino-5-deoxy-D-glucopyranose) (**1**) (Fig. 1). It was first isolated as an antibiotic from *Streptomyces roseochromogenes* but was subsequently isolated from *Streptomyces nojiriensis*, from which the generic name, "nojirimycin," derives. It is a potent inhibitor of α- and β-glucosidases. The corresponding polyhydroxylated piperidine analogs of mannose and galactose, nojirimycin B (**2**) and β-galactostatin (**3**), have also been isolated from *Streptomyces* and are inhibitors of α-mannosidase and β-galactosidase, respectively. The chemical removal of the anomeric hydroxyl group from nojirimycin to form 1-deoxynojirimycin (**4**) does not destroy the inhibitory properties. 1-Deoxynojirimycin, which had been synthesized previously, has been isolated from bacterial cultures and several plant species. It is more stable than nojirimycin and is the model compound for research on iminosugars and has given rise to the trivial nomenclature (e.g. DNJ or 1-deoxynojirimycin; DMJ or 1-deoxymannonojirimycin (**5**); DGJ or 1-deoxygalactonojirimycin (**6**); DFJ or deoxyfuconojirimycin (**7**)). These iminosugars can be described systematically as derivatives of the parent heterocyclic compound or sugar (e.g. DNJ is 2*S*-hydroxymethyl-3*R*,4*R*,5*S*-trihydroxypiperidine or 1,5 dideoxy-1-5-imino-D-glucitol). They all inhibit the corresponding hexosidase.

The first iminosugar to be isolated from plants was fagomine (**8**) (2-hydroxymethyl-3–4,dihydroxypiperidine or 1,2-dideoxynojirimycin), which is inhibitory towards several α- and β-glucosidases and α- and β-galactosidases, showing that only a few hydroxyl groups in the correct configuration may be necessary for inhibition. Derivatives of DNJ with substituents at C-1 have also been isolated (e.g. α-homonojirimycin (**9**) and even β1-*C*-butyl-1-deoxygalactonojirimycin). Glycosides of iminosugars such as the 7-*O*-β-D-glucoside of α-homonojirimycin, which had been synthesized chemically as a drug for diabetes, and the 4-*O*-β-D-glucoside of fagomine have also been isolated from plants. The function of these glycosylated derivatives is unknown.

2.2 Pyrrolidines (Fig. 2)

Polyhydroxylated derivatives of pyrrolidine resemble sugars in the furanose configuration (Fig. 2). The first iminofuranose to be isolated was 2,5-dideoxy-2,5-imno-D-mannitol (DMDP) (**10**), which mimics β-D-fructofuranose and is a potent inhibitor of invertase but does not inhibit the corresponding mammalian enzyme

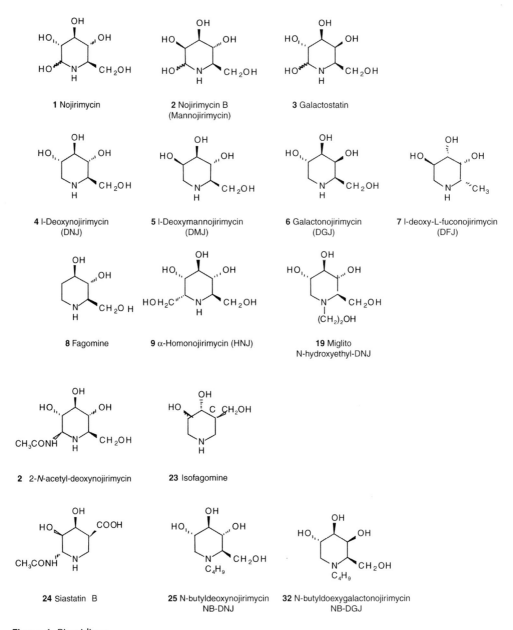

Figure 1. Piperidines.

sucrase. Removal of a hydroxymethyl group from DMDP produces 1,4-dideoxy-1,4-imino-D-arabinitol (DAB-1) (**11**), the 1-deoxy analog of the pentose arabinose, which has been isolated from many plants and is an inhibitor of both the pentosidase α-L-arabinosidase and the hexosidase yeast α-glucosidase. The isomer of DAB (e.g. 1,4-dideoxy-1,4 imno-D-ribitol (**12**)), O-glucosides and derivatives of DAB with substitution of the ring nitrogen, 6-C-alkylated derivatives of DMDP (or 1-C-alkylated derivatives of DAB) (e.g. homoDMDP (13)) have also been isolated. Polyhydroxylated pyrrolidines corresponding to hexoses in the furanose configuration have been synthesized (e.g. 1,4-dideoxy1,4-imino-D-mannitol (DIM) (14)). In general, iminofuranoses can inhibit the corresponding hexosidase that catalyzes the hydrolysis of glycosidic linkages involving the sugar in the pyranose configuration. However, mammalian α-D-mannosidases can be divided into two classes on the basis of their susceptibility to inhibition by the piperidine iminosugar: DMJ (5) or the pyrrolidine iminosugar DIM (14), reflecting differences in catalytic mechanisms [13].

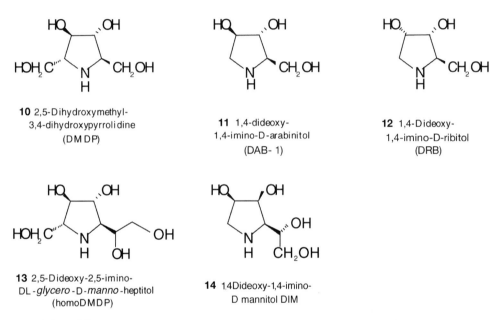

Figure 2. Pyrrolidines.

2.3 Indolizidines (Fig. 3)

The discovery that ingestion of the polyhydroxylated indolizidine swainsonine (15) was the cause of swainsona toxicosis in grazing animals in Australia was a great impetus to research on iminosugars [14]. Swainsonine is a potent inhibitor of lysosomal α-D-mannosidase and causes a phenocopy of the lysosomal storage disease α-mannosidosis, which results from a genetic deficiency of lysosomal α-D-mannosidase. The bicyclic indolizidine can be considered as a fused piperidine and pyrrolidine and it is suggested that the stereochemistry of the hydroxyl groups in swainsonine mimics mannose in a furanose configuration [15]. The inhibitory specificity of swainsonine towards mammalian α-D-mannosidases more closely resembles that of the furanose analog DIM (12) than the pyranose analog

DMJ (5). Swainsonine also inhibits Golgi α-D-mannosidase II, thereby preventing the formation of complex asparagine-linked glycans. It has become a valuable reagent for studying the effect of loss of complex glycans on cellular processes.

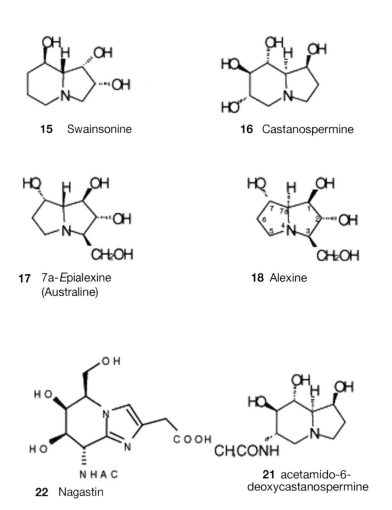

Figure 3. Indolizidines, pyrrolizidines, and nagstatin.

In contrast, castanospermine (16), a trihydroxyindolizidine extracted from another Australian plant, is a potent inhibitor of glucosidases because the stereochemistry of the hydroxyl groups resembles glucose in a pyranose configuration. Castanospermine also inhibits the processing of N-linked glycans and is another widely used reagent. Epimers and derivatives of swainsonine and castanospermine occur naturally and have been synthesized (see [12] and chapter 7 in [4]).

2.4 Pyrrolizidines (Fig. 3)

Pyrrolizidines consist of two fused pyrrolidines and several naturally occurring polyhydroxylated pyrrolizidines that inhibit glycosidase have been identified. Australine (**17**) can be regarded as a ring-contracted form of castanospermine with a strong structural resemblance to DMDP (**10**), which explains its inhibition of α-glucosidases [16]. Alexine (**18**), which is epimeric with australine at the bridgehead position, also inhibits various α-glucosidases [17]. Several epimers of australine have been extracted from *Castanospermum australe* and *Alexa* species.

2.5 *Nor*tropanes or calystegines

The calystegines are polyhydroxylated *nor*tropane alkaloids [18], which were first isolated from the roots of *Calystegia sepium* [19] but are common in potato tubers. They are classified according to the number of hydroxyl groups and most are potent inhibitors of glycosidases. Ingestion of calystegines may lead to inhibition of lysosomal glycosidase and lysosomal storage of incompletely digested glycoconjugates [18].

3. POTENTIAL APPLICATIONS

Glycosidases were the first therapeutic targets for iminosugars because they are involved in a wide variety of essential functions, including nutrition and the turnover and processing of glycoconjugates. In general, iminosugars inhibit exoglycosidases in a competitive and reversible manner with a predictable specificity based on the number and configuration of their hydroxyl groups. However, the effect of the ring nitrogen on the charge and conformation is also important. Subtle changes in specificity and efficiency of inhibition have been achieved by substitution of the lead natural compound. The structural basis and mechanism of action of inhibition of glycosidases by iminosugars have been reviewed extensively [20,21] (see also chapters 3, 8, and 9 in [4]). Endoglycosidases play an important role in the catabolism of polysaccharides for nutrition, defense, and recycling but simple iminosugars are not potent inhibitors of endoglycosidases. The synthesis of oligosaccharides incorporating iminosugars could be a useful strategy for inhibiting this class of enzymes.

Other potential targets are glycosyltransferases [9], the pattern of expression of which determines the repertoire of glycosylation in different cells in normal and disease states. Glycosyltransferases are also important in the biosynthesis of polysaccharides essential for the viability of infectious microorgansims, making them attractive targets for anti-infective drugs. Enzymes involved in the metabolism of nucleic acids, particularly the salvage pathways on which some mammalian cells and microorganism depend, are also targets. *In vivo*, an iminosugar may not act exclusively on one target enzyme. The overall metabolic effect will depend upon the accessibility of the iminosugar to the different susceptible enzymes.

3.1 Intestinal glycosidases, obesity, diabetes, and glycogenolysis

The first application of iminosugars and the cause of their discovery was their ability to inhibit intestinal disaccharidases and consequently absorption of sugars, thereby affecting the blood glucose level. Several α-glucosidase inhibitors (e.g. DNJ (4) and castanospermine (16)) can delay postprandial hyperglycemia, making them potential antidiabetic and obesity drugs. The synthetic derivative of DNJ, N-hydroxyethyl DNJ or Miglitol (19), is approved for use in type 2 diabetes. It decreases moderately postprandial blood glucose, insulin, and glycated hemoglobin levels but causes considerable gastrointestinal discomfort due to undigested polysasccharides [22]. It is contraindicated for patients with gastrointestinal obstruction or irritable bowel disease. It does not cause hyperinsulinemia and therefore could possibly be used in conjunction with insulin in type I diabetes. The pseudotetrasaccharide acarbose, which is not absorbed appreciably, has a similar action.

The blood glucose level can also be regulated by controlling glycogenolysis. DAB-1 (11) is another potential drug for type 2 diabetes because it can decrease glucagon-induced and spontaneous hyperglycemia by inhibition of hepatic glycogen phosphorylase [23], probably by allosteric inhibition of phosphorylase a [24]. Miglitol (19) also inhibits glycogenolysis and markedly decreases the infarct size in ischemia/reperfusion because of reduced myocardial apoptosis [25].

3.2 Inhibition of glycoprotein processing

Iminosugars can inhibit specific processing glycosidases in the endoplasmic reticulum and Golgi apparatus (Fig. 4) and alter the cellular repertoire of N-glycan structures on glycoproteins. This has been exploited extensively to investigate the function of specific glycoproteins and to alter cellular glycosylation for therapeutic purposes (chapter 11 in [4]). Less progress has been made developing inhibitors for processing glycosyltransferases. However, second generation 1-C derivatives of iminosugars, including glycosides, should provide the structural features for inhibiting these enzymes and glycosidases more specifically.

3.3 Immune response

The alteration in N-glycan structures induced by swainsonine (15) stimulates bonemarrow stem cell proliferation after injury but does not stimulate normal cells or hematopoietic tumor cells [26]. It also protects the cells against the cytotoxic effects of cyclophosphamide and the immunosuppressive effects of azido-3′-deoxythymidine (AZT or zidovudine) in vivo. Therefore swainsonine might be useful to stimulate bonemarrow after autologous bonemarrow transplantation or in high-dose chemotherapy or as an adjuvant in AZT treatment of acquired immune deficiency syndrome (AIDS). In contrast, the alteration of glycoprotein processing by castanospermine (16) appears to decrease cell–cell interactions in allorejection and may be useful in prolonging allografts [27]. Recently, several synthetic C-1 substituted piperidine iminosugars have been shown to inhibit the secretion of interleukin 4 (IL-4) and interferon gamma (IFN-γ) by mouse splenocytes and to alter the expression of the immune complexes,

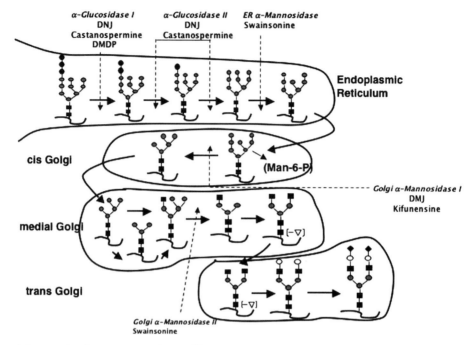

■ N-acetylglucosamine ● Mannose ○ Galactose ▽ Fucose ◆ Neuraminic acid

Figure 4. Inhibition of processing glycosidases by selected aminosugars.

CD3, CD4, CD8, and CD19 to different extents, suggesting cellular specificity [28]. The L-altro derivatives were the most effective and less cytotoxic than the recognized immunosuppressive drug CyA. The precise molecular basis of these effects is unknown but they hold promise for the development of selective immunosuppressive drugs.

3.4 Anticancer drugs (see also Chapter 14)

Many tumor cells display aberrant glycosylation owing to the altered expression of glycosyltransferases. In particular there is increased expression of β-1→6 branched complex *N*-glycans due to increased GlcNAc transferase V activity in many tumor cells. These glycans appear to be required for tumor cell metastasis *in vivo* and their expression correlates well with disease severity. Therefore, prevention of the formation of these glycans by inhibition of the processing glycosidases by iminosugars is a possible therapeutic strategy [29]. Both castanospermine (16) and swainsonine (15), which inhibit α-glucosidase I and α-mannosidase II, respectively, show anticancer activity *in vitro*. Drug development has focused on swainsonine and other α-mannosidase II inhibitors because they allow the formation of hybrid glycans, which are necessary and may be able to substitute for complex glycans in essential cellular processes [30]. Studies in animals showed that swainsonine inhibited metastasis and the growth of tumor cells and had a direct effect on the immune system (see above).

After toxicity and pharmacokinetic studies in animals, a trial of swainsonine in human cancer patients was performed [31]. Several adverse events occurred that were considered to be due to the drug, including neurological symptoms, which may be related to lysosomal accumulation of oligosaccharides in neurons resulting from inhibition of the lysosomal α-mannosidase [14]. The immune system was depressed rather than stimulated. A lower dosing regimen may be needed to avoid some of the side-effects due to the wide biological activity of swainsonine *in vivo*.

Degradation of the extracellular matrix by secreted glycosidases is important in tumor cell invasion and migration [32]. Inhibition of these glycosidases by iminosugars is another possible anticancer strategy [33]. The synthetic "2-*N*-acetyl" derivatives of DNJ (2-acetamido1,2,dideoxynojirimycin) (**20**) [34] and castanospermine (6-acetamido-6-dexoycastanospermine) (**21**) [35] are potent inhibitors of β-*N*-acetylhexosaminidase. A naturally occurring inhibitor of *N*-acetyl-β-D-hexosaminidase is nagstatin (**22**) (Fig. 3), which contains an aromatic ring linking the ring nitrogen to the "anomeric" center of the piperidine [36]. Several analogs have been synthesized with imidazoles, triazoles, or tetrazoles fused to either a pyrrolidine or a piperidine.

Piperidines with nitrogen at the "anomeric" position rather than substituting the ring oxygen are called isoiminosugars or 1-*N*-iminosugars (chapter 6 in [4]) (e.g. the synthetic isofagomine (**23**) (Fig. 1) [37], which is a very potent inhibitor of β-glucosidase). They are better inhibitors of β-glycosidases than α-glycosidase in contrast to the 1-deoxyiminosugars The 1-*N*-iminosugars with an amino group adjacent to the ring nitrogen are called *gem*-diamine1-*N*-iminosugars [38] (e.g. the natural product, siastatin B (**24**) (Fig. 1) [39]). The protonated form of these iminosugars probably mimics the glycopyranosyl cation intermediate in the hydrolysis of glycosides. They resemble natural 2-amino-2 deoxysugars and siastatin B (**24**) is a potent inhibitor of *N*-acetyl-β-D-hexosaminidase. It also has a strong structural resemblance to neuraminic and glucuronic acids and inhibits neuraminidase and β-D-glucuronidase. Synthetic *gem*-diamine1-*N*-iminosugars related to D-glucuronic acid and L-iduronic acid have been shown to inhibit glycosaminoglycan degradation, which is necessary for tumor metastasis and to inhibit tumor sialyltransferases, thereby altering the sialylation of tumor membranes. Preclinical studies in mice and rats have shown these compounds to be promising antitumor metastasis drugs [38].

3.5 Antiviral drugs

The correct glycosylation of viral proteins is necessary for their folding or interaction with receptors [40]. Therefore alteration of the glycosylation of the host cell by iminosugars may abrogate or, perversely, enhance [41] the infectivity of a virus. Inhibitors of the processing α-glucosidase I, such as castanospermine (**16**) or DNJ (**4**) decrease the infectivity of several enveloped viruses *in vitro* and *in vivo* by preventing the correct folding of their envelope glycoproteins (see [42] for references of examples). *N*-Acylated derivatives of DNJ and 6-*O*-acylated derivatives of castanospermine (pro-drugs) are more potent inhibitors than the natural products, because of increased uptake into cells due to the lipophilic substituents. However, the plasma levels of N-butyl DNJ (**25**) achieved in a trial to treat patients with human immunodeficiency virus (HIV) infection were too low to reproduce the antiviral effect observed *in vitro* [43]. Diarrhea due to inhibition of intestinal disaccharidases was also a major complication.

A series of synthetic N-alkylated (C4–C18) DNJ derivatives have been evaluated for antiviral activity and cytotoxicity. Those with side-chains up to C9 were better antiviral agents but longer chain compounds were cytotoxic due to membrane disruption [44]. The N-methoxy-nonyl-DNJ and N-butyl-cyclohexyl-DNJ are even more potent inhibitors of hepatitis B virus and less cytotoxic [45]. Surprisingly the n-(n-nonyl) derivative of deoxygalactonojirimycin (n,n-DGJ), which does not inhibit processing α-glucosidase I, also inhibits hepatitis B virus replication by depletion of the nucleocapsid [46]. Although the iminosugars are potentially very useful antiviral agents, particularly in combination therapy, their mechanism of action is not fully understood.

3.6 Other anti-infective drugs (Fig. 5)

Specific inhibition of an enzyme, which is expressed in an infecting organism but not in the host, is an ideal drug target. Such an enzyme is the elongating α-D-mannosylphosphate transferase of the *Leishmania* parasite. This enzyme is involved in the synthesis of the cell surface phosphoglycans that are essential for the survival and infectivity of the parasite and is not expressed in mammalian hosts. Novel iminosugars (1-oxabicyclic β-lactams) (**26**) (Fig. 5) based on the transition state of the enzyme have been synthesized and shown to be potent inhibitors of the enzyme [47]. They could be the basis of a new type of anti-*Leishmania* drugs. A similar strategy has led to the development of potential iminosugar drugs for combating other infectious microorganisms such as the mycobacteria that cause leprosy and tuberculosis. Pyrrolidine analogs of D-galactofuranose (**27**, **28**) [48] and L-galactofuranose (**29**, **30**) [49] can inhibit *Mycobacterium* galactan biosynthesis and a peptidomimetic of UDP-Gal*f* incorporating agalactofuranose analog has been synthesized [50].

Attempts to find iminosugars with antifungal activity have been less successful. Although some C-2 substituted polyhydroxypyrrolidines do inhibit chitin synthase activity *in vitro* they do not have any antifungal activity towards several pathogenic fungi [51]. The natural fungal product pramanicin (**31**) (a highly lipophilic branched-chain L-arabinonic lactam) is active against *Candida* and has antibacterial activity towards *Bacillus subtilis* [52]. Siastatin B (**24**) (Fig. 1) inhibits mammalian, viral, and bacterial neuraminidases but some derivatives only inhibit the bacterial enzyme. The 3-epimer of siastatin B is a particularly good inhibitor of influenza virus neuraminidase [38].

3.7 Lysosomal storage diseases (see also Chapter 23)

Iminosugars were originally considered to cause lysosomal storage diseases by inhibition of lysosomal hydrolases *in situ* (e.g. swainsonine (**15**) and α-

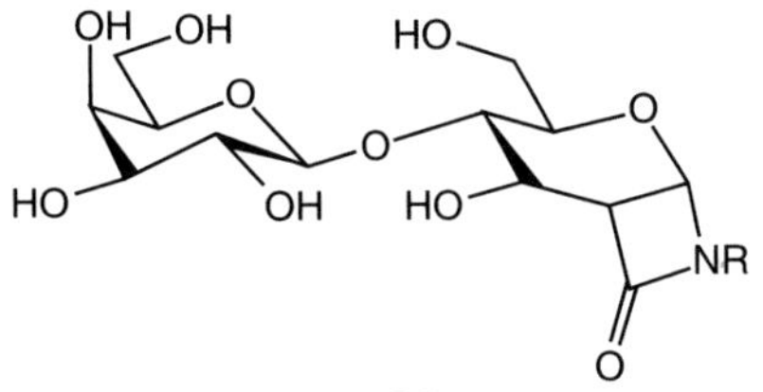

26 1-oxabicyclic β-lactams

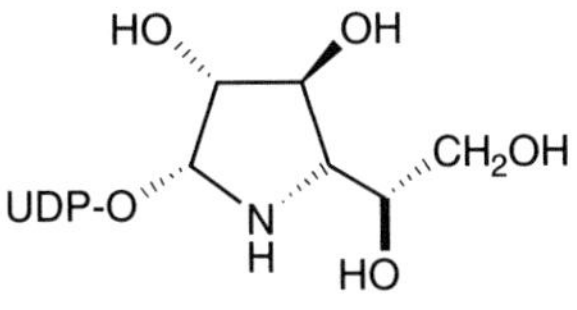

UDP-GAl*f*

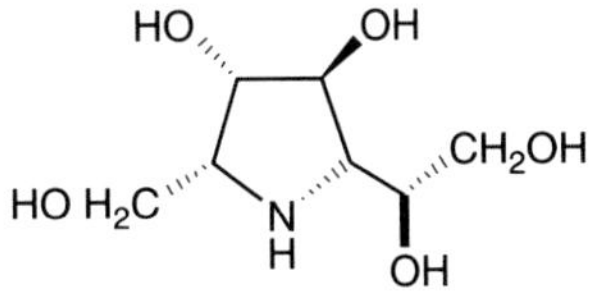

27

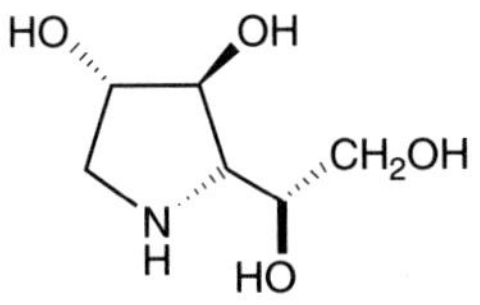

28

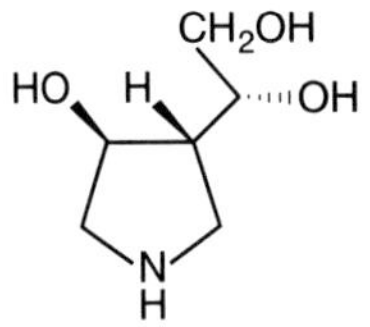

29 D-*galacto*-
furanose analog

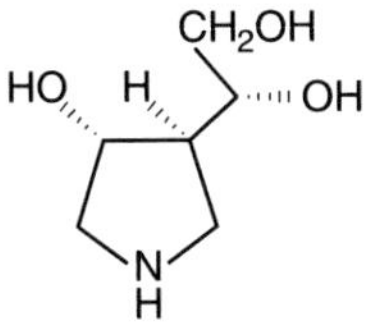

30 L-*galacto*-
furanose analog

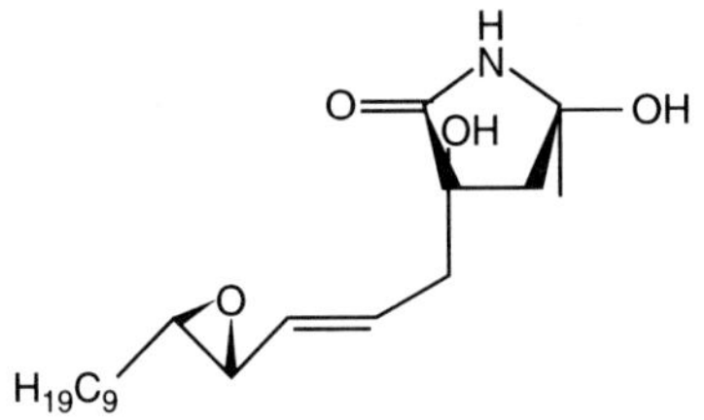

31 Pramanicin

Figure 5. Anti-infective drugs.

mannosidase (**14**)). Swainsonine was used to demonstrate that lysosomal storage was reversible by removal of the inhibitor, providing support for the idea of enzyme replacement therapy [53]. Paradoxically, iminosugars are now being exploited as therapeutic agents in two different ways [54]. N-Butyl-DNJ (**25**) (Fig. 1), which had been developed as an anti-HIV drug, was found to inhibit ceramide glucosyltransferase (or UDP-glucose:N-acylsphingosine-D-glucosyltransferase), the first step in the biosynthesis of most glycosphingolipids at much lower concentrations than required for the inhibition of the processing α-glucosidase. The inhibition was competitive with respect to ceramide and non-competitive with respect to UDP-glucose, indicating that N-butyl DNJ was acting as a mimic of ceramide. It slowed down the rate of synthesis of glycosphingolipids in cells in culture and in normal mice without detriment.

The glycosphingolipidoses are characterized by the lysosomal storage of glycosphingolipids resulting from a defect in their lysosomal catabolism. When N-butyl DNJ was fed to a mouse model of Tay–Sachs disease, accumulation of further lipid was prevented [55]. This suggested that N-butyl-DNJ could be a generic drug for substrate restriction (or deprivation) therapy in all the human storage diseases, in which glucosylceramides were stored [54]. If the patient had sufficient residual ceramide glucosyltransferase activity to cope with the decreased supply of substrate, further storage could be prevented and the accumulated material may even be dispersed. On the basis of an encouraging clinical trial [56], N-butyl-DNJ (also known as OGT 918, Zavesca or miglustat) was licensed for treatment of non-neuronopathic Gaucher disease type 1, in which there is residual activity. Patients showed significant improvement in some clinical features and the initial diarrhea and weight loss due to inhibition of intestinal α-glucosidases decreased with time [57,58]. N-Butyl-DNJ is also being used in trial for several other glycosphingolipidoses [59]. The galactose analog N-butyl-dexygalactonojirimycin (**32**) (Fig. 1) is a more selective inhibitor of ceramide glucosyltransferase because it does not inhibit ceramide galactosyltransferase, lysosomal glycogen catabolism, sucrase, or maltase but it does inhibit lactase [60].

Interestingly, these compounds may have potential as male contraceptives because male mice become reversibly sterile after oral administration of N-butyl-DNJ (**25**) or N-butyl-DGJ (**32**) [61]. The genetic integrity of the germ cells was unaffected, although epididymal spermatozoa were deformed and had lower motility.

As well as acting as an inhibitor, N-butyl-DNJ probably works as a chemical chaperone or stabilizer of the mutant forms of β-glucocerebrosidase in Gaucher type 1 patients [62]. The levels of β-glucocerebrosidase activity in cells expressing wild-type and mutants with missense mutations were increased 1.3–9.9 times when the cells were grown in 10 μM N-butyl-DNJ. In contrast, there was no activation of mutants with null alleles, including some missense mutations, suggesting that an intact active site is necessary for enhancement of activity. The concept of active-site specific chaperone therapy for lysosomal storage diseases using iminosugars to prevent misfolding and premature degradation of mutant enzymes in the endoplasmic reticulum was introduced by Fan [63]. He showed that subinhibitory concentrations of the potent α-galactosidase inhibitor DGJ (**6**) increased the residual activity in lymphoblasts from patients with Fabry disease with two different missense mutations by 7- to 8-fold over 5 days, whereas higher concentrations decreased the activity. The increase in activity was due to accelerated transport to the lysosome and a higher concentration of mature enzyme and not to an increase in mRNA.

Oral administration of DGJ (**6**) to mice expressing α-galactosidase with a missense mutation resulted in increased α-galactosidase activity in some tissues. Comparison of a series of derivatives of DGJ (**6**) indicated a correlation between potency as an inhibitor and effectiveness as a chaperone. A clinical trial of DGJ (**6**) in patients with Fabry disease is in progress. DGJ and similar compounds are water-soluble and can be taken orally. They can cross the blood–brain barrier, making them attractive drugs for lysosomal storage diseases with CNS involvement. Even infusions of galactose, the product of the enzymic reaction and a weak inhibitor of the enzyme led to clinical improvement in a Fabry patient with residual α-galactosidase activity [64]. DGJ (**6**) and *N*-butyl DGJ (**32**) also act as chaperones for mutant β-galactosidases [65].

Several other inhibitory iminosugars, possibly all, have the dual role of inhibitor and chaperone for the corresponding acidic lysosomal glycosidase. Nojirimycin derivatives bind tightly to the enzymes at neutral pH as in the ER, stabilizing the correct conformation and allowing even mutant enzymes to pass the quality control process. The affinity is much lower at the acidic pH of the lysosome and dissociation occurs. Therefore much higher concentrations are required to inhibit the enzyme at acidic pH, explaining the different properties of the iminosugars at different concentrations *in vivo*. Considerable effort is going in to the design of optimal inhibitors/chaperones [66].

4. DNA AND PURINE AND PYRIMIDINE METABOLISM

4.1 DNA polymerase

Eight pyrrolidine iminosugars were tested for their ability to inhibit DNA-metabolizing enzymes [67]. Only 1,4-dideoxy-1,4-imino-D-ribitol (DRB) (**12**) (Fig. 2) showed any appreciable activity. It selectively inhibited eukaryotic but not prokaryotic DNA polymerase and did not inhibit HIV-1 reverse transcriptase, T7 RNA polymerase, or bovine deoxyribonuclease. Kinetic studies and computer modeling suggest that paradoxically the inhibitor binds to the template-primer site rather than the dNTP substrate site. This illustrates the importance of a full understanding of the subtle structure–inhibition relationship for exploitation of these compounds as drugs.

4.2 *N*-Ribosyltransferases

The important and common biological process of transferring purines, pyrimidines, and other heterocyclic bases from ribose or deoxyribose is catalyzed by the *N*-ribosyltransferases. For example, phosphoribosyltransferases are involved in both the de novo and salvage synthetic pathways for nucleotides, nucleoside phosphorylases catalyze the formation of free bases from nucleosides, and DNA glycosylases participate in DNA repair. The *N*-ribosyltransferases are strongly inhibited by imino-C-nucleoside analogs that mimic the transition state of their catalytic mechanism [68]. The inhibition of specific *N*-ribosyltransferases could be useful in the treatment of disorders where nucleotide metabolism is involved in the pathogenesis. A deficiency of human purine nucleoside phosphorylase (PNP) causes a specific immunodeficiency in which activated T-lymphocytes undergo apoptosis. Therefore a drug-induced deficiency of PNP could be effective in treating diseases in which T-cell activation plays a role in the pathology, such as T-cell leukemias, some autoimmune diseases, and transplant rejection.

The iminoribitol-C-nucleoside Immucillin-H (**33**) (Fig. 6), is a potent inhibitor of PNP in the presence of 2′-deoxyguanosine and has been tested in a phase I clinical trial against T-cell leukemia [69]. The synthetic derivatives DADMe-immucillin-H (**34**) and its 9-deazaguanine analog DADMe-immucillin-G more closely resemble the transition state of human PNP and are potent inhibitors *in*

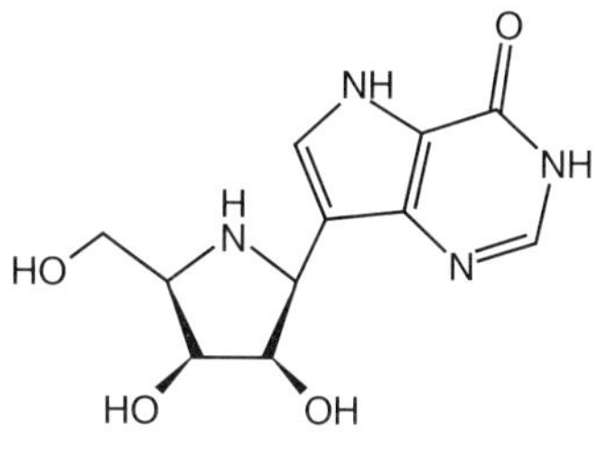

33. Immucillin H

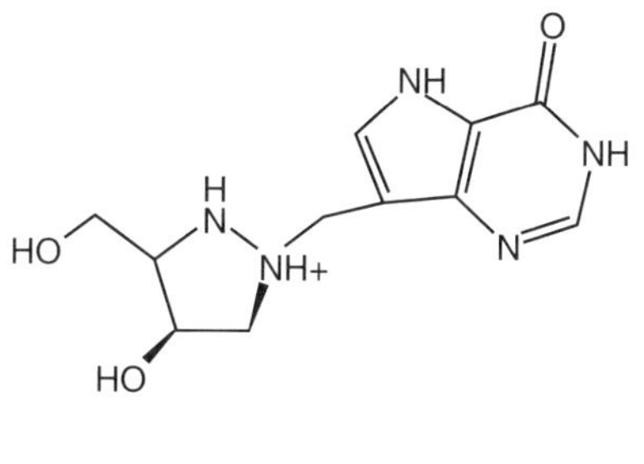

34. DADMe-Immucillin H

Figure 6. Immucillins.

vitro and *in vivo* [70]. Immucillins also inhibit PNP from the malaria parasite *Plasmodium falciparium*, which depends upon the purine salvage pathway for synthesis of RNA and DNA. The parasite is killed by purine starvation when it is cultured in human erythrocytes in the presence of immucillins, indicating their potential as antimalarial drugs [70].

5. CONCLUSION

Iminosugars and their derivatives have a wide therapeutic potential but most compounds do not act exclusively on a single target enzyme. This is illustrated by the diverse properties of *N*-butyl-DNJ (**25**) as an antiviral agent, male contraceptive, inducer of glycogen storage, chaperone, and inhibitor of glycolipid synthesis. Great effort is being made to synthesize or isolate more selective compounds for targeting to specific tissues, cells, and subcellular compartments. Iminosugars lend themselves to use in combination with other drugs because of their metabolic inertness and low toxicity. Other applications for iminosugars will undoubtedly emerge because only a small proportion of the enzymes and proteins involved in the recognition and transformation of carbohydrates have been targeted to date.

Acknowledgments

I would like to thank all my collaborators, who have made this research so interesting and such fun. Their work is especially cited in this review. I also thank the pharmaceutical companies that have supported work in our laboratory.

REFERENCES

1. Winchester B & Fleet GWJ (1992) *Glycobiology* **2**, 199–210.
2. Winchester B & Fleet GWJ (2000) *J. Carbohydr. Chem* **19**, 471–483.
3. Asano N (2003) *Glycobiology* **13**, 93R–104R.
4. Stütz AE (ed.) (1999) *Iminosugars as Glycosidase Inhibitors*. Wiley-VCH, Weinheim, Germany.
5. Fellows LE, Kite GC, Nash,RJ, Simmonds MSJ & Scofield AM (1992) In: *Nitrogen Metabolism of Plants*, pp. 271–284. Edited by K Mengel and DJ Pilbeam. Clarendon Press, Oxford, UK.
6. Watson AA, Fleet GWJ, Asano N, Molyneux RJ & Nash RJ (2001) *Phytochemistry* **56**, 265–295.
7. Asano N (2003) *Curr. Topics Med. Chem.* **3**, 471–484.
8. Fleet GWJ, Fellows LE & Winchester BG (1990) In: *Bioactive Compounds from Plants*, pp. 112–125. Ciba Foundation Symposium 154. Wiley, Chichester, UK.
9. Compain P & Martin OR (2003) *Curr. Top. Med. Chem.* **3**, 541–560.
10. Cipolla L, La Ferla B & Nicotra F (2003) *Curr. Top. Med. Chem.* **3**, 485–511.
11. Michael JP (2004) *Nat. Prod. Rep.* **21**, 625–649.
12. Romero A & Wong C-H (2000) *J. Org. Chem.* **65**, 8264–8268.
13. Winchester B, Al Daher S, Carpenter C et al. (1993) *Biochem. J.* **290**, 743–749.
14. Dorling PR, Huxtable CR, Vogel P (1978) *Neuropath. Appl. Neurobiol.* **4**, 285–295.

15. Cenci di Bello I, Fleet GWJ, Namgoong SK et al. (1989) *Biochem. J.* **259**, 855-861.
16. Molyneux RJ, Benson M, Wong RY et al. (1988) *J. Nat. Prod. Sci.* **51**, 1198-1206.
17. Nash RJ, Fellows LE, Dring JV et al. (1988) *Tetrahedron Lett.* **29**, 2487-2490.
18. Watson AA, Davies DR, Asano N et al. (2000) In: *Natural and Selected Synthetic Toxins Biological Implications*, pp. 129-139. Edited by AT Tu and W Gaffield. American Chemical Society, Washington DC, USA.
19. Tepfer D, Goldmann A, Pamboukdjian N et al. (1988) *J. Bacteriol.* **170**, 1153-1161.
20. Asano N, Kizu H, Oseki K et al. (1995) *J. Med. Chem.* **38**, 2349-2356.
21. Lillelund VH, Jensen HH, Liang X & Bols M (2002) *Chem. Rev.* **102**, 515-553.
22. Drent ML, Tollefsen AT, van Heudsen FH et al. (2002) *Diabetes Nutr. Metab.* **15**, 152-159.
23. Mackay P, Ynddal L, Andersen JV & McCormack JGM (2003) *Diabetes Obesity Metab.* **5**, 397-407.
24. Latsis T, Andersen B & Agius L (2002) *Biochem. J.* **368**, 309-316.
25. Wang N, Minatoguchi S, Chen X et al. (2004) *Br. J. Pharmacol.* **142**, 983-990.
26. Klein JL-D, Roberts JD, Kurtzberg J et al. (1999) *Br. J. Cancer* **80**, 87-95.
27. Grochowicz PM, Bowen KM, Hibberd AD et al. (1992) *Transplant. Proc.* **24**, 2295-2296.
28. Ye X-S, Sun F, Liu M et al. (2005) *J. Med. Chem.* **48**, 3688-3691.
29. Goss PE, Baker MA, Carver JP & Dennis JW (1995) *Clin. Cancer Res.* **1**, 935-944.
30. Fiaux H, Popowycz F, Favre S et al. (2005) *J. Med. Chem.* **48**, 4237-4246.
31. Goss PE, Reid CL, Bailey D & Dennis JW (1997) *Clin. Cancer Res.* **3**, 1077-1086.
32. Bernacki RJ, Niedbala M & Korytnyk W. (1985) *Cancer Metastasis Rev.* **4**, 81-102.
33. Olden K, Breton P, Grzegrorzewski K et al. (1991) *Pharmacol. Ther.* **50**, 285-290.
34. Fleet GWJ, Smith PW, Nash RJ et al. (1986) *Chem. Lett.* 1051-1054.
35. Liu PS, Kang MS & Sunkara PS (1991) *Tetrahedron Lett.* **32**, 719-720.
36. Aoyagi T, Suda H, Uotani K et al. (1992) *J. Antibiot.* **45**, 1404-1408.
37. Jespersen TM, Dong MR, Sierks T et al. (1994) *Angew. Chem.* **106**, 1858-1860.
38. Nishimura Y (2003) *Curr. Top. Med. Chem.* **3**, 575-591.
39. Umezawa H, Aoyagi T, Komiyama T, Komiyama T et al. (1974) *J. Antibiot.* **12**, 963-969.
40. Greimel, P, Spreitz J, Stutz AE & Wrodnigg TM (2003) *Curr. Top. Med. Chem.* **3**, 513-523.
41. Hazama K, Miyagawa S, Miyazawa T et al. (2003) *Biochem. Biophys. Res. Commun.* **310**, 327-333.
42. Whitby K, Pierson TC, Geiss B et al. (2005) *J. Virol.* **79**, 8698-8706.
43. Tierney M, Pottage J, Kessler H et al. (1995) *J. Acquir. Immun. Defic. Syndr. Hum. Retrovirol.* **10**, 549-553.
44. Mellor HR, Platt FM, Dwek RA & Butters TD (2003) *Biochem. J.* **374**, 307-314.
45. Mehta A, Ouzounov S, Jordan R et al. (2002) *Antivir. Chem. Chemother.* **13**, 299-304.
46. Lu X, Tran T, Simsek E & Block TM (2003) *J. Virol.* **77**, 11933-11940.
47. Ruhela D, Chatterjee P & Vishwakarma RA (2005) *Org. Biomol. Chem.* **3**, 1043-1048.
48. Lee RE, Smith MD, Nash RJ et al. (1997) *Tetrahedron Lett.* **38**, 6733-6736.
49. Cren S, Wilson C & Thomas NR (2005) *Org. Lett.* **7**, 3521-3523.
50. Lee RE, Smith MD, Pickering L & Fleet, GWJ (1999) *Tetrahedron Lett.* **40**, 8689-8692.
51. Gautier-Lefebvre I, Behr J-B, Guillerm G & Muzard M (2005) *Eur. J. Med. Chem.* **40**, 1255-1261.
52. Schwartz RE, Helms GL, Bolessa EA et al. (1994) *Tetrahedron* **50**, 1675-1686.
53. Cenci di Bello I, Dorling P & Winchester B (1983) *Biochem. J.* **215**, 693-696.
54. Butters T, Dwek RA & Platt D (2005) *Glycobiology* **15**, 43R-52R.
55. Platt FM, Nieses GR, Reinkensmeir G et al. (1997) *Science* **276**, 428-431.
56. Cox T, Lachmann R, Hollak C et al. (2000) *Lancet* **355**, 1481-1485.
57. Elstein D, Hollak C, Aerts JM et al. (2004) *J. Inherit. Metab. Dis.* **27**, 757-766.
58. Pastores GM, Barnett NL & Kolodny EH (2005) *Clin Ther.* **27**, 1215-1227.
59. Cox TM (2005) *Acta Paedtr.* **94** (Suppl 447), 69-75.
60. Platt F, Jeyakumar M, Andersson U et al. (2003) *Phil. Trans. R. Soc. Lond.* **358**, 947-954.
61. Sugunuma R, Walden CM, Butters TD et al. (2005) *Biol. Reprod.* **72**, 805-813.
62. Alfonso P, Pampin S, Estrada J et al. (2005) *Blood Cells Mol. Dis.* **35**, 268-276.
63. Fan J-Q, Ishii S, Asano N & Suzuki Y (1999) *Nat. Med.* **5**, 112-115.
64. Frustaci A, Chimenti C. Ricci R et al. (2001) *N. Engl. J. Med.* **345**, 25-32.
65. Tominaga L, Ogawa Y, Taniguchi M et al. (2001) *Brain Dev.* **5**, 284-287.

66. **Zhu X, Sheth KA & Li S** (2005) *Angew. Chem. Int. Ed.* **44**, 7450–7453.
67. **Mizushima Y, Xu X, Asano N** *et al.* (2003) *Biochem. Biophys. Res Commun.* **304**, 78–85.
68. **Schramm VL & Tyler PC** (2003) *Curr. Top. Med. Chem.* **3**, 525–540.
69. **Gandhi V, Kilpatrick JM, Plunkett W** *et al.* (2005) *Neoplasia* **106**, 4253–4260.
70. **Schramm VL** (2004) *Nucleosides Nucleotides Nucleic Acids* **23**, 1305–1311.

CHAPTER 26
Carbohydrate biosensors

Raz Jelinek and Sofiya Kolusheva

1. INTRODUCTION

Carbohydrates (also known as oligosaccharides or polysaccharides) constitute a large and diverse class of compounds present in varied materials and which have major roles in applications in chemistry, biology, materials science, and related fields. In the context of biological systems, in particular, carbohydrate research has emerged as the "new frontier" for elucidating fundamental biochemical processes and for identifying new pharmaceutical substances. Along with nucleic acids and proteins, carbohydrates appear to play a critical role in determining biological functions and affecting wide-ranging physiological processes, thus their study and characterization have become increasingly important.

This chapter aims to provide an overview of recent scientific activity pertaining to systems, methods, and devices designed to detect carbohydrates. We have not attempted to cover all aspects of carbohydrate chemistry and biology, carbohydrate detection methods, or issues concerning molecular processes involving carbohydrates; these aspects are broad and prolific fields of study, and the reader is referred to relevant literature [1]. We have also not discussed here the highly technologically and commercially important field of glucose sensing – an active area of research due to the profound health effects of aberrant glucose levels in diabetes – because glucose is technically a monosaccharide rather than a carbohydrate. Similarly, this review does not address the large body of commercially oriented literature (i.e. patents) related to polysaccharide biosensors. Overall, we tried to limit the scope of this chapter to more recent published reports which focus on novel and potentially significant systems or detection methods. Related reviews on the subject have appeared in the literature in the past [2].

2. DECIPHERING CARBOHYDRATE STRUCTURES

Elucidation of the organization and order of the monosaccharide units within oligosaccharides poses one of the most formidable analytical challenges in glycobiology. Varied approaches and generic techniques have been applied to

Glycobiology (C. Sansom and O. Markman, eds.)
© Scion Publishing Limited, 2007

facilitate accurate analysis of the individual monomers in complex carbohydrates [3]. Gel electrophoresis methodologies have been modified for extraction, separation, and analysis of bacterial cell surface (capsular) polysaccharides [4]. Enzymatic processing of carbohydrates and glycoconjugates has been frequently used for determination of carbohydrate structures and sequences because of the overall accuracy of the technique and the requirement for small sample quantities [5]. Recent studies have concentrated on the integration of advanced separation and detection methods for achieving fast and accurate oligosaccharide sequencing. Simultaneous detection by ultraviolet (UV) absorbance and electrospray ionization–mass spectrometry (ESI-MS), for example, provide important structural information on the oligosaccharide components of mixtures [6]. Enzymatic digestion and electrochemical detection of the enzymatic cleavage products have also been widely utilized for determination of oligosaccharide structures [7–9].

Detailed structural analysis of bacterial capsular carbohydrates has been achieved by "enzymatic fingerprinting" procedures combining high-performance anion-exchange/pulsed-amperometric detection liquid chromatography, fluorophore-assisted carbohydrate electrophoresis, and matrix-assisted laser desorption/ionization time-of-flight (MALDI-TOF) mass spectrometry (MS) [10]. This carbohydrate-profiling technique made possible rapid identification of plant cell wall mutants, and was proposed as a viable alternative for more cumbersome genetic or biochemical phenotyping methods [10]. Specifically, Lerouxel et al. explored the advantages and disadvantages of application of the bioanalytical techniques for capsular oligosaccharide analysis, particularly in terms of speed, reliability, and accuracy. The researchers asserted that MALDI-TOF MS offers an efficient and rapid method for carbohydrate analysis [10]. This claim could be somewhat problematic due to the fact that prior knowledge of specific carbohydrate components is necessary for the correct interpretation of MALDI-TOF MS. On the other hand, the technique could indeed serve as an excellent tool for initial fast analysis of cell wall carbohydrates. Combining MALDI-TOF MS with other separation and detection methods and the construction and use of relevant databases could make enzymatic fingerprinting a powerful tool for analysis and sequencing of complex carbohydrates.

A generic and elegant methodology for carbohydrate biosensor design has been the construction of neoglycolipids. These new molecular composites, based on coupling of oligosaccharides to lipid residues, constitute a chemical-synthesis route for deciphering carbohydrate sequences and structures. The attachment of hydrophobic lipid moieties to carbohydrates opens the way for applications of versatile immobilization methods [11,12]. There are several important advantages of the neoglycolipid approach for biosensor purposes. First, neoglycolipids contain preselected single lipid residues rather than the heterogeneity of acyl chains encountered in natural glycolipids – which often adds to the complexity of analysis of saccharide derivatives from natural sources. Another inherent strength of neoglycolipid-based assays is the selective reactivity of different carbohydrates in heterogeneous mixtures following their chemical derivatization, facilitating their separation through varied analytical means. In addition, surface display of carbohydrates immobilized through their lipid chains is well suited to probing directly the biological roles of oligosaccharide sequences as antigens, ligands, or other recognition elements, thus providing valuable information on the "glycome" – the entire spectrum of glycans produced by the cell. Furthermore, neoglycolipids are particularly adaptable for modern micro-array applications for high-throughput evaluation of the specificities of oligosaccharide-recognizing proteins (see below).

In an extension of the original neoglycolipid concept, chemical derivatization techniques utilizing fluorescent glycoconjugates were developed to decipher carbohydrate components in complex mixtures, particularly focusing on ligand discovery within varied mixtures of neutral and acidic oligosaccharides [13]. The important advantage of this approach is that it adds to the detection capabilities for employing neoglycolipids, which by themselves do not contain chromophores other than the saccharides. Further strength of the technique is the analysis of carbohydrates through fluorescence emitted directly from the saccharide-coupled fluorophore (rather than indirect detection of fluorescent substances that bind to the neoglycolipid). A recent report described conjugation of an aminolipid 1,2-dihexadecyl-sn-glycero-3-phosphoethanolamine (DHPE) and the fluorescent label anthracene. This reagent is highly fluorescent and can form neoglycolipids by reaction with diverse oligosaccharides through reductive amination. Such conjugates can be resolved by thin-layer chromatography (TLC) and high-performance liquid chromatography (HPLC), and quantified either spectroscopically or through scanning densitometry.

3. LECTIN-BASED BIOSENSORS

The lectins constitute a broad family of proteins involved in diverse biological processes, occasionally having potent toxic properties [14–16]. Lectins generally exhibit strong binding to specific carbohydrate moieties (glycans) and this property has been extensively exploited as a basis for biosensor design. Furthermore, particular structural profiles of glycans and their recognition by lectins have been attributed to disease progression, making analysis of saccharide–lectin binding processes important as a diagnostic tool [17]. Glucose biosensor designs, for example, have frequently utilized the specificity and high affinity of different lectins to this monosaccharide. Varied detection methods based on lectin–glucose recognition have been reported in the literature, including electrochemical detection of the monosaccharides via immobilization of lectins on electrode surfaces [18], and glucose-sensing based on the competitive reversible binding of a mobile fluorophore-labeled lectin concanavalin A (con A) to immobile pendant glucose moieties within Sephadex beads [19]. Lectins also exhibit high potential in peripheral biotechnology industries, such as food safety; where their unique recognition properties are finding promising applications in detecting microorganisms and carbohydrate additives in foods. Reported data suggest that the use of certain lectins may provide a simple and rapid alternative to traditional methods of bacterial analysis and screening [20].

4. GLYCOPROTEIN AND GLYCOSYLATION BIOSENSORS

Glycoproteins and protein glycosylation (see also Chapters 27 and 28) have attained prominence in recent years as key constituents in varied cellular processes [21]. The exact roles of the carbohydrate moieties in such molecules, however, have not been determined yet. Protein-bound saccharides were suggested to contribute to non-primary functions of proteins, such as non-specific interactions with other carbohydrates or macromolecules, stabilization of protein conformations, or protection from proteolysis. Non-specificity of the expressed saccharides is consistent with both the similarity of carbohydrate structures appearing within diverse glycoproteins and the frequent structural micro-heterogeneity of carbohydrate chains at given sites [21]. This concept is further supported in its overall outline by the viability of cells whose glycosylation processes have been globally disrupted by pharmacological inhibitors [21,22]. Other studies, on the other hand, have revealed the existence of specific receptors for discrete oligosaccharide structures on glycoproteins. Such receptors seem to be either important for compartmentalization of the glycoprotein or for positioning of the cells on which the glycoproteins are located [21,23]. *N*-Linked glycans are believed to play pivotal roles in targeting, transport, and compartmentalization of glycoproteins in cells [24]. Oligosaccharides were also proposed as antigenic determinants of glycoproteins [25].

Diverse glycosylation processes occur on cell surfaces, and elucidating cellular carbohydrate expression and glycosylation pathways is essential for understanding varied cellular events [26,27]. Elegant biochemical techniques were developed for probing oligosaccharide compositions and carbohydrate processes at cell surfaces. Bertozzi and others have expanded upon the concept of "chemical glycobiology" as a generic approach for deciphering biochemical processes in which carbohydrates constitute central components, and for studying structure–function relationships involving surface-expressed oligosaccharides [27,28]. The approach, which was also called "metabolic oligosaccharide engineering," involves chemical modification of specific saccharide units. These unnatural carbohydrates could then be incorporated into various cell compartments and locations via the biosynthetic machinery of the cell [28]. In particular, it was shown that interference with biochemical and metabolic pathways contributing to oligosaccharide biosynthesis could shed light on the progression and significance of such processes [27,29].

5. PATHOGEN DETECTION ASSAYS

Development of biosensors and rapid detection kits for microorganisms such as *E. coli*, *Salmonella typhimurium*, and others are highly desirable due to the adverse and often devastating health effects of pathogen infection [30]. In recent years diverse techniques have been introduced aiming to detect pathogens in shorter times and with maximal potential sensitivity [31]. Varied techniques for pathogen detection are based on the use of antibodies specific to enzymes or other proteins expressed by the microorganism to be examined [32]. Such methods, however, often require prior knowledge of the identity of the pathogenic species to be analyzed. The search for rapid, low-cost diagnostic pathogen techniques has also focused on the use of oligosaccharides, which constitute primary molecular components and markers on pathogen surfaces. The diversity and broad knowledge-base regarding surface-displayed carbohydrates could aid the design of diagnostic tests for specific bacteria. Rapid agglutination assays have been routinely used for detection of microorganisms through binding of their surface carbohydrates to varied external substances, such as antibodies and receptors. The latex agglutination test (LAT), for example, utilizes latex beads coated with polyclonal antibodies against the capsular carbohydrate of particular bacteria. Aggregation of the beads can be observed via the solution turbidity, indicating the presence of bacteria. The technique facilitated, for example, identification of *Mycoplasma* spp. in an early development stage within farm animals [33].

Varied techniques have been developed to facilitate rapid detection of pathogen-displayed carbohydrates that could also be applied in field conditions at high sensitivity. A fluorescence polarization assay (FPA) was successfully applied for serological diagnosis of brucellosis in cattle and other farm animals through antibody binding of the capsular carbohydrate epitopes of several *Brucella* strains [34]. The FPA technology is based on the rotational differences between a solubilized fluorescent-labeled free antigen and the antigen molecule bound to its antibody. In principle, a small molecule will rotate randomly at a rapid rate, resulting in fast depolarization of light, while a larger complex would depolarize light at a reduced rate due to the slower reorientation in water.

Cholera toxin (CT) is the universal marker and binding ligand of the cholera-inducing pathogen [35,36]. The cell surface ligand of CT is the ganglioside GM1 (Fig. 1) and many methods for detection of CT have been based on the multivalent

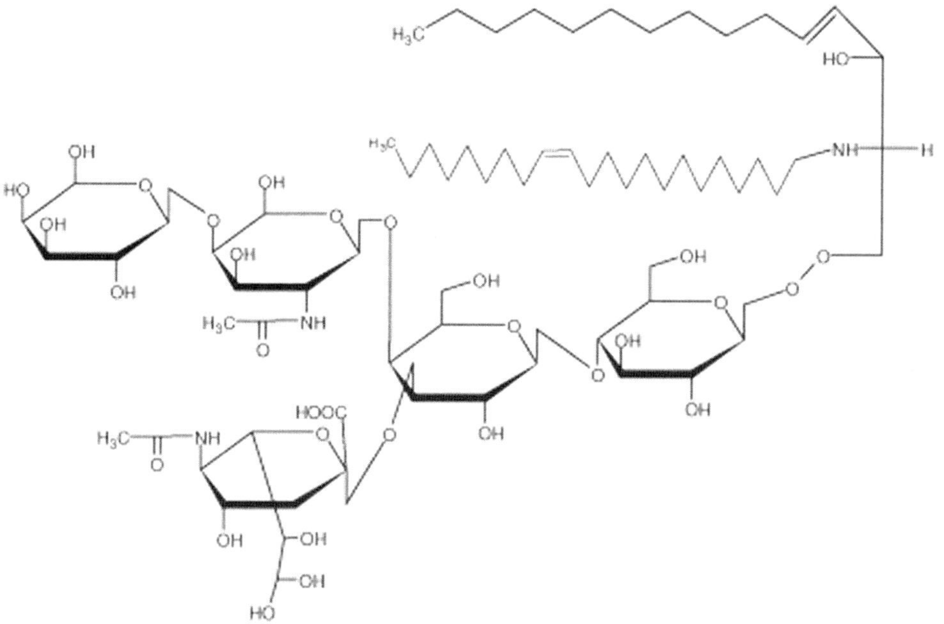

Figure 1. Ganglioside GM1 structure.

binding between the toxin and GM1. Indeed, the strong binding and recognition specificity of this ligand–receptor pair have made the use of this system particularly attractive both as a basis for actual biosensor design and also for demonstrating proof-of-concept for putative biological and pathogen detection schemes. Several representative reports are described herein. Cooper et al. developed a surface plasmon resonance (SPR) sensor chip to which ganglioside-displaying vesicles were attached, facilitating binding of cholera toxin to the chip surface [37]. GM1 was reconstituted within model lipid bilayers in other vesicle-based assays [38]. Several studies presented sensor arrangements in which GM1 molecules were incorporated within phospholipid-covered microspheres, onto which specific binding of CT occurred [39].

Detection of CT using fluorescence resonant energy transfer (FRET, Fig. 2) as the generator of optical signal has been reported in several biosensor schemes [40,41]. One example was a flow cytometry assay based on glass beads coated with phospholipids, which served as the scaffold for the fluorescence-labeled GM1 units [40]. Binding of CT to the GM1-coupled donor and acceptor dyes modified the distance between the fluorophores, and consequently affected the

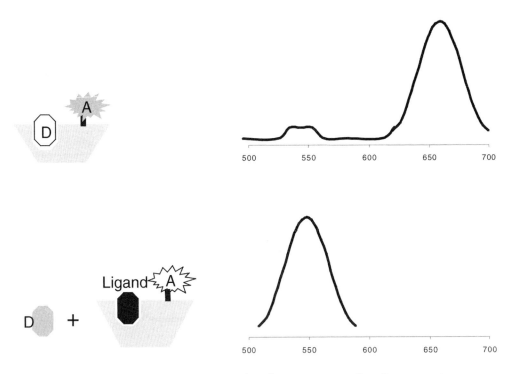

Figure 2. Fluorescence resonance energy transfer (FRET) experiment. Initially, a fluorescent donor molecule (**D**) is located within the binding site of the lectin, leading to a fluorescence energy transfer to an acceptor molecule (**A**) in close proximity, and the observation of an emission spectrum from the acceptor (top). Following binding of the ligand analyte (bottom), the donor molecule is released and fluorescence energy is no more transferred to the acceptor. Thus, the emission spectrum observed (at a lower wavelength) is from the donor.

fluorescence energy transfer. This biosensor arrangement achieved a high detection sensitivity of the toxin – less than 10 pM. Other studies employed the CT–GM1 pair as a model system for construction of biosensors based on FRET pathways [40–43]. Song et al. have presented several sophisticated detection schemes exemplified with the CT–GM1 system. An elegant experiment showed CT detection by FRET where a protein–carbohydrate binding event induced distance-dependent fluorescence self-quenching and/or resonant-energy transfer processes [43]. Another study focused on the design of a "two-tier FRET" biosensor, in which the excitation spectra of the donor and acceptor were sufficiently separated in order to minimize the background fluorescence signal due to indirect excitation of the acceptor fluorescence [41]. Energy transfer in that arrangement was achieved through an intermediate fluorophore, also covalently bound to GM1. These reports point to the feasibility of very high detection sensitivities, specificities, and reliability when advanced fluorescence techniques are employed within an integrated detection system consisting of an appropriate biological recognition system.

Original pathogen colorimetric sensors that respond to molecular recognition phenomena through the occurrence of rapid color transitions have been reported recently [44,45]. Several laboratories have demonstrated that artificial cell membranes made from conjugated lipid polymers (polydiacetylene, or PDA) can, on a simple level, mimic membrane surfaces, allowing both the occurrence and consequent detection of molecular recognition processes (Fig. 3) [44,46–52]. Specifically, the ene-yne conjugated backbones of several polydiacetylene species absorb light at the visible region of the electromagnetic spectrum, thus exhibiting visible colors (in most cases appearing intense blue). Furthermore, it was shown that external perturbations to the polymer induce structural transformations within the conjugated backbone of PDAs, giving rise to dramatic colorimetric transitions (blue–red). In a biological context, it was demonstrated that the blue–red transitions of PDA can be induced by ligand–receptor interactions occurring between soluble molecules and ligands embedded within the PDA matrix. The display of the ligands could be either achieved through covalent binding at the PDA headgroup region (Fig. 3a) [44–46], or through physical incorporation of the recognition element within lipid domains assembled in the PDA framework (Fig. 3b) [49–52]. In PDA-based biosensors, the conjugated polymer backbone essentially acts as a built-in reporter of binding events, measurable by a chromatic change in the visible absorption spectrum. Such assemblies may provide a general approach for direct assays and biosensing devices for varied biological substances and biomolecular recognition events.

Some PDA-based biosensor applications reported on the covalent attachment of the ganglioside GM1 within polydiacetylene liposomes. In this arrangement, specific interactions between GM1 and CT at the interface of the liposomes resulted in a change of the vesicle color (from blue to red) due to conformational changes in the conjugated (ene-yne) polymer backbone induced by the molecular binding [45]. Such "chromatic liposomes" might be used as simple colorimetric sensors for screening of recognition processes involving carbohydrates and other biomolecules. A similar PDA-based colorimetric sensor was constructed in a

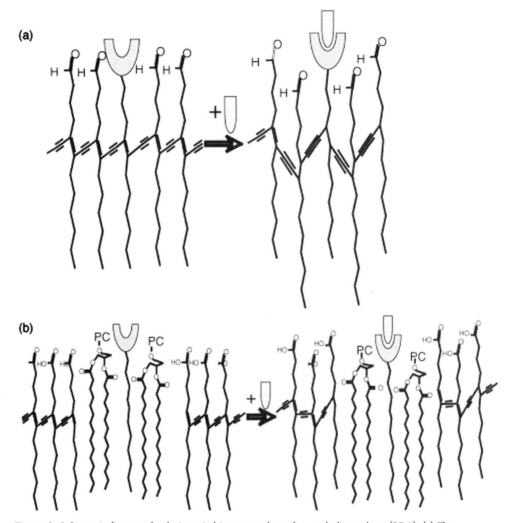

Figure 3. Schematic figures of colorimetric biosensors based on polydiacetylene (PDA). (a) The recognition element is covalently attached at the surface of the PDA framework (which appears blue to the eye). Interaction with the ligand induces a structural transition within the conjugated ene-yne polymer backbone, changing the conjugation length within the polymer network, with a consequent blue–red transition (see text). (b) The recognition molecule is physically incorporated within phospholipids (phosphocholine, PC) domains in the PDA matrix. Ligand–receptor interaction indirectly induces the structural transformation of the polymer (see text).

Langmuir-Blodget film format, rather than the vesicle assemblies discussed above [53]. The film assay exhibited the blue–red transformations induced by biomolecular recognition and by other lipid-perturbing processes occurring at membrane surfaces. Song et al. have similarly reported the incorporation of gangliosides or sialic acid moieties in thin films, which permitted the colorimetric detection of CT or influenza virus, respectively [53].

Carbohydrate-based pathogen biosensors increasingly rely on detection of lipopolysaccharide (LPS) moieties (also denoted endotoxins) on pathogen surfaces. LPS molecules, which consist of carbohydrates covalently attached to a lipid A moiety (Fig. 4) [54], are located on the outer cell surface of various pathogens [55]. LPS plays a major role in conferring resistance of Gram-negative bacteria toward toxic agents, most likely by participating in the formation of an effective permeability barrier at the outer membrane[56,57]. Varied biosensor assemblies have utilized biomolecular recognition between surface-expressed LPS and lectins or other proteins. Ertl et al. described electrochemical biosensor arrays which facilitated E. coli subspecies detection through con A–LPS interactions [58] or through LPS binding to other lectins [59]. In particular, the researchers examined

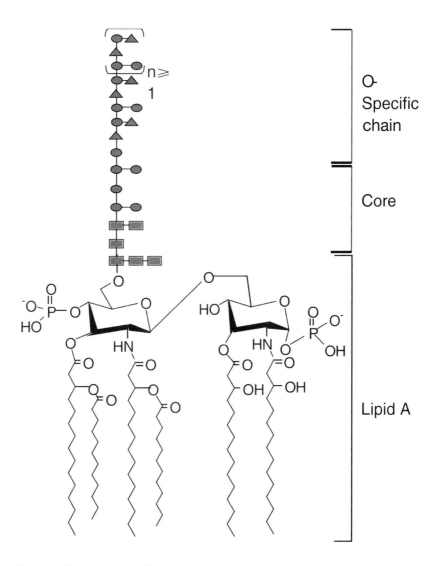

Figure 4. Lipopolysaccharide structure.

whether different lectins could *selectively* bind LPS moieties on surfaces of different bacteria. The construction of devices based on lectin recognition took advantage of the selective and reversible binding between the surface-immobilized lectins and the oligosaccharide groups. Electrochemical detection was facilitated through changes in the redox potential within the bacterial respiratory chain, following pathogen surface immobilization through the lectin–carbohydrate binding [58,59]. It should be emphasized that even though carbohydrate recognition served in these biosensors only as an *indirect* means for bacterial detection, the high affinity between LPS and the lectin could generally assure high sensitivity and fidelity of the sensor. Furthermore, the availability and diversity of known lectins and their carbohydrate ligands could facilitate the construction of sensor arrays for identification of several pathogens through their "signature response" in such arrangements.

6. CANCER DIAGNOSTICS

Modification of carbohydrate expression and glycosylation patterns on cellular surfaces is a common feature of cancer cells [60,61]. A majority of human carcinomas are associated with altered expression of oligosaccharides on membrane glycoproteins, for example in breast cancer [62], adenocarcinomas of the pancreas [63], cervical cancers [64], and others. There have been intensive efforts towards development of diagnostic techniques for tumor identification utilizing carbohydrate markers on cancer cells [61,65]. Dwek *et al.*, for example, have reported an immunohistochemical approach for early tumor detection [61] (see also Chapter 2). Monoclonal antibodies (mAbs) identifying altered glycosylation of specific glycoproteins associated with tumor appearance were used as a diagnostic tool [62,66].

Changes in the localization and relative abundance of carbohydrate species on cell surfaces can be monitored with the aid of specific carbohydrate-binding proteins, such as lectins. Lectin histochemistry has been utilized to identify modulation of the expression of sialic acid on human cervical carcinomas [64]. Plzak *et al.* employed biotinylated galactoside-binding (metal ion-independent) animal lectins (galectins) to detect domains of increased differentiation in human carcinoma tumors [67].

Sialylated Lewis antigens (SLeAs) and their enhanced cell surface expression are recognized markers for various malignancies and metastatic processes [68]. SLeAs (a representative antigen, sialyl Lewis^x, is shown schematically in Fig. 5) have been frequently used as molecular targets in immunohistochemical and serological cancer assays [69,70]. MAbs have been raised and tested against sialylated Lewis antigens with the goal of developing immunoassays for the detection and management of malignancies [71]. An electrochemical biosensor approach for determination of the tumor marker bound sialic acid (b-SIA) has been reported [72]. The sensor consisted of a copolymer-immobilized bilayer containing the enzyme sialidase, placed in contact with an H^+-selective poly(vinylchloride)–poly(vinylacetate) indicator membrane. The release of sialic acid – an α-ketocarboxylic acid with a pK_a of 2.6 – following enzymatic cleavage resulted in a local pH change monitored by the proton-sensitive indicator electrode. This electrochemical sensor was shown to be capable of differentiating between pathological and non-pathological levels of b-SIA within a relatively short detection time (3–5 min) and in reasonable accuracy. The cancer marker sialyl Lewis[x] antigen and its mimetic structures have been comprehensively characterized through analysis of their binding to selectin, a natural lectin [73].

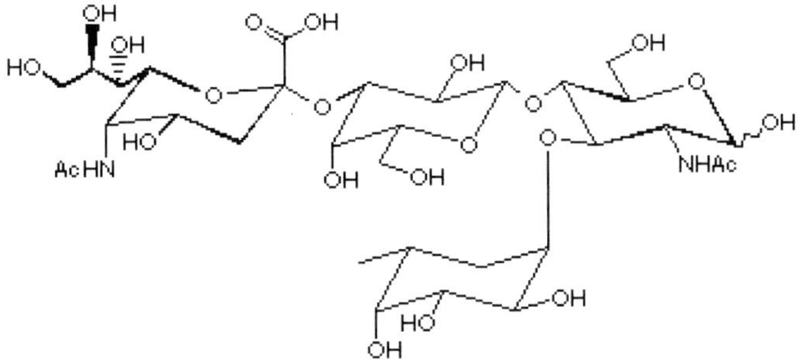

Figure 5. Sialyl Lewis[x] structure.

7. CARBOHYDRATE NANO-BIOSENSORS

The recent emergence of "nanotechnology" as a promising scientific and technological avenue has led to an expanding activity towards development of "nano-biosensors." Some studies have focused on the integration between carbohydrates and nanometer-size systems and devices, while other efforts have attempted to integrate advanced nanotechnology-oriented instrumentation within carbohydrate biosensors. You *et al.* described an amperometric biosensor facilitating high-sensitivity detection of sugar moieties through embedding of nickel nanoparticles within a graphite film electrode [74]. The authors reported that the dispersion of nickel nano-particles within the carbon film yielded an order-of-magnitude improvement in sugar detection limit compared to conventional electrode arrangements. The use of nano-particles was not directly related to the actual detection of the carbohydrate molecules, but rather as a way for improving technical performance of the electrode. Another study has employed nano-size amphiphilic C_{60}-dendrimers for achieving better interactions between the sensor surface and the carbohydrate analytes [75]. Binding was achieved through deposition of ordered Langmuir monolayers of the bucky-ball conjugated with glycodendron headgroups at the air–water interface. The films could be further transferred to solid quartz surfaces, pointing to their potential applications in biosensor design.

Atomic force microscopy (AFM) has been a major driving force in nanotechnology research and development. AFM is conceptually similar to the way old "long-play" records were read by the stylus of a phonograph, where the AFM tip acts like a "stylus" capable to image molecules and atoms on solid surfaces [76]. Among the most widespread applications of AFM in carbohydrate research has been the imaging of single carbohydrate molecules and surface characterization of oligosaccharide assemblies [74–76]. An example of a practical application of AFM is its use for the evaluation of the structure and texture of food carbohydrates [77].

8. BIOSENSORS UTILIZING PROTEIN–CARBOHYDRATE INTERACTIONS

Molecular recognition and interactions between carbohydrates and proteins play key roles in many biochemical processes. The participation of specific oligosaccharide sequences in protein targeting and folding, and in propagating infection and inflammation processes through interactions with receptors and antibodies, have become increasingly apparent [1]. Studying such interactions is also desirable for the development of therapeutic substances that would mimic or interfere with the recognition process. Various approaches have been introduced to probe carbohydrate–protein binding and to utilize such recognition events in the action mechanism of biosensors. However elucidation and understanding of the bioactive domains within oligosaccharides and their protein-binding properties pose distinct bioanalytical and chemical challenges.

From the standpoint of biosensor design, protein–carbohydrate binding has been employed as a platform for the extraction and analysis of various proteins. In most such applications the biosensor operation relies on the immobilization of carbohydrate species, which generally function as the recognition elements, followed by generation of measurable signals induced by association with their complementary macromolecules. Construction of carbohydrate-modified recognition surfaces is synthetically demanding because of the structural complexity of oligosaccharides. Distinct problems have been encountered due to the multiplicity of hydroxyl groups that might make specific binding difficult, as well as the requirement for appropriate linker systems to facilitate display and access to the immobilized oligosaccharides [78]. The two most common carbohydrate immobilization techniques employed in such sensors exploit the high affinity of the biotin–avidin pair [79] or the deposition of alkane thiolate monolayers on gold surfaces [80].

A recent development with potentially significant implications for glycobiology research in general and the study of carbohydrate–protein interactions in particular has been the fabrication of carbohydrate arrays (see Chapters 27 and 28) as a tool for rapid analysis of sugar-binding events and carbohydrate interactions. Examples of such applications include array carbohydrates that are first immobilized on pretreated surfaces, followed by addition of fluorescently labeled carbohydrate-binding proteins; binding occurrence can then be monitored by fluorescence spectroscopy [81]. The

9. CONCLUDING REMARKS

The increasing awareness of the biological importance of oligosaccharide derivatives and growing interest in glycobiology applications have clearly become a major driving force towards development of new techniques for carbohydrate characterization. This review summarized the large body of recent experimental work dedicated to the construction of biosensors and bioassays designed to detect and analyze carbohydrates and glycoconjugates, and sensors utilizing carbohydrates for the detection of other soluble biomolecules.

The complexity and high variability of carbohydrate structures have often placed formidable barriers towards their practical applications, however these properties might well open new avenues to biosensor applications specifically based on the differences among carbohydrate groups and their biological expression. Varied carbohydrate biosensor designs have been based on the molecular recognition and specific binding encountered between polysaccharides and other macromolecules, particularly proteins. Such molecular interactions, including carbohydrate-lectin, carbohydrate-toxin, or saccharide-enzyme affinities, play significant roles in diverse biosensor devices and bioassays, both those aiming to detect and/or analyze oligosaccharides, and those that rely on embedded carbohydrates for detection of other biomolecules.

REFERENCES

1. **Feizi T** (2001) *Glycoconj. J.* **17**, 553–565.
2. **Sturgeon RJ** (1983) *Carbohydr. Chem.* **14**, 5–20.
3. **Welply JK** (1989) *Trends Biotechnol.* **7**, 5–10.
4. **Pelkonen S, Hayrinen J & Finne J** (1988) *J. Bacteriol.* **170**, 2646–2653.
5. **Prime SB, Shipston NF & Merry TH** (1997) *BioMethods* **9**, 235–260.
6. **Thanawiroon C, Rice KG, Toida T & Linhardt RJ** (2004) *J. Biol. Chem.* **279**, 2608–2615.
7. **Ikeda T, Shibata T, Todoriki S, Senda M & Kinoshita H** (1990) *Anal. Chim. Acta* **230**, 75–82.
8. **Sun M & Lee CS** (1998) *Biotechnol. Bioeng.* **57**, 545–551.
9. **Weber PL, Kornfelt T, Klausen NK & Lunte SM** (1995) *Anal. Biochem.* **225**, 135–142.
10. **Lerouxel O, Choo TS, Seveno M et al.** (2002) *Plant Physiol.* **130**, 1754–1763.
11. **Feizi T** (2002) *R. Soc. Chem.* **275**, 186–193.
12. **Feizi T, Stoll MS, Yuen C-T, Chai W & Lawson AM** (1994) *Methods Enzymol.* **230**, 484–519.
13. **Stoll MS, Feizi T, Loveless RW, Chai W, Lawson AM & Yuen C-T** (2000) *Eur. J. Biochem.* **267**, 1795–1804.
14. **Lis H & Sharon N** (1998) *Chem. Rev.* **98**, 637–674.
15. **Vijayan, M & Chandra N** (1999) *Curr. Opin. Struct. Biol.* **9**, 707–714.
16. **Feizi T** (2000) *Immunol. Rev.* **173**, 79–88.
17. **Kim B, Cha GS & Meyerhoff ME** (1990) *Anal. Chem.* **62**, 2663–2668.
18. **Bartlett PN & Cooper JM** (1993) *J. Electroanal. Chem.* **362**, 1.

19. Ballerstadt R & Schultz JS (2000) *Anal. Chem.* **72**, 4185–4192.
20. Patel PD (1992) *Trends Food Sci. Technol.* **3**, 35–39.
21. West CM (1986) *Mol. Cell. Biochem.* **72**, 3–20.
22. Bucala R, Vlassara H & Cerami A. (1994) *Drug Dev. Res.* **32**, 77–89.
23. Sangadala S, Bhat UR & Mendicino J. (1992) *Mol. Cell. Biochem.* **118**, 75–90.
24. Helenius A & Aebi M. (2001) *Science* **291**, 2364–2369.
25. Feizi T & Childs RA. (1987) *Biochem. J.* **245**, 1–11.
26. Yarema KJ & Bertozzi CR. (2001) *Genome Biol.* **2**, 0004.1–0004.10.
27. Bertozzi CR & Kiessling LL. (2001) *Science* **291**, 2357–2364.
28. Dube DH & Bertozzi CR (2003) *Curr. Opin. Chem. Biol.* **7**, 616–625.
29. Charter NW, Mahal LK, Koshland DE Jr & Bertozzi CR. (2000) *Glycobiology* **10**, 1049–1056.
30. Tietjen M & Fung D.Y. (1995) *Crit. Rev. Microbiol.* **21**, 53–83.
31. Leonard P, Hearty S, Brennan J et al. (2003) *Enzyme Microb. Technol.* **32**, 3–13.
32. Kaspar CW, Hartman PA & Benson AK. (1987) *Appl. Environ. Microbiol.* **53**, 1073–1077.
33. March JB, Gammack C & Nicholas R. (2000) *J. Clin. Microbiol.* **38**, 4152–4159.
34. Nielsen K & Gall D. (2001) *J. Immunoassay Immunochem.* **22**, 183–201.
35. Herrington DA, Hall RH, Losonsky G, Mekalanos JJ, Taylor RK & Levine MM. (1988) *J. Exp. Med.* **168**, 1487–1492.
36. Lee YC & Lee RT. (1996) *J. Biomed. Sci.* **3**, 221–237.
37. Cooper MA, Hansson A, Lofas S & Williams DH. (2000) *Anal. Biochem.* **277**, 196–205.
38. Carmona-Ribeiro AM (2001) *Chem. Soc. Rev.* **30**, 241–247.
39. Sicchierolli SM & Carmona-Ribeiro AM. (1996) *J. Phys. Chem.* **100**, 16771–16775.
40. Song X, Shi J & Swanson B (2000) *Anal. Biochem.* **284**, 35–41.
41. Song X, Shi J, Nolan J & Swanson B (2001) *Anal. Biochem.* **291**, 133–141.
42. Song X, Nolan J & Swanson B (1998) *J. Am. Chem. Soc.* **120**, 11514–11515.
43. Song X & Swanson B (1999) *Anal. Chem.* **71**, 2097–2107.
44. Cheng Q, Song J & Stevens RC (2002) *Polymer Preprints* **43**, 128–129.
45. Pan JJ & Charych D (1997) *Langmuir* **13**, 1365–1367.
46. Charych DH, Spevak W, Nagy JO & Bednarski MD (1993) *Materials Res. Soc. Symp. Proc.* **292**, 153–161.
47. Lio A, Reichert A, Ahn DJ, Nagy JO, Salmeron M & Charych DH (1997) *Langmuir* **13**, 6524–6532.
48. Katz M, Tsubery H, Kolusheva S, Shames A, Fridkin M & Jelinek R (2003) *Biochem. J.* **375**, 405–413.
49. Kolusheva S, Kafri R, Katz M & Jelinek R (2001) *J. Am. Chem. Soc.* **123**, 417–422.
50. Kolusheva S, Wachtel E & Jelinek R (2003) *J. Lipid Res.* **44**, 65–71.
51. Rozner S, Kolusheva S, Cohen Z, Dowhan W, Eichler J & Jelinek R. (2003) *Anal. Biochem.* **319**, 96–104.
52. Jelinek R & Kolusheva S (2001) *Biotechnol. Adv.* **19**, 109–118.
53. Song J, Cheng Q, Zhu S & Stevens RC (2002) *Biomed. Microdevices* **4**, 213–221.
54. Hauschildt S, Brabetz W, Schromm AB *et al.* (2000) *Handbook of Experimental Pharmacology*, p. 145. Springer-Verlag, Berlin.
55. Nicaido H & Vaara M (1987) Outer membrane. In *Escherichia coli and Salmonella typhimurium*, pp. 7–22.
56. Wiese A, Brandenburg K, Ulmer AJ, Seydel U & Muller-Loennies S (1999) *Biol. Chem.* **380**, 767–784.
57. Salzer WL & McCall CE (1988) *Mol. Aspects Med.* **10**, 511–629.
58. Ertl P & Mikkelsen SR (2001) *Anal. Chem.* **73**, 4241–4248.
59. Ertl P, Wagner M, Corton E & Mikkelsen SR (2003) *Biosensors Bioelectronics* **18**, 907–916.
60. Feizi T (1988) *Adv. Exp. Med. Biol.* **228**, 317–329.
61. Dwek MV, Ross HA & Leathem AJC (2001) *Proteomics* **1**, 756–762.
62. Ohuchi N, Taeda Y, Yaegashi S, Harada Y, Kanda T & Mori S (1994) *Cancer Mol. Biol.* **1**, 179–192.
63. Ho JJ & Kim YS (1994) *Pancreas* **9**, 674–691.
64. Banerjee S, Robson P, Soutter WP & Foster CS (1995) *Hum. Pathol.* **26**, 1005–1013.

65. Lamerz R (1999) *Ann. Oncol.* **10**, 145-149.
66. Feizi T & Childs RA (1985) *Trends Biochem. Sci.* **10**, 24-29.
67. Plzak J, Smetana K Jr, Betka J, Kodet R, Kaltner H & Gabius H-J (2000) *Int. J. Mol. Med.* **5**, 369-372.
68. Kijima H, Kashiwagi H, Dowaki S et al. (2000) *Int. J. Oncol.* **17**, 55-60.
69. Nakagoe T, Fukushima K, Sawai T et al. (2002) *J. Exp. Clin. Cancer Res.* **21**, 363-369.
70. Davidson B, Berner A, Nesland JM et al. (2000) *Hum. Pathol.* **31**, 1081-1087.
71. Rye PD, Bovin NV, Vlasova EV et al. (1998) *Tumor Biol.* **19**, 390-420.
72. Aubeck R, Eppelsheim C, Braeuchle C & Hampp N (1993) *Analyst* **118**, 1389-1392.
73. Simanek EE, McGarvey GJ, Jablonowski JA & Wong C-H (1998) *Chem. Rev.* **98**, 833-862.
74. You T, Niwa O, Chen Z, Hayashi K, Tomita M & Hirono S (2003) *Anal. Chem.* **75**, 5191-5196.
75. Cardullo F, Diederich F, Echegoyen L et al. (1998) *Langmuir* **14**, 1955-1959.
76. Rief M, Oesterhelt F, Heymann B & Gaub HE (1997) *Science* **275**, 1295-1297.
77. Morris VJ, Kirby AR & Gunning AP (1999) *Scanning* **21**, 287-292.
78. Ohashi E & Karube I (1995) *Fisheries Sci.* **61**, 856-859.
79. Wilchek M & Bayer EA (1988) *Anal. Biochem.* **171**, 1-32.
80. Ulman A (1996) *Chem. Rev.* **96**, 1533-1544.
81. Love KR & Seeberger PH (2002) *Angew. Chem.* **41**, 3583-3586.

CHAPTER 27
Glycoanalysis on a lectin array: applications to the development of biopharmaceuticals and life science research

Ruth Ben-Yakar Maya, Yehudit Amor, Revital Rosenberg, Tamara Byk-Tennenbaum, Albena Samokovlisky, Roberto Olender, Haim Bangio and Rakefet Rosenfeld

1. INTRODUCTION

Analysis of the glycan structure (see Chapter 26) of a glycoprotein is a challenging undertaking, and recognition of its importance is increasing at a fast pace: the diverse roles of glycosylation in biological processes are rapidly growing areas of research [1], with implications that go beyond basic protein research, as the number of protein therapeutics under development is increasing sharply [2]. During the manufacture of protein therapeutics, glycosylation is highly sensitive to the process used: the type of host cell, the particular clone chosen, and the growth conditions all affect the glycosylation of the products [3]. Therefore, there is a growing need for characterization and monitoring of the glycan structure at all stages of discovery, development, and manufacturing of protein therapeutics. The currently used methods for glycoanalysis, however, are complex, typically requiring high levels of expertise and days to provide answers, and are not readily applicable to samples at all stages.

Procognia has developed a lectin array-based method for rapid analysis of glycosylation patterns of glycoproteins. The binding of a glycoform mixture to the lectin array results in a characteristic fingerprint that is highly sensitive to changes in the glycoform composition. Although specificity towards monosaccharides or disaccharides is exhibited by most lectins [4], other features of the glycan in which these epitopes reside often affect the binding affinity of the lectins. For example, mannose-recognizing lectins may bind high mannose-type glycans with high affinity, and the core mannose residues of a biantennary complex glycan with lower affinity. To take advantage of this feature of lectin binding, we use a set of 20–30 lectins with overlapping specificities. This set was chosen from over 70 lectins characterized, based on the level of specificity and activity towards glycans attached to intact glycoproteins, when the lectins are bound to nitrocellulose membranes. We have characterized these lectins using both a large collection of carefully chosen, well-characterized glycoproteins, and a set of enzymatically remodeled glycosylation variants of these proteins. These reference proteins and their variants were fully characterized using high-performance liquid chromatography (HPLC) [5] and mass spectrometry [6]. Based on the binding of this set of glycoproteins to the lectin array, a database of lectin–glycan recognition rules was compiled and computational tools to deconvolute the fingerprint data were developed. The deconvolution provides quantitative information on the overall structures of *N*-linked glycans (e.g. antennarity), and additional epitopes (e.g. antennae termini) present in the sample.

The technology is rapid, enabling dozens of samples to be analyzed within hours with very small quantities of starting material, and requires no specialized expertise. The analysis can be performed on samples in fermentation broth, eliminating the need for time-consuming purification and sample preparation. Complex protein mixtures, such as serum and cellular extracts can also be analyzed with this method.

2. THE TECHNOLOGY

A set of 20–30 lectins is arrayed on a nitrocellulose membrane-coated glass slide. Each lectin is arrayed at several concentrations and in replicates on each slide; the concentration ranges are tailored for each of the lectins, and calibrated to provide a linear response within the same range, regardless of the affinity of the lectin. A sample of intact glycoprotein is applied to the array, and its binding pattern is detected by a fluorescent label. The label can be either on the glycoprotein itself, or on a probe that is directed toward either the protein moiety or a carbohydrate moiety. Common labeled probes are polyclonal antibodies, lectins, or streptavidin for biotinylated samples. The technology provides two levels of data: a fingerprint – a histogram of the dose–response slope measured for each printed lectin, and an interpretation table, which results from deconvolution of the fingerprint. For single proteins, the deconvolution results in a glycan profile – a list of glycan epitopes and relative abundance of each of these in the tested sample. For complex mixtures, the deconvolution results in a comparative interpretation between samples – the relative abundance of particular epitopes, relative to a standard run within the same experiment.

3. THE FINGERPRINTS

Human milk lactoferrin (hmLF) is a glycoprotein with a relatively simple glycosylation pattern [7]. It has two *N*-linked glycosylation sites that are occupied by any of five major glycans, which can result in up to 25 major glycoforms. All five of these glycans are complex biantennary structures containing a core fucose. They differ in their levels of sialylation (0, 1 or 2 sialic acids per glycan) and the presence or absence of antennary fucose. A fingerprint of hmLF is depicted in Fig. 1a. It was produced using a labeled anti-lactoferrin polyclonal antibody probe. Each bar in the histogram represents the total signal from the binding of the hmLF glycoform mixture onto one of the array-bound lectins. The lectins on the array are grouped by their specificities. The fingerprint of hmLF is typical of biantennary complex glycans. There are few very low signals on lectins in the complex glycan specificity group, which recognize mainly tri- and higher order antennary structures. There are significant signals on lectins in the glucose/mannose specificity group, which generally recognize the core structures of biantennary complex N-linked glycans. There are no signals on the lectins from the mannose specificity group, which generally recognize high mannose N-linked glycans. The binding pattern observed on these lectins indicates that all of the lactoferrin glycans are of the complex biantennary type. There are significant signals on lectins of both the Gal/GalNAc (galactose/N-acetyl-galactosamine) and the Galβ specificity groups (lectins Gal/GalNAc 2 and 8, and lectins Gal beta 1 and 2, respectively). Their patterns and intensities indicate that they recognize the terminal galactose of non-sialylated antennae. Additional signals are observed on the fucose-recognizing lectin, and on the sialic acid-recognizing lectin that is specific for (2-6)-linked sialic acid (sialic acid4). A signal is also observed from one of the lectins recognizing Galα(1,3)Gal (Gal alpha3). This lectin also recognizes terminal Galβ, but with lower affinity. The absence of signals from the other two Galα-recognizing lectins indicates that the sample lacks this epitope.

The terminal sugars of hmLF were successively trimmed enzymatically, and the resulting glycoform mixtures were applied to the lectin array (Fig. 1b). Enzymes were added successively to the same initial protein solution, and samples were taken for analysis on the lectin array after each enzymatic reaction. Following desialylation, the signal on the sialic acid-recognizing lectin disappears, and signals on the glucose\mannose-recognizing lectins increase slightly. Our data suggest that this signal increase is due to enhanced accessibility of their concomitant epitopes. Signals on the Gal/GalNAc and Galβ-recognizing lectins increase to different degrees. This reflects their differential sensitivity to the presence of sialic acid on neighboring antennae. Cleavage of the terminal galactose from the desialylated glycans results in an additional increase in the signals on two lectins from the glucose/mannose specificity group, Glc/Man1 and 3, which recognize the core of the biantennary structures. This result further supports the conclusion that better accessibility of the core structures results in enhanced binding to these lectins. Removal of terminal galactose also results in appearance of a signal on the lectin that recognizes the newly exposed terminal GlcNAc, lectin terminal GlcNAc1. Signals on galactose-recognizing lectins decrease considerably, but remaining low signals from these lectins suggest that removal of galactose was not complete. The signal on the lectin that recognizes the antennary fucose, fucose6, changes very little, consistent with the inability of the galactosidase to cleave galactose in the presence of the antennary fucose. When GlcNAc is cleaved from

(a)

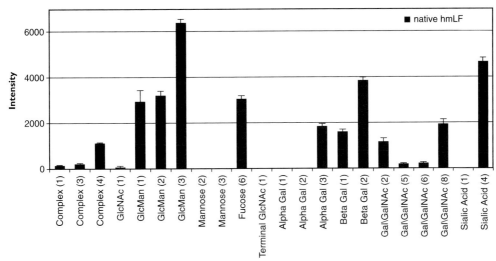

(b)

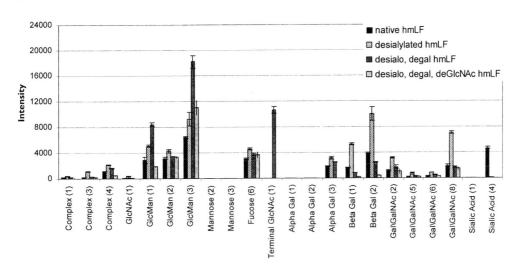

Figure 1. Fingerprints of human milk lactoferrin (hmLF). A solution of 0.1 µM hmLF was applied to the lectin array. The fingerprints were detected by using a labeled anti-lactoferrin antibody as a probe. Twenty-two array-bound lectins were used in these experiments, and are grouped by their specificities on the abscissa.

the de-sialylated, de-galactosylated glycans, the signal on the lectin that recognizes the terminal GlcNAc disappears. There is significant reduction in signal on lectins Glu/Man1 and Glu/Man3, because both of these lectins require the antennary GlcNAc residues for high-affinity binding. Additional signal decrease is observed from the lectins that recognize galactose. This result can be explained by continued cleavage by the galactosidase, since enzymes were neither removed nor inhibited throughout the successive enzymatic steps.

These fingerprints demonstrate the sensitivity of the lectins in the array to changes in the glycan structures. Yet due to the complexity of lectin binding specificities and the differing binding affinities, the fingerprint cannot provide a direct readout of the relative abundance of the glycan epitopes within the sample. For example the level of sialylation of native hmLF cannot be directly deduced from the relative signals on the sialic acid and the galactose recognizing lectins. Quantification of the glycan epitopes requires computational algorithms that are built on a comprehensive understanding of glycan recognition patterns for each of the lectins.

4. FINGERPRINT DECONVOLUTION

Many lectins recognize the same or similar epitopes, but they recognize these epitopes with different affinities and have different dependencies on other features of the glycan in which the epitope resides [8]. For example, some lectins that recognize complex tri- and tetra-antennary structures are inhibited by the presence of sialic acid at antennae termini, and therefore their signals do not directly correlate with the abundance of tri/tetra-antennary glycans in the sample. Other lectins, such as those that recognize terminal saccharide residues, may be sensitive to the number of antennae present on the glycan. These intricate details of the specificities and the affinities of the lectins towards specific epitopes when presented on intact glycoproteins are often unknown. Since samples of glycoproteins are never single glycoforms [9], directly measuring these affinities is not feasible. To overcome this limitation, we have used the lectin arrays to analyze a large set of carefully chosen, well-characterized proteins and multiple enzymatically produced glycosylation variants of several of these. The proteins were chosen to represent a broad range of N-linked glycans produced by mammalian cells under both physiological and cell culture conditions. These fingerprints, along with reference data on each sample produced with conventional technologies, constitute the benchmark on which the deconvolution algorithms were built.

In order to develop fingerprint deconvolution algorithms using the information from the benchmark, a bioinformatics infrastructure was constructed. This infrastructure contains three major components: a Quantitative Fingerprint Interpretation Relation (QFIR) tool that finds lectin signal-based parameters that optimally correlate with the glycan epitopes being analyzed; a set of mathematical toolboxes for combining these parameters into predictive functions; and an automatic training/testing module that divides the benchmark into various training and test sets for evaluation of each of the predictive functions found. This latter module is very important for ensuring generality of the calculated functions, as it prevents overfitting of the functions to the fingerprints within the benchmark.

The QFIR module works by calculating all ratios between all pairs of lectins and lectin groups, defined by their specificities, and various functions of these ratios (such as log or square-root). Then, for each of these parameters, the tool searches for the set that best correlates with each of the glycan epitopes that are analyzed. Typically, we find that the best correlation between a specific lectin and its concomitant epitope is between 0.5 and 0.7 (dependent on the particular epitope), whereas the QFIR tool combines lectin signals and is able to identify parameters with correlations above 0.9. Prior to the development of interpretation functions from the identified parameters a set of heuristic rules is applied to choose the parameters that will be used. Although statistically the best parameters to choose are those that show the highest correlation with a specific epitope, improved robustness is achieved by applying heuristic rules such as favoring parameters that rely on groups of lectins rather than single lectins, and favoring parameters that are consistent with known specificities of the lectins.

Following the choice of parameters with high correlations towards each of the glycan epitopes being analyzed, these parameters are combined into a predictive function to produce the interpretation function for each epitope. This is achieved by testing various mathematical toolboxes with each set of parameters. These toolboxes contain various interpolation algorithms, as well as non-linear tools such as decision trees.

The final interpretation function combines the interpretation functions for each of the epitopes in such a way as to provide the most accurate interpretation on the various test sets. It is noteworthy to point out that when building interpretation functions for a benchmark that contains a large set of different glycoproteins, the obtained accuracy is not as high as when constructing subsets of benchmarks using specific proteins or protein families. For example an interpretation function that was constructed from a benchmark of a large number of monoclonal antibody samples results in high-resolution interpretations that are all within 10% of the reference data.

5. GLYCOANALYSIS OF HUMAN MILK LACTOFERRIN GLYCOVARIANTS

To demonstrate the full output of the lectin array technology, Fig. 2 presents fingerprints of multiple glycosylation variants of hmLF. Gradual degalactosylation was performed on a fully desialylated sample, and was achieved by incubating the sample with limiting amounts of galactosidase (the sample in which 55% of the galactose was removed is presented in the figure); gradual addition of Galα was achieved by limiting substrate during the enzymatic reaction. Detection of the Galα(1-3)-Gal epitope is important for manufacturers of biopharmaceuticals, because it is antigenic in humans [10,11] and is sometimes found on therapeutic proteins produced in rodent cell cultures.

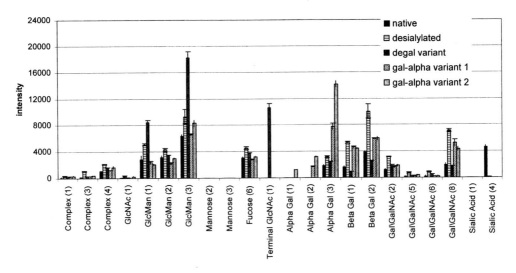

Figure 2. Fingerprints of glycovariants of hmLF. Fingerprints were obtained as in Fig. 1. The desialylated sample was synthesized with *Arthrobacter ureafaciens* neuraminidase; the degalactosylated sample was synthesized with *Streptococcus pneumoniae* beta 1,4-galactosidase (Gala); addition of Gala was performed using recombinant (*E. coli*) porcine α-1,3 galactosyltransferase; different levels of galactosylation were obtained by varying the amounts of substrate.

The fingerprints depicted in Fig. 2 were automatically deconvoluted, and resulted in the interpretations presented in Table 1. HPLC data are presented in parentheses. In the benchmark used for development of the deconvolution algorithms there is limited diversity in the abundance levels of core and non-core fucose: for all samples, the complex *N*-linked glycans have either >90% or <10% core fucose, and antennary fucose was neither cleaved nor synthesized to create variable abundance levels. Therefore, both core and non-core fucose are reported as "detected" vs. "not detected." The algorithms assign 97–100% of the structures as complex biantennary glycans, consistent with the 100% known abundance. The antennae termini predictions are within 10% of the reference data, with the exception of the GlcNAc in the degalactosylated variant, which is overestimated by 12%. These results demonstrate the ability of the algorithms to produce accurate data from the lectin fingerprints.

6. GLYCOANALYSIS OF COMPLEX PROTEIN MIXTURES

The lectin array technology can be used to analyze complex protein mixtures. In the case of mixtures, deconvolution of the fingerprints is more complex, and

Table 1 Deconvolution of fingerprints of human milk lactoferrin (hmLF)

Glycan structure	Abundance (%)[a]				
	Native	Desialylated	De-gal variant	α-gal variant 1	α-gal variant 2
Complex N-linked glycans					
Biantennary	100 ±1 (100)	98±1 (100)	99±0 (100)	98±1 (100)	97±1 (100)
Tri/tetra-antennary	0 (0)	2±1 (0)	1±0 (0)	2±1 (0)	3±1 (0)
High-mannose	0 (0)	0 (0)	0 (0)	0 (0)	0 (0)
Antennae termini[b]					
GlcNAc	0 (0)	0 (0)	67±8 (55)	0 (0)	0 (0)
GalNAc	0 (0)	0 (0)	0 (0)	0 (0)	0 (0)
α-Gal	0 (0)	0 (0)	0 (0)	9±3 (13)	41±4 (50)
Sialic acid (2–6)	52±10 (48)	0 (0)	0 (0)	0 (0)	0 (0)
Sialic acid (2–3)	0 (0)	0 (0)	0 (0)	0 (0)	0 (0)
Additional epitopes[c]					
Core fucose	det. (100)	det. (100)	det. (100)	det. (100)	det. (100)
Out-of-core fucose	det. (30)	det. (30)	det. (30)	det. (30)	det. (30)

[a]Data in parentheses represent reference glycoanalysis.
[b]Abundance of all non-β-gal terminating antennae are calculated and reported by the deconvolution algorithms.
[c]Epitopes for which the deconvolution provides a binary result: det., detected; n.d., not detected.

comparing fingerprints produces valuable insight into the differences in glycosylation patterns between samples.

Figure 3 presents fingerprints of bovine serum from different development stages (fetal, newborn, and calf). Serum was depleted of albumin, immunoglobulin G (IgG), and lipids prior to application to the array, and directly labeled with 5′-fluorescein isothiocyanate (FITC). The fingerprints in Fig. 3a show complex N-linked type glycans with decreasing levels of antennarity with development stage. The fetal serum shows very high level of antennarity, both by the signals of the complex glycan-recognizing lectins, and on the sialic acid lectins, and the calf serum shows the lowest level of antennarity and sialic acid. Both 2-6- and 2-3-linked sialic acids are present. The fingerprints of the serum samples do not show large numbers of signals, probably due to the fact that we are measuring a small number of proteins whose abundance levels dominate the analysis. Neither high mannose glycans nor O-linked glycans are detected in any of the samples, although their presence cannot be ruled out. Our experience shows that O-linked glycans are often masked by the bigger and bulkier N-linked glycans, and in these cases signals for O-linked glycans (on the Gal/GalNAc lectins) become apparent only when N-linked glycans are removed by peptide N-glycosidase F (PNGaseF). The fingerprints of desialylated serum (Fig. 3b) confirm the above conclusions. The same pattern is observed on the complex glycan-recognizing lectins, and the levels of galactose are highest for the fetal calf serum.

Interestingly, the levels of galactose are very similar for the newborn and calf serum, yet significant signals on one of the sialic acid lectins are still evident in newborn serum sample, suggesting incomplete desialylation of this sample.

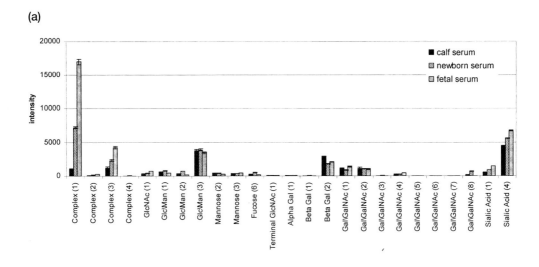

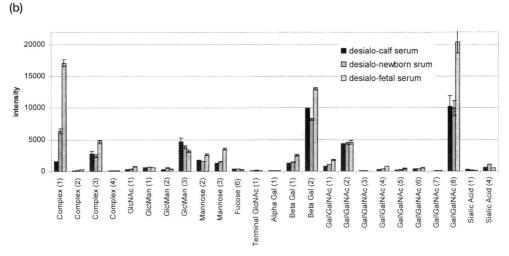

Figure 3. Fingerprints of bovine serum from different developmental stages. Serum samples were depleted of albumin, IgG and lipids, and labeled with 5′-fluorescein isothiocyanate (FITC).

Figure 4 depicts fingerprints of cell surface glycoproteins from the mouse 3T3L1 cell line. These are preadipocyte cells, which, following induction, differentiate into adipocytes (fat cells) that take up glucose in an insulin-dependent mode. This cell line is widely used in diabetes research. Cells were labeled with a sulfated biotin derivative that does not penetrate cells, and therefore only cell surface glycoproteins were labeled. Following the labeling, membrane proteins were extracted using a commercial cellular fractionation kit (Qproteome Kit, Qiagen), and dialyzed prior to application to the array. Fluorescently labeled streptavidin was used as a probe to detect the bound glycoproteins. The fingerprints in Fig. 4 demonstrate the presence of both N- and O-linked glycans (the latter evident from the signals on the Gal/GalNAc-recognizing lectins). Following differentiation, which is morphologically defined by the appearance of fat drops in the cell body, large differences in fingerprints on the lectin array are observed: there is a significant decrease in signals on the lectins belonging to the complex group, indicating a decrease in the level of antennarity with differentiation. This decrease is further supported by the strong decrease in signals on the Galβ-recognizing lectins beta Gal2. In addition, differentiation is accompanied by an increase in the level of fucosylation, which is evident by the signal from the fucose lectin (fucose6). This lectin recognizes exclusively antennary fucose and can potentially recognize Lewis y and Lewis b groups.

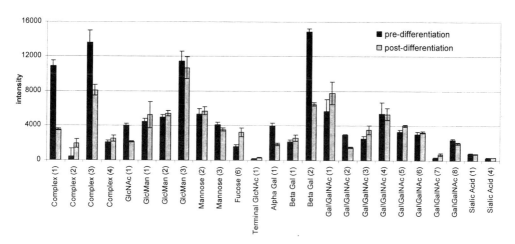

Figure 4. Fingerprints of cell surface proteins from 3T3L1 cells before and after differentiation. 3T3L1 cells were induced to differentiate to adipocytes. Cells before and after differentiation were labeled with a membrane impermeable sulfo-biotin reagent. Total membrane proteins were extracted using Qproteome cell compartment fractionation kits (Qiagen), and extracts were dialyzed prior to application to the lectin arrays. The fingerprints were detected using fluorescently labeled streptavidin

7. DISCUSSION

Glycosylation, one of the most important modifications of proteins, has significant impact on the biological properties of the proteins, and aberrant glycosylation has been observed in various disease states such as inflammatory diseases and cancer [12,13]. Despite numerous studies that unequivocally demonstrate that oncogenic transformation is accompanied by changes in glycosylation, elaborate structural characterization of these changes and understanding of their precise roles in the transformation process remain elusive. A key obstacle limiting such studies is the lack of a technology that can analyze the glycan composition of glycoproteins at high-throughput, directly from *in vivo* samples. The lectin array technology can be used as a first-line tool for the characterization of global glycosylation patterns of biological samples, and for identifying changes that accompany biological processes.

Recently, several reports of lectin arrays have appeared in the literature [14–17]. In these, lectins are arrayed onto glass slides, and the binding of either intact glycoproteins [14–16] or whole cells [17] is measured. The major difference between the technology presented herein and these reports is the ability to deconvolute the lectin fingerprints to provide quantitative information on the glycosylation of the sample. Our comprehensive characterization of the lectin binding on the arrays has demonstrated that the fingerprint obtained on the arrays is not a quantitative readout of the glycan structures present in the sample, and various knowledge-based tools, along with heuristic rules, are required in order to process the large number of low-specificity signals obtained on the array to produce an accurate profile of the glycan structure.

During the manufacturing of protein therapeutics, glycosylation is difficult to monitor and control. The number of glycoprotein therapeutics under development is increasing rapidly [18,19], and during the recent years, production capacity for some recombinant protein drugs has been a bottleneck in meeting market need. This capacity shortage led to significant efforts to develop cell lines and culture conditions capable of producing large quantities of protein. Optimization of such cell lines and growth conditions solely for protein quantity often results in altered glycosylation patterns of the expressed glycoproteins, which may affect product quality. Obtaining clones and optimizing growth conditions that will retain desired glycosylation patterns, along with other product attributes, is a key to relieving the capacity shortfall for therapeutic protein manufacturing. The lectin array technology, when applied to single proteins, can provide accurate glycan profiles for various glycan epitopes, rapidly and without sample purification.

In summary, the lectin array platform presented herein can provide reliable glycosylation analyses of tens of samples within hours, and can be applied at all stages of discovery, development and manufacturing of protein therapeutics, as well as in basic research.

REFERENCES

1. **Dove A** (2001) *Nat. Biotechnol.* **19**, 913–917.
2. **Anscom A** (2004) In *Pharmaceutical outsourcing opportunities post launch*. Edited by S Heffner. Kalorama Information Market Intelligence, New York.
3. **Elliott S, Lorenzini T, Asher S** *et al.* (2003) *Nat. Biotechnol.* **21**, 414–421.
4. **Sharon N & Lis H** (2003) *Lectins*. Kluwer Academic, Netherlands.
5. **Guile GR, Rudd PM, Wing DR, Prime SB & Dwek RA** (1996) *Anal. Biochem.* **240**, 210–226.
6. **Harvey DJ** (2003) *Int. J. Mass Spectom.* **226**, 1–35.
7. **Spik G, Coddeville B & Montreuil J** (1988) *Biochimie* **70**, 1459–1469.
8. **DiVirgilio S** (1997) In *Lectins, Biology, Biochemistry, Clinical Biochemistry*, Vol. 12. Edited by Edilbert van Driessche, Sonia Beeckmans and Thorkild C Bøg-Hansen. TEXTOP, Denmark.
9. **Dell A & Morris HR** (2001) *Science* **291**, 2351–2356.
10. **Galili U, Rachmilewotz EA, Peleg A & Flechner I** (1984) *J. Exp. Med.* **160**, 1519–1531.
11. **Galili U, Shohet SB, Kobrin E, Stultz CIM & Macher BA** (1988) *J. Biol. Chem.* **263**, 17755–17762.
12. **Hakomori S** (2002) *Proc. Natl Acad. Sci. USA* **99**, 10231–10233.
13. **Dube DH & Bertozzi CR** (2005) *Nat. Rev. Drug Discov.* **4**, 477–488.
14. **Pilobello K, Krishnamoorthy L, Slawek D & Mahal L** (2005) *ChemBioChem.* **6**, 985–989.
15. **Kuno A, Uchiyama N, Koseki-Kuno S** *et al.* (2005) *Nat. Methods* **11**, 851–856.
16. **Angeloni S, Ridet JL, Kusy N** *et al.* (2005) *Glycobiology* **15**, 31–41.
17. **Hsu K-L, Pilobello KT & Mahal LK** (2006) *Nat. Chem. Biol.* **2**, 153–157.
18. **Brekke OH & Sandlie I** (2003) *Nat. Rev. Drug Discov.* **2**, 52–62.
19. **Andersen DC & Krummen L** (2002) *Curr. Opin. Biotechnol.* **13**, 117–123.

CHAPTER 28

Lectin chips and CarboDeep™: a rapid carbohydrate fingerprinting technology and its application for the food scientists

Gabriel Faiman and Ofer Markman

1. INTRODUCTION

Microarrays have evolved to answer the need to carry out tens and hundreds of reactions simultaneously for a small number of samples [1–9]. They were originally developed for nucleic acids due to their relatively simple synthesis, stability, hybridization properties, and straightforward binding to solid supports. An example of a microarray application in glycobiology is described in [1]. Protein arrays are much more complex, not only because there are no simple solutions to synthesize proteins *in vitro* but also because the tertiary structure of proteins is essential to their binding activities. Therefore, special care must be taken in order to bind proteins to solid matrixes and keep their functionality [8].

Since lectins were discovered they have been used as analytical tools in order to detect the presence of specific glycan groups. This use was limited initially to chromatography and histochemistry. However, the use of lectins as analytical tools developed rapidly [7] as the importance of glycosylation was established and with it the glycogenomics era.

Lectin arrays [3–6] are an innovative technology utilized to analyze glycan contents based on the biochemical recognition of sugar moieties by an array of natural proteins. The lectins in the microarray are carefully chosen such that the specificities are overlapping on one side, so that signals can be reinforced, yet the space of recognition is large enough to obtain a meaningful fingerprint of lectin binding signals. In several cases of known glycoproteins the fingerprint can be translated into an approximate relative carbohydrate composition. In general, however, the fingerprint can only be used as a basis of comparison between different products, such as in the analysis of complex mixtures.

Lectin arrays have a large number of advantages compared to conventional glycoanalysis methods such as chromatographic techniques and mass spectrometry. The conventional glycoanalysis methods are based on the physical and chemical properties of the glycomolecules, whereas lectin arrays are based on the biological binding properties. Due to this fact, lectin arrays can analyze samples that are not pure and/or that have not yet been cleaved from proteins. The sensitivity of the chemical analytical methods is limited by the chemical characteristics of the molecules analyzed: fluorescence, ionization ability, charge, mass, magnetic polarization, etc., while the sensitivity of the lectin arrays is limited by the magnitude of the binding constants of the lectins to the glycans. Importantly, the ambient analyte theory proves the enhanced sensitivity that protein microarrays have. Some of the chemical analysis methods available are high-performance anion exchange chromatography with pulse amperometric detection (HPAEC-PAD) [15], fluorescence chromatography [14,17], capillary electrophoresis (CE) [18], fluorophore-assisted carbohydrate electrophoresis (FACE)[19], mass spectroscopy [22] with its various applications such as electrospray ionization (ESI) [26], matrix-assisted laser desorption/ionization (MALDI) [23] and nuclear magnetic resonance (NMR) imaging (see Chapter 7) [32,33]. Several methods utilize two orthogonal methods such as chromatography and mass spectroscopy [13,25,29,30] or even two orthogonal mass spectroscopy [27]. These methods or combinations of them are commonly used for analysis of carbohydrates such as starch [20], bacterial saccharides [21], and glycans from proteins [16,24,31] as well as many food applications.

Compared with the 20 amino acids in proteins, glycans have only a few monosaccharide building blocks. However, due to the numerous branching positions of each monosaccharide, the diversity of glycans possible is far more complex than the diversity found in proteins, which are defined by a linear sequence. Furthermore, the same glycoprotein can have different glycan compositions defining a wide range of glycoforms, which pose a great challenge to the analytical methods used.

This complexity is reflected in the initial output of the lectin arrays, which is a fingerprint that characterizes the glycan content of the sample. This output can be used straightforwardly as a sample specific pattern, which can be standardized and compared for quality assurance purposes. Clustering and comparison of fingerprints can be used to detect differences and similarities between samples without knowing the specific glycan contents. Further deconvolution of the fingerprints and analysis using empiric knowledge-based databases can lead to fingerprint interpretation, which looks at the relative amounts of glycans descriptors detected in the sample.

The food industry needs techniques that will enable easy identification and analysis of its ingredients. Complex carbohydrates or polysaccharides are ubiquitous within the world of foods, both as natural components and as added ingredients. They are found in the form of starch, cellulose, gums, and pectins with extremely varied and versatile functions, ranging from stabilizers, thickeners and binders to emulsifiers, fillers, solubilizers, and lubricants. Several food-related health hazards are related to the complex sugars of ingredients, such as lactose intolerance, lactalbumin allergies, or general allergic responses. Changes in complex carbohydrates during food processing are extremely frequent; the Maillard reaction (browning) for example. Although some are welcomed, others affect the quality of the final product. The unique structure of each complex carbohydrate determines its properties as a food ingredient. Therefore, alterations in the polysaccharides will modify the overall properties of the food material. The characterization of the complex carbohydrate patterns of food ingredients is an excellent way of following upon general changes of food ingredients. Complex carbohydrates can also be found attached to proteins and lipids in the form of constitutive polysaccharides, and their structure varies depending on the organism from which the proteins and the lipids are extracted. Overall, there is a wealth of information about glycoproteins and glycolipids as food constituents that has hardly been explored with existing analytical technologies.

We have developed a lectin array that can be used to analyze the glycan analysis of purified compounds, such as lactoferrin and glycomacropeptides from milk. This lectin array can be used to identify samples, as seen in Figs 1 and 2. Furthermore, the lectin arrays can differentiate between different species, and can thus detect adulterations in dairy products (Fig. 3). The lectin array can also be used, albeit without interpretation, to analyze mixtures and detect abnormalities. Moreover, industrial processes such as fermentation can be followed and analyzed since they usually involve changes in glycosylated species (not shown).

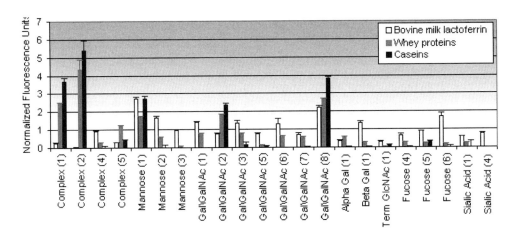

Figure 1. Whey and casein fractions were prepared from 3% fat fresh bovine milk and 0.25% fat bovine milk powder. Samples were defatted by centrifugation. Monosaccharides and small molecules were removed by dialysis (MWCO 3500). Casein and whey fractionation was carried out by acidification and centrifugation according to standard procedures. The fractions were quantified for protein content by the Bradford assay and analyzed by sodium dodecyl sulfate polyacrylamide gel electrophoresis (SDS PAGE). The samples were analyzed by the CarboDeep technology at protein concentrations of 0.2–0.5 mg/mL. Bovine milk lactoferrin was a commercially available sample and was analyzed using the CarboDeep platform at a 0.1 μM concentration.

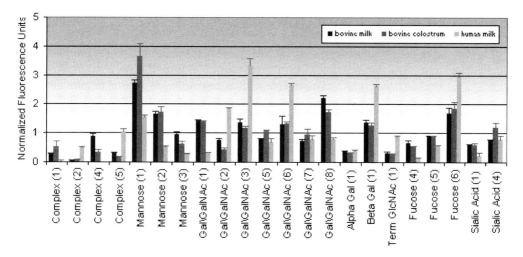

Figure 2. Human, bovine milk, and bovine colostrum lactoferrin samples were from commercially available sources and were analyzed using the CarboDeep platform at 0.1 µM concentrations.

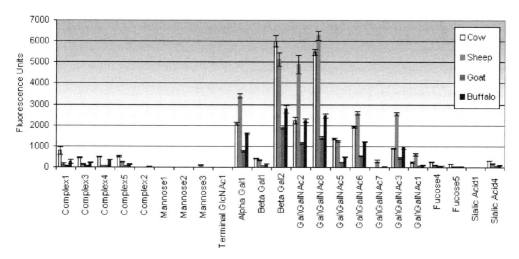

Figure 3. Glycomacropeptide (GMP) samples were prepared from fat fresh milk from four different species. Samples were defatted by centrifugation. Monosaccharides and small molecules were removed by dialysis (MWCO 3500). The milk plasma proteins were treated with rennet for 15 min at 37° C. Proteins were precipitated by TCA acidification and centrifugation. The supernatant containing the GMP was dialyzed and lyophilized. Peptide concentration was determined by the bicinchoninic acid method (BCA). Samples were analyzed using the CarboDeep platform at 1 mg/mL concentrations.

2. TECHNOLOGICAL APPLICATIONS

The use of lectin arrays, as the use of microarrays in general, has revolutionized the way test reactions are carried out. The use of this technology allows researchers to perform hundreds and thousands of reactions and tests on a minute amount of sample simultaneously. It also provides a tool for high-throughput reactions carried out in minimal time, almost real time, which could help in the production line of food materials and in safety issues. Yet, issues of quality follow the field to maturity, and are handled in a few ways and protocols [34–36].

The lectin arrays are a "new microscope" in the food industry [37] – they allow us to look into carbohydrates in timescales and with a simplicity never before possible. They thus have the potential to have a widespread impact on the industry (compare [37] and [38]). Some industry issues are close to application and we expect these to pioneer this process and become a suitable replacement to some of the complex and advanced but less comprehensive methodologies [38–42]. Heat treatment in food processing and food drying is critical in today's industrial food-processing environment, and it is important to monitor the time and temperature of the treatment as well as reaction times flow and other conditions to control heat-related reactions, the most critically of which is browning (the Meillard reaction). In the sections below we expand and focus on some other applications of lectin arrays in the food industry.

2.1 Safety

To help reduce contamination at all points in the food chain we need a better understanding of the way that microorganisms interact with their environment through the glycosylated proteins on their surface walls. It is important to be able to not only detect but also characterize pathogens and develop an understanding their physiological responses.

2.2 Identity, authenticity and adulteration

Identity is gaining power in modern quality industrial processes. Important questions include: Is this molecule the same as the one I produced last week? Is the process stable? Am I using the same molecules and processes as my competitors? Is this safe? Is this product stable upon shelf life?

The global economy today requires the manufacturer and customer to know the exact content and origin of any food produced: there are many reasons for this, the most important of which are safety and quality/reputation (regional or corporate). There are also important reasons for manufacturers to protect their customers from false claims and from adulteration as these reflect highly on the products and brand reputation. Lectin arrays have potential to be useful in authenticating cheeses, milk, and whey powder, especially in view of predicted requirements to label the origin of caseins in the cheese industry.

2.3 Taste and feel

In the food industry the way our food tastes is critical to success and several of the players in taste and feel are carbohydrates. We expect lectin arrays to help in the major efforts being made to reduce the need for food sensorial testing and in the development of an artificial mouth. This will reduce the need for expert panels and tasting tolls in the industry.

2.4 Formulation and food treatment

The sensitivity of lectin arrays to a variety of molecules already on the market as ingredient co-products (such as starch, gums, and alginates) suggest that we will soon see the incorporation of lectin arrays and a panel of predefined (so called "inert") co-ingredients in highly monitored mix-and-process processes in the food industry, such as those used in the production of powdered food for small children and the elderly and as feed for young animals.

Biofilm is another product related to the carbohydrate–bacterial interaction with immense economic impact on the food industry, both through the negative impact on filters and membranes and through several positive impacts (e.g. for probiotics delivery). Again we can predict a vast impact of lectin arrays both in basic research and in the applied technologies.

2.5 Composition

Food composition is becoming crucial to the new way the community looks at food because of issues of food allergenicity and sensitivity. Food composition and food labeling is very much part of quality of life in the twenty-first century. The ability to identify the metabolites from food, including complex carbohydrates, is more important than ever before. Lectin arrays have the ability to provide answers on a molecular level about the type of monosaccharides and type of bonding the sugar is made of, the type of glycans, their abundances and their relatedness to health and safety of food.

2.6 Mixtures

The determination of the composition of a food is often hindered by the fact that food is a mixture of many materials. This is true for sugars in food as well and in particular for complex sugars. Most complex carbohydrates in the food industry are in the form of inhomogeneous mixtures. Often these mixtures have important properties and sometimes it is the inhomogeneous nature of the mixture that gives them these properties. At the same time the reproducibility of the nature of the mixture is critical for process repeatability and reproducibility. In such cases lectin array fingerprints can be a useful tool.

2.7 Food forensics

The ability to use fingerprints to define identity is critical in discriminating foods for the forensic and epidemiological applications. Various pathogens and toxins are related to specific sugars, as are allergens and other sources of intolerance such as lactose.

2.8 Speciation

Milk speciation is important not only because of dairy allergies but also because of economic and nutritional issues. The ability to discriminate cow's milk from other milks is critical both for economical and health reasons. Cow's milk is the most abundant and least expensive type of milk. Buffalo's and goat's milk as well as sheep's milk are considered alternative and higher class milks in most industrial countries. The ability to identify the milk species is relevant for labeling issues, economics, and safety. In the European Union several cheese labels can only be used for certain milk species (such as mozzarella for buffalo's milk)

3. SPECIFIC MARKETS

3.1 Lactofact™

NutriCognia has developed a semi-quantitative rapid glycoanalysis assay for milk lactoferrin (bmLF) named LactoFact, which can control the production process and product development as well as improving quality control procedures. Lactoferrin is widely used as a supplement in baby formulas and other milk beverages. As seen in Fig. 2, a characteristic fingerprint is obtained for different samples of lactoferrin such as bovine milk, human milk, and bovine colostrums. The proprietary algorithms that analyze the fingerprint provide a glycoanalysis which includes the level of high mannose and biantennary glycans as well as the glycan termini: sialic acid, galactose, and GalNAc (not shown).

3.2 GMPact

The glycoanalysis kit for glycomacropeptides (GMPs), named GMPact, is another product developed by NutriCognia. GMPs, a product of the cleavage of caseins, are used in the dairy industry as a supplement as an amino acid source for patients with phenylketonuria, as a stimulant of cholecystokinin release that increases the satiety sensation, and as a source of sialic acid, which is important in the development of the central nervous system. Until now there has been no easy way of analyzing the quality of GMPs or their glycosylation, which is relevant to their functionality. NutriCognia's CarboDeep technology changes this picture since the glycoanalysis and comparison of different GMP samples can be performed in a couple of hours. This gives insight to the *O*-glycans of GMPs which provide evidence of their production, purification, and formulation methods. Furthermore, analysis of GMP can be used in adulteration detection, since many differences are observed between GMPs from different species (Fig. 3).

3.3 Probiotics and prebiotics

Probiotics are microorganisms used as dietary supplements to provide a natural flora, whereas prebiotics are supplements that stimulate the growth of the existing physiological flora. The analysis of probiotics and prebiotics can be carried out using lectin arrays due to the ubiquitous glycan structure, both in the form of membranal glycoproteins and as secreted exopolysaccharides in probiotics and complex carbohydrates, such as inulin in prebiotics.

Their effects on nutrition and health and their protective effects are all part of the probiotic evolution. It is now clear that the ability of bacteria in the gut to recognize specific sugar epitopes, the polysaccharides secreted by the bacteria, and the bacterial modification of the gut environment all contribute to the health effects of probiotics as well as prebiotics and symbiotic health effects. We see a great future for lectin and glycan arrays in this arena.

3.4 Dairy research

Lectin arrays promise innovative analysis to dairy research areas such as: (1) ingredients such as glycolipids, glycoproteins, prebiotics, etc.; (2) follow-up of new chemical and physical processes, such as heat treatment-based lactose adducts of proteins and amino acids; and (3) differentiation of different milks (species) milk fractions and concentrates [43].

3.5 Milk processing

The development of lectin array-based analysis brings to the dairy industry a new window in processing. The ability to monitor the carbohydrates as the process progresses allows a variety of new opportunities in milk fermentation, cheese making, and heat treatment of milk as well as in milk fractionating (into milk fat globules, casein concentrates, whey, and whey derivatives).

3.6 Whey

There are a number of areas in the whey industry that lectin array technology can help with: (1) the identification of glycoprotein whey ingredients, such as lactoferrin and GMPs; (2) whey lines – lectin arrays can discriminate rennet ("sweet") whey from the common ("sour") whey because of the increase in O-glycans due to casein glycopeptides in rennet whey; and (3) lectin arrays can help define specific fingerprints for specific whey lines.

3.7 Beer

There are a number of applications used in the dairy industry which could help the fast analysis of complex sugars in the beer industry. These are related to beer authenticity, beer fermentation, monitoring of malt processing, production of the malt and the incoming ingredients such as starch, dextrans, and pentosans.

3.8 Baby food and feed

Several ingredients in baby food are glycomolecules, such as whey proteins, milk powder, glycolipids/gangliosides, and lactoferrin, which is added to baby food in East Asia and the Pacific region. The ability to gain high value information on these in a short time frame will allow better control of the content of baby food. Several issues related to young animal nutrition and protein availability are related to their sources of complex carbohydrates, to the source of whey added to their nutrition and to casein in their milk powder protein sources.

3.9 Pastry egg and milk

A key to the usability of lectin arrays in food is the ability to define each of the ingredients by specific carbohydrates; there are marked differences between the carbohydrate species found in eggs and those found in milk or milk powders, butter, and flour.

Other applications in the food industry will arise as the technology matures, and applications will be developed in area such as hydrocolloids, meat, and more.

4. CONCLUSIONS

There are already a variety of analytical applications for carbohydrate analysis of foods in general and for lectin arrays in particular. The time and span of such processes is more academic driven in the current situation due to the fact that the field of lectin arrays in food is only in the pioneering stage, as well as due to the immaturity of complex carbohydrate analysis in food technology and nutrition. In Chapter 27 Dr Rosenfeld explored the applications of lectin arrays in more traditional biotech applications.

REFERENCES

1. Comelli EM, Head SR, Gilmartin T et al. (2006) Glycobiology 16, 117–131.
2. Angeloni S, Ridet JL, Kusy N et al. (2005) Glycobiology 15, 31–41.
3. Zheng T, Peelen D & Smith LM (2005) J. Am. Chem. Soc. 127, 9982–9983.
4. Kuno A, Uchiyama N, Koseki-Kuno S et al. (2005) Nat. Methods. 2, 851–856.
5. Hirabayashi J (2004) Glycoconj. J. 21, 35–40.
6. Pilobello KT, Krishnamoorthy L, Slawek D & Mahal LK (2005) Chembiochem 6, 985–989.
7. Jelinek R & Kolusheva S (2004) Chem. Rev. 104, 5987–6015.
8. Stoll D, Templin MF, Bachmann J & Joos TO (2005) Curr. Opin. Drug Discov. Dev. 8, 239–252.
9. Uttamchandani M, Walsh DP, Yao SQ & Chang YT (2005) Curr. Opin. Chem. Biol. 9, 4–13.
10. Templin MF, Stoll D, Bachmann J & Joos TO (2004) Comb. Chem. High Throughput Screen. 7, 223–229.
11. Kusnezow W & Hoheisel JD (2003) J. Mol. Recognit. 16, 165–176.
12. Kusnezow W, Syagailo YV, Goychuk I, Hoheisel JD & Wild DG (2006) Expert Rev. Mol. Diagn. 6, 111–124.
13. Bruggink C, Maurer R, Herrmann H, Cavalli S & Hoefler FJ (2005) Chromatogr. A. 1085, 104–109.
14. Lilla Z, Sullivan D, Ellefson W, Welton K & Crowley R (2005) J. AOAC Int. 88, 714–719.
15. Corradini C, Bianchi F, Matteuzzi D, Amoretti A, Rossi M & Zanoni S (2004) J. Chromatogr. A. 1054, 165–173.
16. Siemiatkoski J, Lyubarskaya Y, Houde D, Tep S & Mhatre R (2006) Carbohydr. Res. 341, 410–419.
17. Xia B, Kawar ZS, Ju T, Alvarez RA, Sachdev GP & Cummings RD (2005) Nat. Methods 2, 845–850.
18. Dutta U & Dain JA (2005) Anal. Biochem. 343, 237–243.
19. Gao N (2005) Methods 35, 323–327.
20. Yao Y, Guiltinan MJ & Thompson DB (2005) Carbohydr. Res. 340, 701–710.
21. Li SY, Holtje JV & Young KD (2004) Anal. Biochem. 326, 1–12.
22. Zamfir AD, Bindila L, Lion N, Allen M, Girault HH & Peter-Katalinic J (2005) Electrophoresis 26, 3650–3673
23. Harvey DJ (2005) Proteomics 5, 1774–1786.
24. Peter-Katalinic J (2005) Methods Enzymol. 405, 139–171.
25. Siemiiatkoski J & Lyubarskaya Y (2005) Dev. Biol. 122, 69–74.

26. Brivio M, Oosterbroek RE, Verboom W, van den Berg A & Reinhoudt DN (2005) *Lab. Chip.* **5**, 1111–1122.
27. Clowers BH & Hill HH Jr. (2005) *Anal. Chem.* **77**, 5877–5885.
28. Zamfir AD, Bindila L, Lion N, Allen M, Girault HH & Peter-Katalinic J (2005) *Electrophoresis.* **26**, 3650–3673.
29. Guignard C, Jouve L, Bogeat-Triboulot MB, Dreyer E, Hausman JF & Hoffmann L (2005) *J. Chromatogr. A.* **1085**, 137–142.
30. Liu Y, Urgaonkar S, Verkade JG & Armstrong DW (2005) *J. Chromatogr. A.* **1079**, 146–152.
31. Harvey DJ (2005) *Expert Rev. Proteomics* **2**, 87–101.
32. Huckerby TN, Nieduszynski IA, Giannopoulos M, Weeks SD, Sadler IH & Lauder RM (2005) *FEBS J.* **272**, 6276–6286.
33. Hricovini M (2004) *Curr. Med. Chem.* **11**, 2565–2583.
34. Kreil DP & Russell RR (2005) *Brief Bioinform.* **6**, 86–97.
35. Burgoon LD, Eckel-Passow JE, Gennings C *et al.* (2205) *Nucleic Acids Res.* **33**, e172.
36. Altman N (2005) *Appl. Bioinform.* **4**, 33–44.
37. Hsu KL, Pilobello KT & Mahal LK (2006) *Nat. Chem. Biol.* **2**, 153–157.
38. Gilboa-Garber N, Lerrer B, Lesman-Movshovich E & Dgani O (2005) *Electrophoresis* **26**, 4396–4401.
39. Ferranti P (2004) *Eur. J. Mass Spectrom.* **10**, 349–358.
40. Frazier RA & Papadopoulou A (2003) *Electrophoresis* **22–23**, 4095–4105.
41. de la Fuente MA & Juarez M (2005) *Crit. Rev. Food Sci. Nutr.* **45**, 563–585.
42. Guy PA & Fenaille F (2006) *Mass Spectrom. Rev.* **25**, 290–326.
43. Berlin CM, Crase BL & Furst P (2005) *J. Toxicol. Environ. Health A.* **68**, 1803–1823.

Afterword

Nathan Sharon

This book is the most precious gift that I ever received. I am extremely grateful to the many friends and colleagues who contributed to the book, in particular to Clare Sansom, whose idea it was, and to Ofer Markman, co-editor. I am also thankful to Clare for the excellent summary of my scientific activities in the first chapter of the book. In addition my thanks are due to the many colleagues and friends who contributed the other 26 chapters of the book that provide a view of the current state of the fast-growing area of glycobiology. As pointed out by Clare in her chapter, the idea to prepare this book evolved in connection with the scientific symposium held at the Weizmann Institute in November 2005 to celebrate my 80th birthday. I wish to use this opportunity to thank again Timor Baasov from Technion, Haifa Institute of Technology, and Ed Bayer from our Department, for planning the symposium and for so ably organizing it.

Having expressed my deep gratitude to those who made this book possible, I will allow myself to indulge in reminiscences about some aspects of my career hardly touched upon by Clare or not mentioned by her at all.

1. EARLY TRAINING

In 1950 it was my great fortune to be taken by the late Aharon Katzir, as one of his early graduate students. This was just after I completed my studies for the MSc degree in biochemistry at the Hebrew University, Jerusalem. It was he who introduced me to carbohydrates. Another happy event was my acceptance in 1954, as a young postdoc, by Aharon's brother Ephraim Katchalski-Katzir into his fledgling Biophysics Department.

Ephraim expected me to devote my time to polypeptides, a subject to which he was then making monumental contributions and in which all other members of his department were engaged. Still, he gave me quite early on the freedom to follow some chance observations I made on carbohydrates and then on lectins, a completely neglected subject at the time, and supported me fully in my endeavors. I then spent two years in the lab of Roger Jeanloz at Harvard Medical School, from whom I learned much about carbohydrates. In the early 1960s I became involved with bacillosamine, a rare amino sugar I discovered, and with lysozyme, subjects I left in the middle 1970s, and of lectins, in which I am interested to the present time. So this is in a nutshell how it came about that I spent essentially all my scientific life between carbohydrates and proteins.

Both Aharon and Ephraim infused me, as their other students, with the love of science, and with the recognition that engaging in scientific research, especially in a small country like Israel, is a special privilege, justified only if it is done at the highest standards. They taught us that to succeed, one has to choose an important problem, not necessarily in the mainstream of scientific research, to stick to it and strive to become a foremost expert on the subject. Eventually, if one persists, works hard, has a prepared mind and is favored by luck, one may make seminal discoveries, and one's name becomes associated with a particular subject or even a whole field.

Aharon and Ephraim also taught us that scientists in Israel have the duty to go out of their way to serve their country, and in this vein inspired me to devote many years to the popularization of science. I was therefore doubly thrilled that when I was awarded the Israel prize in biochemistry in 1994, the citation referred also to my activities in presenting science to the general public.

2. SPREADING THE GOSPEL

Along with my research work proper, I was engaged in other activities, which I believe served to attract scientists young and old to carbohydrates, glycoproteins and lectins, as well as to science in general. One of these was teaching an extended course on complex carbohydrates since 1963, primarily at the Feinberg Graduate School of the Weizmann Institute, and also at other institutions in Israel and abroad. In the aftermath of the October 1973 war I prepared a series of lecture notes for the students who were unable to attend the course because they were on military service. The notes were published in 1975 as a book, '*Complex Carbohydrates*', which was well received. One reviewer wrote that the book 'is an excellent, readable introduction to those topics which [the author] has chosen'. Another said 'the contents are extremely clear, and students fortunate enough to have sat through such lectures cannot fail to have benefited enormously from such a lucid account.'

The book was translated into Japanese by my friend Toshiaki Osawa, and was in wide use for nearly two decades. In the introductory chapter I stated my belief 'that the specificity of many natural polymers is written in terms of sugar residues, not of amino acids or nucleotides.' During the years the emphasis of my course changed and it focuses now on the molecular biology of glycoproteins and glycolipids. More recently the course, which I renamed *Molecular and Cellular Glycobiology*, started to attract increasing numbers of students.

I also wrote a large number of reviews, book chapters and encyclopedia entries on glycoproteins and lectins, most of them with Halina Lis. They appeared in advanced treatises, such as *The Antigens* (edited by Michael Sela) and *The Proteins* (edited by Hans Neurath and Bob Hill), in prestigious serials such as *Annual Reviews of Biochemistry* (three times) and *Advances in Immunology*, and in a variety of first rate journals. Our first review on lectins that appeared in *Science* in 1972 is still high on the list of lectin papers cited, as is the second one published in the same magazine journal in 1986. Several of our articles appeared in magazines used for teaching, such as *Scientific American* (five in number between 1969 and 1993) and its foreign language translations, and in other widely circulated journals, among them *Chemical and Engineering News*; the latter article, because of the interest it created, was reprinted in *Molecular and Cellular Biochemistry* and also translated into Japanese. Together with Halina I prepared in 1989 a monograph on lectins and in 2003 a second edition of the same, both of which have been translated into Japanese. Some 20 years ago I co-edited a treatise on lectins, to which we contributed several chapters. I have no doubt that these publications helped to spread the glycoprotein (and lectin) gospel among wide audiences, as acknowledged, for example, by Ron Schnaar, editor of *Glycobiology*, in his introduction to the review on the history of lectins which Halina and I published in 2004 in this journal: 'Along with his longtime collaborator Dr. Halina Lis, Dr. Sharon ... has been a tireless and highly effective advocate for glycobiology worldwide'.

3. SCHOOLS, WORKSHOPS AND SYMPOSIA

Over the years I organized several schools, workshops and symposia on carbohydrates, glycoconjugates and lectins. Among them was the tenth International Symposium on Glycoconjugates that convened in Jerusalem in September 1989, with almost 450 participants from all over the world. Regular biannual meetings of these symposia, which serve as important catalysts to research on glycoproteins and related subjects, started in 1973 with the *Colloque International des Glycoconjugués* (later designated as the second symposium in the series) convened by Jean Montreuil in Lille in 1973. It was an exciting event, which attracted about 140 scientists, including several of the leaders of carbohydrate research at the time, among them David Aminoff, Ernst Buddecke, Jean Emile Courtois, Zacharias Dische, Fujio Egami, Alfred Gottschalk, Sen Hakomori, Roger Jeanloz, Elvin Kabat, Rex Montgomery, Ward Pigman and Ikuo Yamashina.

The most memorable of the events I hosted was the Edmond de Rothschild School on Glycoproteins. This two-week school was scheduled for October 1973, but had to be postponed for one year because of the war. The lecturers included leading glycoprotein researchers, among them Gilbert Ashwell, John Clamp, Leon Cunningham, Victor Ginsburg, Mary C. (Susie) Glick, Roger Jeanloz, Torvard Laurent, Albert Neuberger, Elizabeth Neufeld, Garth Nicolson and Harry Schachter, as well as some who were beginning to make their mark on the subject like Jeremiah Silbert and Roland Schauer. Several of the students are now also well known, for instance Kurt von Figura, Hans Kresse and Ralph Schwarz. Alfred Gottschalk, the pioneer glycoprotein researcher, who was to give the opening lecture, unfortunately passed away in October 1973. On the program were topics such as the sialic acids, chemical synthesis of glycoprotein glycans, membrane glycoproteins of normal and malignant cells, blood type determinants, glycoprotein biosynthesis, mucopolysaccharidoses and the biological role of the carbohydrate, especially in hepatic recognition of circulating glycoproteins. These topics are still timely and, in spite of the enormous progress achieved during the last three decades, would fit well into the program of a present day glycobiology course or conference on the subject.

4. EPILOG

Looking back at my scientific career, it is clear that it was shaped more than once by fortuitous encounters and chance observations, and was powered by the twin drives of curiosity and a desire to contribute to society's well-being. Like most of my colleagues, I immensely enjoyed my work that was also my favorite pastime. I was fortunate in having outstanding mentors and highly talented graduate students and coworkers. Of the latter I wish to single out Halina Lis, with whom I have been collaborating closely since the early 1960s. I owe a debt to the many agencies that supported my research throughout the years, among them the National Institutes of Health, the US Department of Agriculture, the

Volkswagen Foundation and the US-Israel Binational Foundation and the Israel.

It was my good luck to spend a major part of my scientific life in the conducive atmosphere of the Weizmann Institute, and also to benefit from the hospitality of a number of leading laboratories at several other world class institutions. Among others, it gave me the opportunity to meet countless interesting people worldwide, and to strike lasting friendships with quite a few. I continue to watch with much interest and excitement the rapid development of glycobiology, with its increasing impact on medicine and industry, which I have helped to advance.

In closing, I wish to express my profound gratitude to my wife Rachel, to our daughters Esty and Osnat, and to our son-in-law Danny who were a constant source of loving support to my efforts. This is an opportunity to acknowledge my debt to them publicly.

Index

Adipocytes, 349
Amoebapores, 302–303
Angiogenesis, 274, 280
Antigens, 261
   blood group, 218, 250
   Hanganutziu-Deicher (H-D), 283
   Lewis, 218, 227
   T, 226, 227
   Thomsen Friedenreich (TF), 261, 262, 263, 264, 267, 270
   Tn, 224, 226, 227, 249
Antisense RNA, 299
Apoptosis, 117, 118, 161, 162, 165, 221
Arrays, *see also* Microarrays
   carbohydrate, 336
   gene, 258
   glycan, 84, 86, 145, 174–175, 256
   lectin, 340–350, 352
Arthritis, 171
   osteoarthritis, 45
Atomic force microscopy (AFM), 335
Autoimmunity, 169

ß-cell, 244
B-cell receptors, 166
Biantennary glycans, 63, 358
   complex, 63
Bioinformatics, 83, 86–88, 344
Biosensors, 325–337
   arrays, 333
   nano-biosensors, 335
   PDA-based, 331

Blastogenesis, 237
Blood clotting factors, 217
Blood group,
   ABO, 257
   ABO (H), 278
ß-secretase, 58

Ca^{2+} homeostasis, 291, 292
Cancer, 4, 5, 10, 26, 78, 226, 231, 284–286, 288, 350
   anticancer drugs, 315
   colon, 227, 230, 261–271, 278, 282
   diagnostics, 334–335
   gastrointestinal, 270, 278
   immune therapy, 283
   immunity, 172
   lung, 286
   markers, 249
   progression, 280
   prostate, 108, 118
   tumor cell invasion, 207
   vaccine, 10
Capillary electrophoresis (CE), 81
Capsular polysaccharides (CPS), 180, 185–186, 326
Carbohydrate,
   active enzymes (CAZy), 84, 222, 225, 226
   arrays, *see* Arrays
   recognition domain (CRD), 2, 108, 110, 112, 113–115, 128, 129, 162, 299
Carbohydrate-protein interactions, 14, 336

Cartilage, 42, 44
Cell,
    adhesion, 94, 109–110, 113–115, 162, 207, 217, 218, 270, 275, 289, 298
    aggregation, 270
Cell-cell,
    interactions, 238
    recognition, 191
Cerezyme®, see GlcCerase
CFTR, 19
Chaperones, 126
    Cosmc, 226
Chemokine, 194
Chemotaxis, 168
Chitin, 15
Cholera toxin (CT), 329
Chondroitin sulfate, 38, 42
Chromatin, 279
Circular dichroism (CD), 53, 59
    synchrotron radiation CD, 52, 53, 54, 57
Coagulation factors, 129
Collagen, 107, 217
Confocal laser microscopy, 277
Consortium for Functional Glycomics (CFG), 6, 85, 87, 88
Core 1 ß3-gal-transferase, 226–227
Crystallography, 5
    jacalin, 150
    X-ray, 5, 52, 150, 203, 213
Cyclin-dependent kinases (CDKs), 115
    inhibitors (CDKIs), 116
Cysteine proteinases, 299–302
Cystic fibrosis, 19
Cytokines, 147, 161, 166, 167, 183, 217, 221, 265
    interleukin-4, 147
    NFκB, 269

Dermatan sulfate, 38, 42
Diabetes, 314
DNA microarrays, see Microarrays
DNA repair, 321
DNJ, 317
Dystroglycan, 240–241, 243

Endoplasmic reticulum (ER), 8, 123, 124, 126–128, 131, 209, 288
    ER-associated degradation (ERAD), 127, 130
Endosulfates, 99, 100
Entamoeba histolytica, 297–305
Epidermal growth factor, 111
Epimers, 42, 43, 312
Erythrocytes, 3, 142, 144
Evolution, 155
Exoglycosidase, 70, 74
Extracellular matrix (ECM), 45, 107, 119, 203, 298

Fertilization, 237
Fibronectin, 94, 102, 107, 109, 304
Fluorescence *in situ* hybridization (FISH), 266
Fluorescence resonant energy transfer (FRET), 330, 331
Fourier transform infrared (FTIR), 52, 54, 55–58
Fucose, 39, 65, 67, 74, 126, 258

G protein-coupled receptors (GPCRs), 191, 194, 304
Galactose, 40, 63, 65, 126, 142, 151, 240, 265
Galactosyltransferase, 212
Gangliosides, 141, 144, 146, 236, 243, 283, 288, 291, 329
Gel electrophoresis, 326
Gene arrays, see Arrays
Gene microarrays, see Microarrays
Gene silencing, see Silencing
Gene therapy, 18, 294–295
GlcCerase (Cerezyme®), 293
Glucoronic acid, 39

Glucose, 39, 65
Glucosidase, 9
Glycan arrays, see Arrays
Glycan-binding protein (GBP), 85
Glycocalyx, 251, 254, 255, 256, 266
Glycoconjugates,
  cell surface, 303–304
Glycomacropeptides (GMPs), 358
Glycome, 49, 52, 84, 86–88, 258
Glycopolymers, 14, 16
Glycosaminoglycan (GAG), 5, 38, 43, 49, 78, 79, 83, 95, 241–242
  sugar ratios, 83
Glycosidases, 266
Glycosphingolipids (GSLs), 242–243, 288–295
  Glycosphingolipid storage diseases, 5, 10
Glycosulfatases, 266
Glycosyltransferases (GTs), 90, 129, 217, 219, 222–224, 313, 315
  Asp-x-Asp (DxD) motif, 223
  core 3 ß3GlcNAc-transferase, 227–228
  inverting, 223, 231
  polypeptide GalNAc-transferases, 224–226
  retaining, 223
  structures, 222–224
Golgi, 108, 123, 124, 126, 131, 141, 203, 208, 209, 220, 251, 262, 263, 288
GPI anchor, 101, 303
Greek keys, 150, 151, 155

Hemicelluloses, 214
Heparan sulfate (HS), 40, 49, 54, 56, 81, 94–104, 207
  composite sulfated region (CSR), 95
  HS saccharides, 58–59
Heparan sulfate proteoglycans (HSPGs), 100–104, 242
  glypicans, 103–104
  syndecans, 100–103

Heparin, 40, 46, 52, 54, 56, 88, 94, 97, 207–208
  anticoagulant activity, 97
  biosynthesis, 98–99
  low molecular weight heparin (LMWH), 89
Hepatitis B, 9
Hepatitis C, 8
High mannose chains, 129, 347, 358
High-performance liquid chromatography (HPLC), 62, 66, 70, 81, 327, 341, 346
Human immunodeficiency virus (HIV), 8, 127, 145, 289, 317, 319
  HIV-1, 10, 164, 207
Hyaluronan, 44, 241
Hypoxia, 283
  hypoxia inducible factor (HIF), 280, 282

Iminosugars, 5, 8, 9, 308–322
  indolizidines, 311–312
  nojirimycin, 309
  piperidines, 309
  pyrrolidines, 309–311
  pyrrolizidines, 313
Immune response,
  adaptive, 163
  innate, 168, 178–186
Implantation, 238
Inflammation, 160, 162, 271, 288
Inflammatory bowel disease, 166, 171, 261–271
  Crohn's disease, 261, 267, 269, 271
  ulcerative colitis, 261, 264, 266, 269
Inflammatory diseases, 350
Influenza, 3, 25, 31, 317, 332
Integrins, 109, 110, 119, 165

Lactoferrins, 341–344, 358
Laminin, 107
Lectin arrays, see Arrays

Lectins, 2, 14, 15, 17, 128–132, 161, 264, 327
  adhesins, 146, 183, 298
  artocarpin, 153
  ß-prism 2 lectin fold, 150, 155, 156, 158
  C-type, 85, 178, 179–180, 185, 186
  calnexin, 8, 126, 128, 129, 130
  calreticulin, 8, 126, 128, 129
  collectins, 179, 182
  concanavalin A, 327
  folds, 4
  galectins, 4, 85, 107–119, 160, 162, 270
  hemagglutinins, 3, 144
  integrins, 190, 191, 194–196
  P-type lectins, 162
  pentraxins, 162
  phytohemagglutinin (PHA), 264
  selectins, 107, 162, 190–194, 196–197, 228, 238, 275, 278, 284, 286
  siglecs, 85, 147, 162, 277
  soya bean agglutinin (SBA), 264
  tachylectins, 162
Leishmaniasis, 211
Leukocytes, 190, 196, 239
Lipophosphoglycans (LPGs), 211–212
Lipopolysaccharides (LPS), 180, 182, 185, 333
Lysosomal storage disorders, 206, 311, 317–320
Lysosomes, 109

M cells, 268
Macrophages, 180, 182, 183
Major histocompatibility complex (MHC), 127
Malaria, 146
Mannans, 157
Mannose, 20, 63, 65, 124, 131, 151, 185
  receptors, 180, 185
  6-phosphate, 209–211

Mass spectrometry (MS), 44, 62, 64–66, 81, 82, 326, 341, 352
  electrospray ionization (ESI), 62, 64, 69–70, 75, 81–82, 326, 353
  Fourier transform ion cyclotron resonance MS (FTIR-MS), 83
  matrix-assisted laser desorption/ionization (MALDI), 62, 64, 67–69, 75, 81–82, 326, 353
  MS/MS spectra, 73
Membrane blebbing, 109
Metastasis, 270, 275, 284–286, 315
Microarrays, 83, 352, see also Arrays
  DNA, 282
  gene, 84
Migration, 94, 109, 162
Mimetics, 33
Mitochondria, 108
Molecular modeling, 52, 54, 55, 153, 302, 320
Monoclonal antibodies (mAbs), 334
Motifs, 141
  DxD, 223
  von Willebrand, 249
MUC genes, 248, 252, 253, 254
Mucin-like glycoproteins, 228
Mucins, 138, 144, 218, 248–258, 262
  variable number of tandem repeat (VNTR) regions, 218
Mucopolysaccharidoses (MPS), 208
  MPS II (Hunter syndrome), 208
  MPS III (Sanfilippo syndrome), 208
Mycobacteria, 212–213

N-acetyl-D-glucosamine, 32, 74, 126
N-acetylgalactosamine, 39
N-acetylneuraminic acid, 139
  see also Sialic acid
NB-DNJ (Zavesca®), 11, 293, 319, 322
Neoglycoproteins, 14, 15, 16, 17, 18
Neuraminidases, 25, 144, 317

N-glycolyl neuraminic acid (NeuGc), 282
N-glycosylation, 123, 263
N-linked glycans, 62–63, 64, 66–67, 73, 192, 238–240, 244, 344, 347
  biosynthesis, 63
  motif, 63, 123
  release, 66
Nodulation (Nod) factors, 202, 208–209
Notch receptors, 240, 243
Notch signaling, 236, 237, 240, 243
N-sulfation, see Sulfation
Nuclear magnetic resonance (NMR), 52, 53, 54, 55, 57, 58, 81, 83
  NMR spectroscopy, 44, 98, 302, 353

O-antigen, 182, 185
O-fucose glycans, 240
O-linked glycans, 63, 73, 153, 164, 192, 219–222, 230–232, 238–240, 248, 249, 252, 347
  mucin, 217, 238
O-mannose glycans, 240–241
Oogenesis, 235
Opsonin, 181
Osteoarthritis, see Arthritis
O-sulfation, see Sulfation

p53, 116
polymerase chain reaction (PCR), 85
  reverse transcriptase PCR (RT-PCR), 85
Polymers,
  glycosylated, 16
Protein–carbohydrate interaction, 154
Proteoglycans, 5, 51, 217, 241–242, see also Heparan sulfate proteoglycans (HSPGs)

Schiff's base, 15
Sharon, Nathan, 2, 14, 248, 258, 288

Shear flow, 190–197
Sialic acids, 24, 62, 63, 65, 68, 69, 126, 136–148, 358
  N-acetyl neuraminic acid (NeuAc), 24, 34, 137, 138
Sialidases, 25, 30, 141, 144
Sialin, 141
Sialyl Lewis,
  SL^a, 268, 274,
  SL^x, 27, 143, 147, 219, 237, 239, 250, 261, 274, 275, 278, 334
  SL$^{x/a}$, 274, 279, 282, 284
Sialylmimetic, 25, 27, 35
Sialyltransferase, 26, 33–34, 139
Signal,
  peptide, 123
  recognition particle (SRP), 123
  sequence, 249
  transduction, 4, 110
Signaling pathways, 256
Silencing,
  epigenetic, 279, 280
  gene, 274, 278–279
Size exclusion chromatography, 42
Sodium dodecylsulfate polyacrylamide gel electrophoresis (SDS PAGE), 67
Spermatogenesis, 236
Storage disorders, 8, 289
  Gaucher disease, 11, 89, 289, 290, 293, 294, 319
  glycosphingolipidoses, 319
  MPS I, 89, 289
  MPS II, 208
  MPS III, 208
  N-butyl DNJ, 319
  Niemann–Pick disease, 289
  Sandhoff disease, 289
  Tay–Sachs disease, 319
Sulfatases, 203, 206–207
  carbohydrate, 206–207
  extracellular, 206
  lysosomal, 206

Sulfation, 42, 203–209
　N-sulfation, 96, 99
　O-sulfation, 40, 45, 96
　of CS, 42
　of DS, 42
Sulfotransferases, 44, 203–209, 278
　carbohydrate sulfotransferases, 203–205
Sulfs, see Endosulfates and also Sulfatases
Surface plasmon resonance (SPR), 16, 330
Systems biology, 78, 85

T-cell, 10, 161, 163–165, 243, 321
　receptors, 166
Thin-layer chromatography (TLC), 327

T-lymphocytes, 146
Transferases,
　inverting, 226
　retaining, 225
　sialyltransferases, 231

UV absorbence, 326

Virulence factors, 180, 298, 304
VNTR domains, 250, 255, 256
Von Willebrand factor, 248

Warburg effect, 280

Xylose, 40